ENVIRONMENTAL HEALTH - PHYSICAL, CHEMICAL AND BIOLOGICAL FACTORS

DUST

SOURCES, ENVIRONMENTAL CONCERNS AND CONTROL

Environmental Health - Physical, Chemical and Biological Factors

Additional books in this series can be found on Nova's website under the Series tab.

Additional e-books in this series can be found on Nova's website under the e-book tab.

Public Health in the 21st Century

Additional books in this series can be found on Nova's website under the Series tab.

Additional e-books in this series can be found on Nova's website under the e-book tab.

ENVIRONMENTAL HEALTH - PHYSICAL, CHEMICAL AND BIOLOGICAL FACTORS

DUST

SOURCES, ENVIRONMENTAL CONCERNS AND CONTROL

LAURENT B. WOUTERS
AND
MICHEL PAUWELS
EDITORS

Nova Science Publishers, Inc.
New York

Library of Congress Cataloging-in-Publication Data

Dust : sources, environmental concerns, and control / [edited by] Laurent B. Wouters and Michel Pauwels.
p. cm.
Includes bibliographical references and index.
ISBN 978-1-61942-547-7 (hardcover)
1. Dust. I. Wouters, Laurent B. II. Pauwels, Michel.
TD884.5.D87 2011
620'.43--dc23
2011049554

Published by Nova Science Publishers, Inc. † New York

CONTENTS

PREFACE

In this book, the authors present current research in the study of the sources, environmental concerns and control of dust. Topics discussed in this compilation include the characterization of heavy metals content in attic dust from copper ore and ferronickel smelter processing plants in Macedonia; dust control in the mining industry of Australia; the health and environmental effects of dust storms in the Arabian Peninsula; dust fallout and its potential hazard on public health in Kuwait; photochemical and climate implications of airborne dust and assessing the quality of the urban environment by the elemental concentrations of foliage dust.

Chapter 1 - The chemical composition of urban dusts in Slovenia (Europe) is the topic of this contribution. Urban dusts are important substances in the environment because they can be an important pathway for toxic metals into the human body. The goal of this work is the presentation of the chemical composition of selected urban deposits (dusts) and their relation to spatial macrolocation (rural/urban environments), geological background, topsoil composition, dominant natural/anthropogenic factors and other influential factors. The evaluation was done on the basis of 83 sampling locations, where attic dust, household dust and topsoil were sampled. Twenty-three of them were placed in towns; others ware placed in natural environments in Slovenia. In the work, distributions of 41 chemical elements were evaluated. According to the multivariate statistics, the dominant geochemical associations of elements were recognized. Their areal distribution in sampled materials across the countryside and larger towns in Slovenia was determined using the universal kriging method. The geochemical properties of household and attic dust were compared with topsoil and evaluated in terms of their elemental contents and correlation coefficients. The proportion of anthropogenic impact to toxic metal concentrations in urban sediments was assessed. As based on comparisons of household dust, attic dust and topsoil using the multivariate statistical method (factor and cluster analysis), four patterns of elemental distributions were established. The two natural geochemical associations, Ti-V-Al-Th-Sc-Fe-Y-Nb-Co-Mn-La and Ba-Na-La, are mainly influenced by the weathering of crust or soil. The association Pb-Zn-Cd-Sb-Mo-Hg-Sn-Cu represents the chemical elements anthropogenically introduced into the environment. Distribution of this association represents the consequences of the influence of Pb and Zn smelters in the past. High Hg is a consequence of centuries of lasting operation of the Idrija mercury mine and smelters and of military activities during the First World War, known as Soča (Insonzo) front line. The fourth association of Ni-Cr-Co is influenced by natural and anthropogenic factors. Distribution is mainly influenced by lithology, but it can also be attributed to the anthropogenic influence as a result of iron processing.

An alarming fact is that high contents of Cd, Cu, Hg, Pb, Sn and Hg were measured in household dust. These concentrations exceed their corresponding values in natural sediments by more than twenty times. The urban sediments, especially household dust, are substances to which the humans are exposed on a daily basis. Dust particles containing toxic metals can enter human organisms by being swallowed or inhaled. Several authors established significant associations between the heavy metal contents in household dust and concentrations in body liquids (blood, urine). Thereof direct hazards to population in Slovenia may be derived. High contents of toxic metals in urban dust are potentially dangerous, especially to children. Small infants are the most endangered group because of their higher intake of dust compared to adults and owing to their higher sensitivity to the influence of toxic metals.

Chapter 2 - A comprehensive monitoring was applied to assess the environmental pollution in the Republic of Macedonia. Large amounts of fine dust are generated during blasts and excavations of mining minerals and from highly technological industrial processes, whereas they are distributed in the air by the winds. For that issue characterization of heavy metals content in attic dust and total deposited matter was performed. The significant emission sources that contribute to atmospheric pollution with heavy metals on the territory of the Republic of Macedonia appear to be all mines and smelter plants. In one of the region three collection stations were placed for monthly monitoring of heavy metals content in total deposited dust is near a copper mine in the Radoviš region and the second location is in the vicinity of the smelter plant known for its ferronickel industrial activity in the Kavadarci region. In both locations two most affected villages and the nearest town was monitor within one year. Characterisation of contents showed higher content of heavy metals in deposited dust, which is due to distribution of fine particles from mine and smelter plant respectively, in air and their deposition in settlements in the close vicinity. Attic dust samples were also used in order to determine and monitor long-term air pollution with heavy metals around the copper mine and the ferronickel smelter plant. The total contents of about 20-40 elements were analysed by inductively coupled plasma - mass spectrometry (ICP-MS), inductively coupled plasma - atomic emission spectrometry (ICP-AES) and electrothermal atomic absorption spectrometry (ETAAS). Basic statistical methods and multivariate exploratory (factor and cluster analysis) techniques were applied for data processing. The results of the study of spatial distribution of different trace elements over these regions show some areas with critically high content of some elements. It was found that high contents of those anthropogenic elements are deposited in the close vicinity of the mine and the smelter plant.

Chapter 3 - Dust control is a widespread problem affecting the mining industry in Australia and most parts of the world where the mining of natural resources is taking place. Due to the ever increasing focus on the health and safety of personnel, as well as environmental concerns, the problem of dust control is certainly an important one that needs total attention. Within the mining industry, the majority of dust problems arise from the transportation of the material being mined. However there are also problems associated with product that is exposed to the environment during storage, such as in large stockpiles. In this chapter current technology is presented and discussed that addresses both the characterisation of dust and different avenues for classifying a material in terms of "dustiness". In addition, analysis techniques currently available for use in the design of materials handling equipment are also presented. The particular emphasis is on passive dust control measures with the focus on transfer chutes, as it is well known the transfer of material from one conveyor to another is a major cause of dust problems. The work presented herein includes a series of case studies that seek to highlight the challenges involved in the design of plant equipment to handle dusty materials.

Chapter 4 - Dust lifted into the atmosphere can be carried far afield and, apart from disruption to human activities, can have serious and life threatening implications. The Arabian Peninsula is a major dust producing region. This chapter briefly discusses the health and environmental effects of dust storms and, in more detail, the dynamics and the two atmospheric mechanisms that cause dust storms and allow dust to be raised in suspension for prolonged periods in the atmosphere. Particular reference is made to blowing dust and dust storms during the summer and winter Shamals, as well as those caused by the Kaus wind, the less frequent Nashi wind and the usually short lived thunderstorm associated dust storm. Seasonal and diurnal distribution and frequency is discussed. Attention is drawn to the importance of numerical weather prediction models and a simple dust forecast index proposed. The latest advances in satellite imagery for dust detection are discussed and a dust storm forecasting methodology is presented.

Chapter 5 - Dust and dust storms are always associated with widespread consequences, hazards and complications on the environment, public health, soil quality, economy … etc. Rates of dust fallout deposition on multiple large scale areas throughout the world significantly indicates the adverse effect of dust storms on the deposition of soil particles in arid and semi-arid areas. Soil erosion, sediment suspension and sediment deposition are considered to be the major processes that are involved in a dust storm and could contribute in a diversity of environmental hazards. Soil quality, particularly its fertility, can be directly affected by soil erosion due to the deterioration of its nutrient contents. Deflation of clay and silt particles by the wind action is the major contributor to the high sand content of soil. Suspended particles during a dust storm directly affect the atmospheric environmental quality and greatly reduce the visibility in some severe cases to less than 0.5 meter, which results in traffic accidents and their consequent complications. Sediment deposition and sand approach results in considerable economics losses in the urban areas, different utilities' plants, field crops and cultivated areas, and disruption of communications. The factors playing a significant role in dust storms and dust fallout could be the environment type, geomorphology and relief differences.

Chapter 6 - Airborne mineral dusts suspended in earth's atmosphere can have dramatic effects on climate, regional photochemistry and human life. In this chapter, both the direct and indirect climate forcing effects of dust will be discussed. Physical models for treating light scattering and absorption by dusts will be considered. Since dust absorbs light in the ultraviolet and blue portion of the electromagnetic spectrum, its presence can reduce actinic flux and alter atmospheric photochemistry. This phenomenon will be considered from the perspective of common atmospheric pollutant gases. Finally, the chemistry occurring at / on dust surfaces will be considered and their significance discussed.

Chapter 7 - It is well established that dust explosions represent a serious hazard in many process industries and silo facilities. Many catastrophic cases have been reported, and statistics show that the number of incidents per year is still significant. Venting is the most common protective measure used in silos in order to prevent the appearance of unacceptably high internal pressures. However, the installation of venting devices is not always simple, and technical complications frequently arise. Furthermore, the cost of protection by venting must be taken into account. Vent areas in silos should be large enough to prevent damaging overpressures, but not so large that the use of venting becomes impracticable. Thus, vent area sizing is a critical issue. To calculate vent area size, standards EN 14491 (2006) and NFPA 68 (2007) are commonly used in Europe and North America, respectively, but in certain situations, they are contradictory. The aim of this study was to analyze dust explosion venting in silos by comparing the above venting standards and also conducting CFD simulations. The pressures and associated vent areas in these numerical simulations were compared to those contemplated in the standards. In addition, some calculations were carried out for venting devices with inertia. The simulated explosion pressures showed the expected trends for the associated vent areas and agreed reasonably well with the values contemplated in NFPA 68 (2007). For low overpressure values, the differences were significant compared to those vent areas predicted by standard EN 14491 (2006). Numerical simulations could serve as a powerful tool for helping engineers to design explosion protection and calculate the structure of vented silos. However, the results of numerical simulations should be extrapolated with caution since they are absolutely dependent on the flow characteristics and the dust concentration of the initial dust cloud.

Chapter 8 - The dust composition due to the impact of industry and traffic in the some Albanian cities are subject of this investigation. 42 dust samples from the house lofts or green house gutters, where undisturbed dust deposition has occurred for a period of about 30 years, were investigated from the vicinity of three industrial sites (i.e. Burrel: a ferrochromium smelter; Elbasan: a metallurgical complex; Porto-Romano: a former chemical plant). The characterization of the dust samples consisted in assessing the elemental and speciation analysis, by means of atomic absorption spectroscopy (AAS) and X-ray diffraction. The elemental concentration of the dust samples from these industrial sites, were compared with local background dust concentrations. The geochemical signature indicated that the contamination of dust by Cr, Fe, Ni, Co, Mn and to a lesser extent by Zn and Pb in the samples from the industrial sites (42 in total) is apparent when total concentrations are compared with the background values (three samples). The time these industrial sites have been active, played an important role in the geochemical signature of the dust accumulated, especially for chromium, iron, cobalt, nickel. An additional source dominated of aluminium, potassium, and sodium indicates that a natural source, not linked to the industrial processes, but most probably from pulverized clay-rich soils also contributes to the composition of the dust particles in the investigated sites. Additionally the composition of the dust indicates that the industrial activities are the main source of the contamination especially with the heavy metals. The composition of dust particles is ruled by the intensity of the industrial activity as confirmed by XRD and SEM-EDX, while the distribution is dominated from the prevailing winds. In the dust particles collected outside the ferrochromium smelter/stalk chimney, was seen that ferrochromium was the main species relating to the industrial activity, chromite and quartz were also detected most likely relating to the technological process use in the industrial plants as well as to its performance.

Maps for the spatial distribution show the spreading of the contaminants in relation to the pollution sources and the influence of the prevailing winds in transportation and deposition in distance from these investigated industrial sites. Although with a limited data set of samples, it is clear that the prevailing winds are the dominant factor in the transportation of the dust particles in distance from the industrial site. Moreover, from the air contamination history, as recorded by the age of the dust samples it can be concluded that the air quality has been severely affected, mainly during the 1980's, due to the intensive industrial activity in these industrial sites. Knowing that the health related issues, due to the quality of air are strongly related to asthma, these studies might be an initial step in applying some on-site risk management measures minimise the dust release and to filter the gases and dusts released from these industrial sites. Obviously these conclusions may be useful in the long term health studies in these industrial sites as well as to determine and regulate the air quality in the Albanian cities.

Chapter 9 - The sources and pathways of dust play an important role for the global climate and biogeochemical cycles. In that context, the assessment of the elemental concentration of foliage dust elevated to an interesting topic of many scientific studies. Foliage dust may indicate heavy metal contamination and the source of contamination from atmospheric deposition in urban areas. Foliage dust contains heavy metals that may have harmful effects on human health; thus, elemental contents of foliage dust are useful to assess air pollution. The concentrations of heavy metals in dust samples may vary across cities depending on the density of industrial activities. The aim is to summarize the sources of dust and heavy metal concentration in dust samples. In a case study we demonstrated that dust is a useful indicator to assess the level of pollution in urban environment. The authors studied the elemental concentrations in foliage dust along an urbanization gradient in Vienna, Austria. Samples were collected from urban, suburban and rural areas. The authors analyzed 10 metals in samples from all three areas: Al, As, Ba, Co, Cu, Fe, Pb, S, Sr and Zn. They found that the elemental concentrations of foliage dust were significantly higher in the urban area than in the rural area for aluminium, barium, iron, lead, phosphor and selenium. Urbanization changed significantly the elemental concentrations of foliage dust and the applied method proved itself as a useful way to monitor the environmental load.

In: Dust: Sources, Environmental Concerns and Control
Editors: Laurent B. Wouters and Michel Pauwels
ISBN: 978-1-61942-547-7

Chapter 1

CHEMICAL COMPOSITION OF URBAN DUSTS IN SLOVENIA

Robert Šajn, Gorazd Žibret* and Jasminka Alijagić
Geological survey of Slovenia, Ljubljana, Slovenia

ABSTRACT

The chemical composition of urban dusts in Slovenia (Europe) is the topic of this contribution. Urban dusts are important substances in the environment because they can be an important pathway for toxic metals into the human body. The goal of this work is the presentation of the chemical composition of selected urban deposits (dusts) and their relation to spatial macrolocation (rural/urban environments), geological background, topsoil composition, dominant natural/anthropogenic factors and other influential factors. The evaluation was done on the basis of 83 sampling locations, where attic dust, household dust and topsoil were sampled. Twenty-three of them were placed in towns; others ware placed in natural environments in Slovenia.

In the work, distributions of 41 chemical elements were evaluated. According to the multivariate statistics, the dominant geochemical associations of elements were recognized. Their areal distribution in sampled materials across the countryside and larger towns in Slovenia was determined using the universal kriging method. The geochemical properties of household and attic dust were compared with topsoil and evaluated in terms of their elemental contents and correlation coefficients. The proportion of anthropogenic impact to toxic metal concentrations in urban sediments was assessed. As based on comparisons of household dust, attic dust and topsoil using the multivariate statistical method (factor and cluster analysis), four patterns of elemental distributions were established.

The two natural geochemical associations, Ti-V-Al-Th-Sc-Fe-Y-Nb-Co-Mn-La and Ba-Na-La, are mainly influenced by the weathering of crust or soil. The association Pb-Zn-Cd-Sb-Mo-Hg-Sn-Cu represents the chemical elements anthropogenically introduced into the environment. Distribution of this association represents the consequences of the influence of Pb and Zn smelters in the past. High Hg is a consequence of centuries of lasting operation of the Idrija mercury mine and smelters and of military activities during the First World War, known as Soča (Insonzo) front line. The fourth association of

*E-mail: gorazd.zibret@geo-zs.si; +386-1-2809-757.

Ni-Cr-Co is influenced by natural and anthropogenic factors. Distribution is mainly influenced by lithology, but it can also be attributed to the anthropogenic influence as a result of iron processing.

An alarming fact is that high contents of Cd, Cu, Hg, Pb, Sn and Hg were measured in household dust. These concentrations exceed their corresponding values in natural sediments by more than twenty times. The urban sediments, especially household dust, are substances to which the humans are exposed on a daily basis. Dust particles containing toxic metals can enter human organisms by being swallowed or inhaled. Several authors established significant associations between the heavy metal contents in household dust and concentrations in body liquids (blood, urine). Thereof direct hazards to population in Slovenia may be derived. High contents of toxic metals in urban dust are potentially dangerous, especially to children. Small infants are the most endangered group because of their higher intake of dust compared to adults and owing to their higher sensitivity to the influence of toxic metals.

Keywords: top soil, attic dust, house dust, toxic metals, Slovenia

1. Introduction

The name dust is used in a variety of ways, and with different meanings. These range from the material that accumulates on the earth's surface, such as on streets and in living and working environments, to the particulate material suspended in the atmosphere.

Atmospheric dust, or atmospheric particulate matter, originates from a wide variety of natural processes and anthropogenic activities, including volcanism, forest fires, rock/crust degassing, combustion of fossil fuels, agricultural practices, industrial manufacturing, and construction activities (Fergusson, 1992; Ozaki et al., 2004; Tasdemir and Kural, 2005; Wilson and Pyatt, 2007).

The discrete airborne particles that compose atmospheric dust consist of a complex mixture of metals, acids, biogenic material, and other organic and inorganic compounds that may represent health risks. Through several exposure routes (such as inhalation, skin contact, absorption in mucosal membranes of eyes and airways, swallowing and ingestion) the particles may reach into humans, especially children (Cizdziel and Hodge, 2000; Molhave et al., 2000). The health effects of toxic metals distributed with air-dust depend on their mass concentration and affect the respiratory tract. Due to fact that children breathe deeper and faster, the particles penetrate deeper in their lungs. Moreover, because they spend more time outdoors and are more active, children belong to the group in the population most exposed to the dust. Another group with high risk is senior citizens, especially those with a weakened cardiovascular and respiratory system (Gauderman et al., 2004; Vallero, 2008).

Atmospheric dust undoubtedly contributes to household dust by permeating people's homes, primarily through vents, doors, open windows and tracking. Because a typical resident spends 65-90% of the time inside the home, household dust has been used to assess human exposure to variety of pollutants, including those originating from outside the home (Cizdziel and Hodge, 2000). Eventually much of the atmospheric dust becomes surface dust, either through dry deposit or washed out by precipitation. These surface dusts will, in most cases, contain contributions from material that had not been airborne, or at least that had been for only a short time, such as weathered buildings materials, soil, deposited litter, and large particles from traffic emissions (Fergusson, 1992).

The main purpose of this study is the comparison of chemical compositions of different sampling media simultaneously, such as soil, house dust and attic dust. This study addresses the influence of many parameters on the composition of household dust and attic dust. Based on the sampling scheme, which was done over the whole territory of Slovenia (approximately 20.000 km^2), several dominant influences on the household and attic dust composition can be recognized, such as population influence, mining and smelting influences and the influence of the natural factors (lithology). All of them are discussed in this contribution. Such comparisons will allow the reader to have insight about the scales of the different influential factors. Differences between towns and the countryside, between lithological settings, and between household and attic dusts will be presented according to the basic parametric and nonparametric statistical variables. Relatively large amounts of analyzed samples allow for making multivariate statistics over large sets of variables (elements). Factor and cluster analyses show us the connections between different elements and their spatial distribution.

1.1. Emission and Deposition of Metals in Dust

Many authors recognize the great importance of studying chemical compositions of atmospheric particles and their interaction between soils, street sediment, and attic and household dusts. Toxic metals are ubiquitous in environmental compartments as low natural concentrations, and they were always present in a minute amount in our environment.

There are so many sources for releasing dust into the environment. The most important human activities that emit significant amounts of toxic metals are transportation and industrial production. Mining activities, ore processing and the processing of waste are found as significant emitters of metals in air, especially open pit metal mining. Open pits and ore tailings are quite often unprotected, and fine grain particles are carried away by wind or water. Transportation distance might be very long under favorable conditions. These elements get distributed among soils, air, surface dust and water. As metals can not be degraded or decomposed, they are usually accumulated in the environments (Banerjee, 2003; Imperato et al., 2003; Pacyna et al., 2007; Agarval, 2009; Šajn et al., 2011). As a matter of fact, metal pollution has become a global environmental problem, owing to the large-scale atmospheric transport. It is noticeable that human-induced metals have been detected even in snow samples in Greenland and Antarctica (Gorlach and Boutron, 1992; Candelone et al., 1995; Hong et al., 1996; Boutron et al., 1998; Van de Velde et al., 2005; Hur et al., 2007).

Deposition is the process by which aerosol particles collect or deposit themselves on solid surfaces and thus decrease the concentration in the air. The rate of deposition or the deposition velocity depends on particle size. Very large particles will settle out quickly through the sedimentation process. Sedimentation is much faster for larger particles than for smaller particles.

Two different types of deposition exist: dry and wet (Godish, 2004; Vallero, 2008). Dry deposition is characterized by the direct transfer of gas phase and particulate matter from the air to the ground, vegetation, water bodies and other surfaces. In the absence of precipitation, dry deposition plays a major role in removing pollutants from the atmosphere (Finlayson-Pitts and Pitts, 2000). Wet deposition is a deposition of solid particles by water media, such as rain, snow, fog, cloud, dew formation or similar (Godish, 2004; Vallero, 2008). Two different processes are important: condensation and wash-out. In the condensation process, water condenses around solid (dust) particles, which act as a condensation nuclei. In this way, disposal of pollutants by rain, snow, etc. occurs. On the other hand, "wash out" is the process by which pollutants are removed from the atmosphere by being absorbed in or desorbed on water droplets.

1.2. Household and Attic Dust

In geochemical studies, the term "dust" usually refers to street dust and household dust. Household dust represents an important vector for the ingestion and consequent accumulation of toxic substances by humans, particularly by young children. Household dust has many internal and external sources, including garden soil, road dust, human hair and skin, carpet and clothing fibers, paint chips and fungi, resulting in a heterogeneous matrix of organic matter and inorganic and metallic particles. Trace metals, like Cd, Cu, Pb and Zn, exist in many of these sources and therefore exhibit considerable enrichment in the household environment relative to their crustal abundances (Culbard et al., 1988; Fergusson and Kim, 1991; Fergusson, 1992; Rasmussen, 2004). Toxic metals find their way into residential homes either as airborne dust (e.g., leaded gasoline emissions from motor vehicles) or through items used or activities carried out within the house (e.g., renovating or types of heating).

However, contaminated residential dust provides a critical link in the exposure pathway for most young children. Through their hand-to-mouth actions, many children are inadvertently ingesting the metal toxins. In some cases, a child may not exhibit conspicuous pica activity, yet he/she may still be ingesting over 50 mg of lead from traces of dust, dirt or soil (Mielke et al., 1988). In Birmingham, England, the daily uptake of lead for a 2-year-old was estimated to be 36 mg (Davies et al., 1990). Rasmussen et al. (2001) showed that ingestion of house dust is the main exposure pathway for Pb (69%) for children living in contaminated areas.

Atmospheric particles are tiny particles of solids or liquids suspended in air. These particles vary in size and density (Finlayson-Pitts and Pitts, 2000). Particles 0.005–0.1 μm in diameter are primary particles produced from high temperature combustion processes and gas condensation. Metals emitted by those processes into the atmosphere have high solubility and reactivity, especially under the low pH, and can be carried far away from the sources by the atmosphere. Such processes could contribute particle matter to the atmosphere, pedosphere and hydrosphere under certain dynamic conditions (Hršak et al., 2003; Avila and Rodrigo, 2004; Hou et al., 2005). The particles within the size range 0.1-1.0 μm are the result of the accumulation of smaller particles. The largest particles with diameters >1 μm are called coarse particles and are the result of mechanical abrasion (from soil, plants, sea salt, volcanoes, windblown material and abrasion) (Fergusson, 1992).

A particular type of household dust is attic dust. It represents dust deposited in the attics abandoned by inhabitants, so that tenant influence is minimized. The attic dust is derived predominantly from external sources, such as aerosols deposit and soil dusting, and less from household activities. While household dust is a material to which we are exposed daily, attic dust clearly shows the size and shape of the anomaly produced by atmospheric pollution (Šajn 2005, 2006). The attic dust as sampling material has the advantage that its composition remains constant, i.e., chemically unchanged, with time. Investigations of attic dust chemistry therefore reveal the average historical state of the atmosphere (Šajn, 1999, 2000, 2003). The use of undisturbed attic dust has the advantage of being a measurement, albeit indirect, of air pollution. An attic dust measurement provides an integrated measure based upon the above variables over time; it is, therefore, closer to the endpoint in the process continuum from sources to exposure and, ultimately, effects (Lioy, 1990). The use of attic dust was also successful in tracing plutonium aureole in Nevada, which was a result of atomic bomb experiments (Cizdziel et al., 1998; Cizdziel et al., 1999).

Several investigations of attic dust as an important indicator of pollution have also been studied in West Balkan, mostly around pollution sources such as ironworks, smelters, mines, ore deposits, and waste dump. Systematic studies were done all over Macedonia, especially around the places where mining and ore processing had significant consequences for the environment. Baseline data regarding trace metal levels have been established by comparing different sampling media: soil, attic dust, moss and lichen (Bačeva et al., 2011; Balabanova et al., 2010, 2011).

The use of undisturbed attic dust as a tool for reconstructing historical air pollution was evaluated in many geochemical studies in Slovenia. The first studies were mostly limited to individual industrial facilities, concerning health effects of particulate matter (Drev et al., 1986; 1992; Lukan and Roter, 1990; Lukan et al., 1990; Gspan in Hrašovec, 1993; Gabor, 1994). According to the Hydrometeorological Institute of Slovenia, measurements of suspended particulate matter were performed in Ljubljana, Maribor in Celje during 90s in several monitoring sites. However, the first systematic studies of soil, street sediment and attic dust were done for whole territory in 1999 (Sajn, 1999; 2000; 2003). Attic dust was also used for tracing the mercury halo in the Idrija area (Gosar at al., 2002; 2006) and pollution of heavy metals in Celje area (Šajn, 2005; Žibret, 2008; Žibret and Šajn, 2008b;), Mežica area (Šajn at al., 2000; Šajn, 2006) and Litija area (Jemec and Šajn, 2007).

2. Description of Study Area

2.1. Geographical Description of the Area

General climatological description is made according to the data from the book *Enciklopedija Slovenije* (engl. Encyclopedia of Slovenia, Zupančič, 1991) and according to the book *Klimatografiji Slovenije I in II* (Clymatography of Slovenia, HIRS, 1989a, 1989b). Slovenia, or, officially, The Republic of Slovenia, is a central European country (Figure 1) stretching on latitudes between 45 and 47°N and longitudes between 13 and 17°E. Slovenia borders Austria, Italy, Hungary and Croatia. It is a small country with an area of a little more than 20,000 km^2 and a population of 2 million. It is a member of the European Union and NATO. The lowest point is at the sea level in the southwestern part on the Adriatic coast, and the highest point is Mt. Triglav (2864 m) in the northwestern part of Slovenia in the Julian Alps. The climate varies depending on and according to the position relative to the sea and elevation. In the southwestern part near the Adriatic Sea there is a Mediterranean climate with mild winters and hot and dry summers. In the alpine area, in the northern and western parts, alpine climate dominates, and in the eastern part the continental climate prevails. These three dominant climatic systems interact with each other, and this is the reason for Slovenia's rich biodiversity. Pecipitation also shows a similar pattern, having an Alpine precipitation maximum in the mountains in western Slovenia with more than 3500 mm of rain annually, and dropping to 800 mm in the eastern part, which is situated in the Panonian basin. Ground inversions on calm and clear nights represent a common phenomenon within the inland regions during winter time, especially from November to February, persisting sometimes all day long or even for several days. A high relative humidity commonly accompanied by a fog appears to be a typical characteristic for the air below and inside the inversion layer. Strong temperature inversions also sometimes occur in the coastal region of the country (Primorje), when the valley of the river Po and the northern part of the Adriatic are covered with cold and moist air mass.

Population density in Slovenia is around 100 inhabitants per square kilometer. A characteristic of the population distribution is that population is spread across almost whole territory. One quarter of the population in Slovenia lives in the seven biggest towns. Ljubljana is the largest town with a little less than 300,000 inhabitants. Only the second largest town of Maribor also has more that 100,000 inhabitants; all other towns have less than 100,000 inhabitants. Population is dispersed in 6,029 settlements with an average of 337 inhabitants (SURS, 2009). The most densely populated areas are central Slovenia, the northeastern part, and the southwestern part (near the sea). Seven towns have more than 20,000 inhabitants: Ljubljana, Maribor, Celje, Kranj, Velenje, Koper and Novo Mesto. These seven towns have 24.8% of the total population.

Data from 2001 (Skumavec and Šabić, 2005) shows that 63.3% of the territory is forest and 30.5% is the agricultural land. Most of it is meadows, and only a small amount is arable land. Urbanized areas cover 2.8% of the territory. The GDP per capita is currently around $24,000 (depending on the exchange rate between euro and US$). Activities which contribute much to the GDP are manufacturing (16.8%) and wholesale and retail (11.0%). The mining sector contributes 0.4% and electricity, gas and other energy sector 2.3% (SURS, 2011).

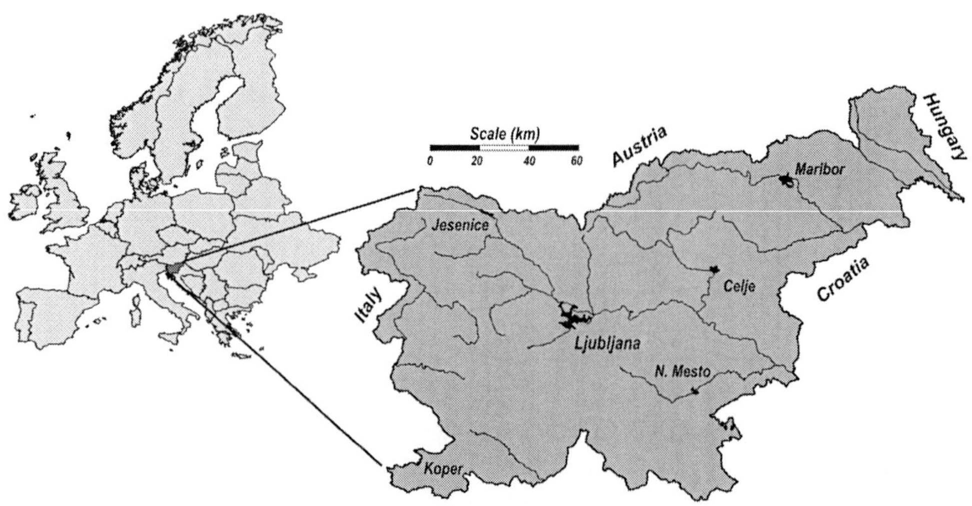

Figure 1. The position of Slovenia.

2.2. Geological Description of the Area

Lithological description of the area is made according to the data from *Enciklopedija Slovenije* (Encyclopedia of Slovenia, Buser et al. 1989), geological map of Slovenia in the scale 1:500,000 (Buser in Draksler, 1989) and according to the chapter in the book of *The Geology of Slovenia* (Vrabec et al., 2009).

Slovenia is placed on the junction between four major Alpine structural units: the Dinarides, the Southern Alps, the Eastern Alps and the Pannonian Basin (Vrabec et al., 2009). Each of these units has its own geological history. Paleogeographically speaking, the majority of present-day Slovenia belonged to the Adriatic continental microplate, which detached from the African plate at the beginning of Mesozoic and drifted north towards the European plate. The collision began in the beginning of Mesozoic and still exists today. The Alpine mountain chain and Dinaridic orogen belt are the consequences of this collision. This area is also seismically active in recent times. This is why the geological composition of the territory is very complex and every generalization is subjected to a certain degree of error (Figure 2).

Speaking from the sedimentological and petrological point of view, the oldest rocks in the territory are of the Silurian age. Paleozoic sedimentary rock formations are not abundant, and are mainly composed of different clastic rocks (claystones, sandstones, quartz sandstones, conglomerates, etc.), which are found in central Slovenia. Other clastic rocks, formed as the consequence of the Alpine orogenic processes, are younger and belong to the different basins. Flysch and flyshoid sediments of the Cretaceous, Paleocene and Eocene age are found here. Sedimentary rocks also dominate on the eastern part of Slovenia.

They are mainly of the Neogene age and are formed as a consequence of sedimentation in the post-orogenic Panonian basin between Oligocene and Pliocene. Inside these post-orogenic sedimentary rocks, grains which originate from felsic igneous rocks prevail and are mixed with the eolian post-glacial deposits. Clastical deposits are also found in the central part of Slovenia in the more densely populated basins as Quaternary alluvial deposits mainly of carbonatic composition.

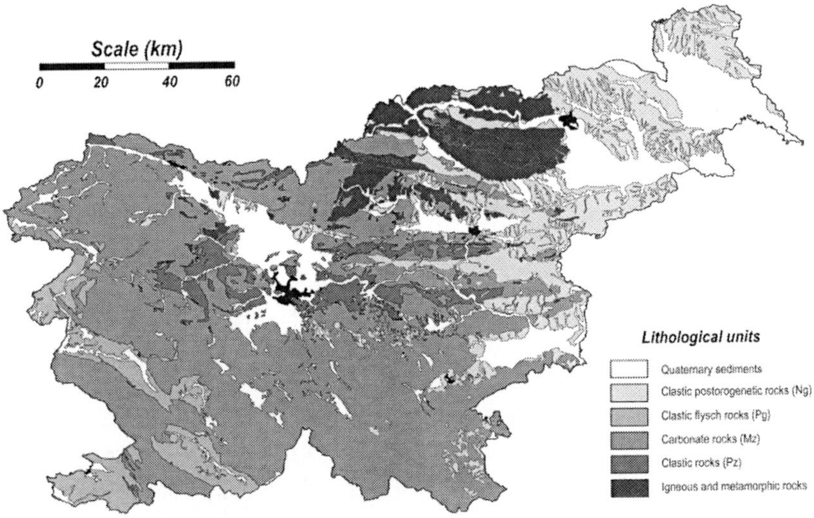

Figure 2. Simplified lithological map of Slovenia.

The majority of the territory of Slovenia is composed of carbonatic rocks, limestones and dolomites. These types of rocks dominate the central and western parts of Slovenia. Karstic landscape is developed here. The carbonatic sedimentation was mainly dominate in Mesozoic age (mainly Triassic), but carbonates of Paleozoic and Paleogene age can also be found.

Igneous and metamorphic rocks are found in the northern part of Slovenia. Granites, granodiorites, keratophires and porphirites are found among igneous rocks, and filites, gneisses, amphibolites, eclogites, marbles and quarcites are found among metamorphic rocks. Tuff deposits are the consequence of the Oligocene andesitic volcanic activity.

For the purpose of this contribution, the whole geology of Slovenia was divided into five geological units:

- Molasse deposits, which dominated the eastern part of Slovenia from Neogene until recently;
- Flysch deposits in basins of the western parts of Slovenia, mainly of Paleogene age;
- Carbonates, which dominate the central and western parts of Slovenia, mainly of the Mesozoic age;
- Clastic rocks in central Slovenia, mainly of the Paleozoic age
- Igneous and metamorphic rocks of northern Slovenia.

2.3. Pollution of the Study Area with Heavy Metals

Slovenia has long been known for its numerous mines and ore processing. From the times of the Roman Empire to present, 49 mines and open pits are opened; among them, four are large (Idrija, Mežica – Topla, Litija and Žirovski vrh – Figure 3). There are also 25 ore processing plants and smelters that are mostly operating in the vicinity of larger mines (the largest ones existed in Idrija, Žerjav and Celje).

There were 33 iron works operating in the vicinity of mines and open pits; three large ones have further developed and are still operating today (Jesenice, Ravne na Koroškem and Štore; Figure 3.; Budkovič at al., 2003).

In the last decade, thc focus of the geochemical research was put on the polluted areas. In Slovenia, several polluted areas exists (Žibret and Šajn, 2008a). The main source of metals was ironworks, metal mines and smelters. Among them, we have to mention: Mežica area (Šajn at al., 2000; Vreča at al., 2001; Šajn, 2006), Idrija area (Gosar at al., 2002, 2006; Žibret and Gosar, 2006; Gosar and Žibret, 2011;), Celje area (Šajn, 2005; Žibret, 2008; Žibret and Šajn, 2008b; Žibret and Rokavec, 2010;), Jesenice area (Šajn at al., 1999), Litija area (Jemec and Šajn, 2007; Šajn and Gosar, 2007), Ljubljana area (Šajn at al., 2011) and Drava watershed area (Peh at al., 2008; Šajn at al. 2011).

In the area of Celje, a town with about 40,000 inhabitants, very high contents of Cd, Cu, Pb and Zn are found. Their source was the smelting of zinc ore between 1873 and 1970. Concentrations of toxic metals in topsoil exceed the official limit of critical concentration on 18 km^2. For example, the average content of Cd in the downtown area (7.5 mg/kg) is 15 times above the Slovenian average. In the area of Jesenice, which has about 15,000 inhabitants, the impact of centuries of long-lasting ironworks activities in a narrow alpine valley has been investigated.

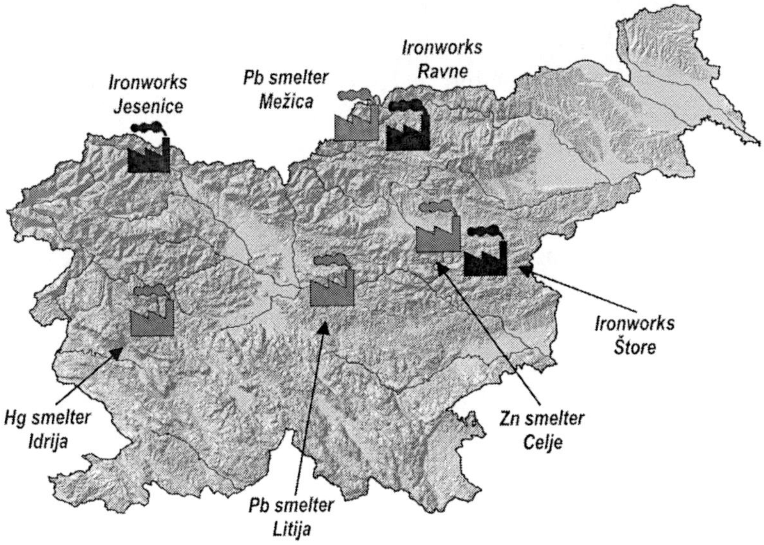

Figure 3. Abandoned and active metal smelters and ironworks in Slovenia.

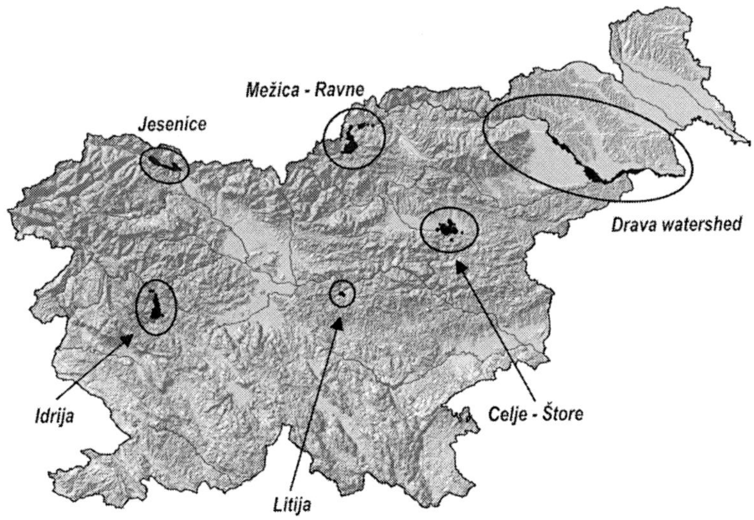

Figure 4. Critically polluted areas in Slovenia, as a consequence of past mining and smelting activities.

The concentrations of metals in topsoil exceed the official limit of critical concentration on 13 km^2. In the Mežica valley, 300 years of lead and zinc ore mining and smelting left a very negative impact on the environment.

The area is strongly polluted with Ag, As, Cd, Cu, Hg, Mo, Pb, S, Sb, Sn and Zn. The concentrations of toxic metals in topsoil exceed the official limit of critical concentration on 24 km^2 of the research area. In Idrija and its close surroundings, the influences of a half millennium of mining and smelting dominate over the rich natural enrichment and dispersion of mercury. Mercury concentrations in topsoil exceed the critical values for topsoil on 21 km^2.

The Drava watershed, with confluents, is an important area of mining and smelting activities, which began in the Antic period, developed in Middle Ages and reached its peak in the middle of the last century. Numerous mines and smelters, Bleiberg-Kreuth, Cave del Predil and Mežica, have left great consequences on the chemical composition of Drava alluvial sediments. On the territory of recently flooded lowland, averages of Cd, Pb and Zn exceed the Slovenian average by approximately 10-26 times with regard to particular elements and location of sampling. In the whole studied area, 85 km^2 is critically polluted with metals, especially with Zn, according to the Slovenian and Croatian legislations.

On the basis of the results of our investigations in the vicinity of larger mines and smelters, we estimate that the critical limit for toxic metals in topsoil is exceeding on 160 km^2 (Figure 4).

3. Materials and Methods

The sampling strategy was made in order to determine the dominant influential factors on the composition of household dust and attic dust. These factors are:

- soil composition;
- population;
- mining and smelting activities;
- regional trends;
- lithology,

3.1. Sampling

Samples were collected in natural areas, covering the whole territory of Slovenia (countryside) and larger towns. The territory of Slovenia was covered with 60 sampling points. In Slovenian towns, 23 sampling points were selected: 7 samples in Ljubljana (290.000 inhabitants), 4 samples in Maribor (95.000 inhabitants), 4 samples in Celje (40.000 inhabitants), 3 samples in Jesenice (15.000 inhabitants), 3 samples in Novo Mesto (35.000 inhabitants) and 2 samples in Koper (24.000 inhabitants). Figure 5 shows the positions of the sampling points. Three samples were taken from each sampling point: topsoil (0-5 cm), attic dust and household dust.

Soil sampling was conducted on natural areas. A horizon was sampled on the areas of automorphic (natural) soil from a maximum depth of 5 cm. Possible organic horizons (Oh) were excluded from sampling. Anthropogenic soils, like soils on fields, constructions or other similar areas, were not sampled. Moreover, sampling of forest soil was also avoided in this research. Most suitable soil samples were at the locations where there were visible washing/accumulation horizons (A-B horizons), which were found most often on meadows. A topsoil sample (0-5 cm) was represented by 15 subsamples in the area of 50 m radius. Approximately 2 kg of soil was taken in situ. Subsamples were mixed and roots and grass removed. 1 kg of soil was sealed in the plastic bag for further processing. Other material was discarded. Topsoil sampling was conducted according to the European guidelines for soil pollution studies (Theocharopoulos, et al., 2001; Salminen et al., 2005).

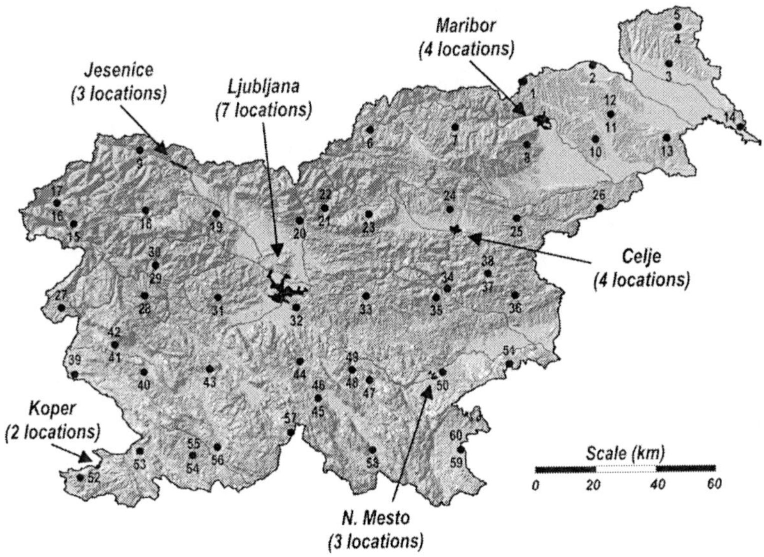

Figure 5. The position of sampling points in the countryside (N=60) and in towns (N=23).

Attic dust was sampled in the attics of old houses. The condition regarding the age of the house was that the house needed to be built before the industrial revolution in Slovenia; this is before the year 1900. The age of the house was determined according to conversations with the owners of the house and according to the architectural elements presented. Two conditions also needed to be met: attics were not used for storing hay or crops and attics were not used for dwelling purposes. After obtaining the permission of the owner of the property, dust was collected from the wooden roof-carrying construction with hard plastic brushes prior the removal of tiles, sand, remains of vegetation, or other debris. Only electrostatically connected dust to the wood was collected from several places in the attics. In total, approximately 100 g of attic dust was collected and sealed in the plastic bag for further processing (Šajn, 1999, 2000).

Household dust: The sample of the household dust was represented by at least 3 full vacuum cleaner bags from the houses at the sampling point. Bags where the vacuum cleaner was used for cleaning the car, garage, workshop or any other possible areas that might contain the particles of non-local origin were not collected. Vacuum cleaner bags were sealed in the plastic bag for further processing (Šajn, 1999; 2000).

3.2. Laboratory Preparation

The purpose of the samples' pre-processing was to dry the samples and to extract only fine-grained material for chemical analysis. Topsoil samples were air-dried at 40°C until a constant weight was achieved. After that the soil was sieved through a 2 mm sieve. Stones and other coarse grained material were discarded. Other material was gently crushed in a ceramic mortar. Special attention was paid so that only soil aggregations were crushed, while sand particles remained intact. The material was then sieved under 0.125 mm sieve and 5 g of the material <0.125 mm represented the substance for chemical analysis (Salminen et al., 2005).

Attic dust was prepared in a similar manner. Dust was dried; possible dust aggregations were gently crushed. The material was sieved under 0.125 mm sieve and 5g of the material <0.125 mm represented the material for chemical analyses.

Household dust was prepared in a different manner. The purpose of the preparation was to extract as much atmospheric deposit as possible and to avoid analyzing fibers or other coarse particles. The material inside the vacuum cleaner bag was rubbed on the 2 mm sieve. Hair, textile and other material was then discarded. Material which fell under the 2 mm sieve was then dried at 40°C for one day. After that, the material was sieved several times, first under 0.5 mm sieve, then under 0.25 mm sieve and finally under 0.125 mm sieve. During each process, sieving was performed 3 or 4 times. Sieving was done with several strong hand strokes, without the shaking of longer duration machine sieving. In this way we were able to discard the majority of fibers, which would otherwise have fallen under the sieve. Finally, 5 g of the material <0.125 mm represented the material for chemical analysis (Šajn, 1999, 2000).

3.3. Chemical Analysis

All samples were analyzed at the ACME, Ltd. laboratory in Vancouver, Canada. Determination of 41 chemical elements (Al, Ca, Fe, K, Mg, Na, P, S, Ti, Ag, As, Au, Ba, Be, Bi, Cd, Ce Co, Cr, Cu, Hf, La, Li, Mn, Mo, Nb, Ni, Pb, Rb, Sb, Sc, Sn, Sr, Ta, Th, U, V, W, Y, Zn and Zr) was performed by inductively coupled plasma mass spectrometry (ICP-MS) following a four-acid digestion (mixture of $HClO_4$, HNO_3, HCl and HF at 200°C), which is considered to be a total digestion method. Mercury was determined by cold vapor atomic absorption spectrometry AAS-CV after aqua regia digestion (mixture of HCl, HNO_3 and water at 95°C). All samples, replicates and geological standards were submitted to the laboratory in random order. This procedure assured unbiased treatment of samples and random distribution of possible drift of analytical conditions for all samples. The sensitivity of the analysis, in the sense of the lower limit of detection, was adequate for 41 out of 42 determined elements. Only Au was below the lower detection limit. The accuracy of the analytical method for the remaining 41 elements was estimated by calculation of the relative systematic error between the determined and recommended values of the geological standard materials: GXR-2, GXR-5 and GXR-6 (Govindaraju, 1989; http://georem.mpch-mainz.gwdg.de). Most of the elements showed very low deviations from the recommended range of the values, i.e., the mean of all determined elements in the standards generally differed by less than 30% of the recommended values. Large negative deviations were observed for P and Nb. Precision is a measure of repeatability in the determination of a parameter in the same sample, regardless of deviation from the true value (Rose et al., 1979). The precision for each element was tested by considering relative differences between pairs of analytical determinations on the same sample and by drawing Thompson-Howarth plots (Analytical Methods Committee, 2002). Overall precision for all elements was better than 15

% at the 95% confidence level. The reliability of the analytical procedures and results produced for 41 elements was considered satisfactory for further statistical analysis.

3.4. Data Processing

Spatial data acquired by digitization of existing topographic databases and geological maps were used, with additional reambulation of the data. All field observations, analytical data, and measurements were entered into a data matrix. Au was removed from the final database used in the statistical analysis.

Data of 249 samples of topsoil, attic dust and household dust (the countryside and larger towns) were used in the statistical analysis. Parametric and nonparametric statistical methods were used, and normality tests of data distributions were performed (Snedecor and Cochran, 1967; Hollander and Wolf, 1999). Multivariate cluster and R-mode factor analyses (FA) were used to reveal associations of chemical elements (Davis, 1986; Facchinelli et al., 2001; Micó et al., 2006; Zhang, 2006; Luo et al., 2007; Templ et al., 2008). Due to the relatively small number of sampling points, all available samples of topsoil, attic dust and household dust were used for performing the multivariate statistic analysis. In this way we assured enough cases for making multivariate statistics with large number of elements.

The universal kriging interpolation method with the uniform linear variogram model (Lado et al., 2008; Beelen et al., 2009) was applied for the construction of maps showing the spatial distribution of factor scores, as well as maps displaying the distribution of metals in sampling materials. The basic grid cell size for the interpolation was 1 x 1 km. The percentile values of distribution of interpolated values were chosen for class limits. Seven classes at the following percentile values were selected: 0-10, 10-25, 25-50, 50-75, 75-90 and 75-100.

The data analysis and production of maps were performed on a PC using the Statsoft Statistica software package and Golden software (Surfer, MapViewer).

4. Results

4.1. Basic Statistics and Normality of Distribution

Basic statistical parameters and the histograms are showing the elemental distribution, which can be a normal (Gaussian) distribution or log-normal distribution. Normal distribution is regarded as an indication of uniform natural environment, while a log-normal distribution might indicate of a highly variable environment, especially due to anthropogenic impact.

Distribution, which can not be regarded as a normal (Gaussian) distribution, even after a logarithmical transformation, is usually produced by diverse and very strong anthropogenic influences. The examination of the distribution histograms and monovariate statistical parameters shows that most natural elements have a log-normal distribution. Normal distribution is observed with naturally distributed chemicals such as Al, K, La, Mg, Nb, Sc, Sn, Th and V in topsoil. In attic dust, only Ca shows a natural distribution, while in the household dust this characteristic can be found for Ca and Mg.

Other elements in topsoil, attic dust and household dust have a log-normal distribution.

Figures 6, 7, and 8 compare the concentrations of elements in topsoil, attic dust and household dust, respectively. Tables 1, 2, and 3 show the basic statistics for topsoil, attic dust, and household dust for the countryside (non-polluted areas, 60 samples), while tables 4, 5, and 6 show the same values for samples collected in towns (23 samples).

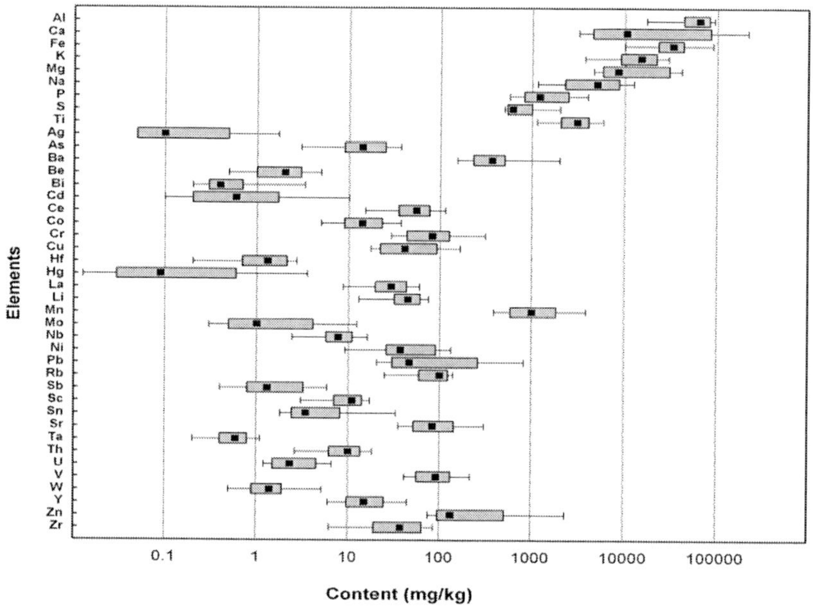

Figure 6. Distribution of chemical elements in topsoil (N=83). Whiskers - range (minimum and maximum value); Box - 10th and 90th percentile values; Square – median value.

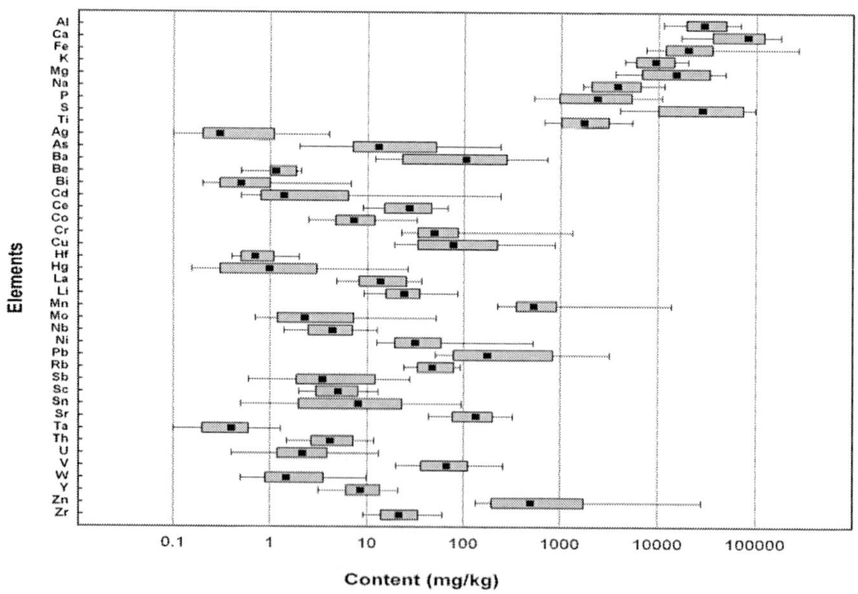

Figure 7. Distribution of chemical elements in attic dust (N=83). Whiskers - range (minimum and maximum value); Box - 10th and 90th percentile values; Square – median value.

Table 1. Descriptive Statistics of Elemental Concentrations in Topsoil – Countryside (N=60)

	unit	X	Xg	Md	S	Sx	CV	Min	Max	P10	P90
Al	%	6.8	6.7	6.9	1.2	0.15	17	3.1	9.3	5.4	8.3
Ca	%	1.3	0.89	0.68	1.5	0.20	120	0.32	9.5	0.44	3.2
Fe	%	3.5	3.5	3.5	0.61	0.079	17	2.2	4.8	2.8	4.3
K	%	1.7	1.6	1.6	0.45	0.058	27	0.84	3.0	1.2	2.2
Mg	%	0.89	0.81	0.73	0.50	0.065	56	0.46	3.1	0.55	1.4
Na	%	0.52	0.46	0.52	0.24	0.031	46	0.11	1.2	0.20	0.85
P	%	0.12	0.11	0.11	0.041	0.0053	35	0.057	0.26	0.075	0.17
S	%	0.067	0.062	0.050	0.029	0.0037	43	0.050	0.20	0.050	0.10
Ti	%	0.32	0.32	0.33	0.080	0.010	25	0.15	0.59	0.23	0.41
Ag	mg/kg	0.074	0.070	0.050	0.025	0.0033	34	0.050	0.10	0.050	0.10
As	mg/kg	16	15	15	6.6	0.85	41	6.0	37	9.0	24
Ba	mg/kg	350	330	350	99	13	28	150	690	220	470
Be	mg/kg	2.3	2.0	2.0	1.0	0.13	45	0.50	5.0	1.0	3.5
Bi	mg/kg	0.42	0.40	0.40	0.12	0.015	28	0.20	0.70	0.30	0.60
Cd	mg/kg	0.64	0.50	0.45	0.47	0.061	75	0.10	2.0	0.20	1.3
Ce	mg/kg	60	58	60	18	2.4	30	24	110	37	79
Co	mg/kg	17	16	16	6.0	0.77	35	5.0	37	11	24
Cr	mg/kg	90	86	91	24	3.1	27	29	140	62	120
Cu	mg/kg	41	35	31	29	3.7	69	17	170	20	71
Hf	mg/kg	1.4	1.2	1.4	0.55	0.072	41	0.20	2.7	0.70	2.2
Hg	mg/kg	0.081	0.063	0.065	0.057	0.0074	71	10	260	25	160
La	mg/kg	33	31	31	10	1.4	32	12	60	21	47
Li	mg/kg	50	49	50	10	1.3	21	31	75	36	63
Mn	mg/kg	1200	1100	1000	470	60	40	380	2200	670	1900
Mo	mg/kg	1.5	0.99	0.80	2.2	0.28	140	0.30	12	0.50	3.7
Nb	mg/kg	8.8	8.4	8.7	2.4	0.31	27	3.5	16	6.1	12
Ni	mg/kg	55	48	50	28	3.6	51	9.2	130	27	93
Pb	mg/kg	41	40	41	12	1.6	29	20	87	28	54
Rb	mg/kg	110	100	110	16	2.1	15	64	140	87	120
Sb	mg/kg	1.2	1.1	1.1	0.53	0.068	44	0.40	3.7	0.70	1.5
Sc	mg/kg	12	12	12	2.0	0.26	17	7.0	17	9.0	15
Sn	mg/kg	3.7	3.3	3.1	3.9	0.50	110	1.8	33	2.3	4.5
Sr	mg/kg	81	77	77	30	3.8	37	35	170	50	130
Ta	mg/kg	0.60	0.58	0.60	0.18	0.023	29	0.20	1.1	0.40	0.85
Th	mg/kg	11	10	10	2.7	0.35	25	4.7	18	6.8	14
U	mg/kg	2.5	2.3	2.1	1.1	0.14	42	1.2	6.1	1.5	4.1
V	mg/kg	110	100	100	31	4.0	29	41	220	74	130
W	mg/kg	1.4	1.4	1.4	0.41	0.054	29	0.80	3.0	0.90	1.9
Y	mg/kg	18	16	17	7.3	0.94	41	7.0	44	9.1	27
Zn	mg/kg	120	120	120	29	3.8	23	75	220	89	170
Zr	mg/kg	40	36	39	18	2.3	44	6.2	86	19	66

X – mean; Xg – geometrical mean; Md – median; S – standard deviation; Sx – standard error of mean; CV – coefficient of variance; Min – minimum; Max – maximum; P10 – 10th percentile; P90 – 90th percentile; Data rounded to two digits.

Table 2. Descriptive Statistics of Elemental Concentrations in Attic Dust – Countryside (N=60)

	unit	X	Xg	Md	S	Sx	CV	Min	Max	P10	P90
Al	%	3.3	3.0	3.0	1.3	0.17	40	1.1	6.9	1.7	4.9
Ca	%	7.5	6.6	7.6	3.5	0.45	47	1.7	18	3.1	12
Fe	%	2.0	1.9	1.9	0.78	0.10	39	0.75	4.3	1.1	2.9
K	%	1.1	1.1	1.1	0.32	0.041	29	0.45	2.0	0.73	1.5
Mg	%	1.7	1.4	1.3	1.1	0.15	67	0.36	4.8	0.60	3.5
Na	%	0.42	0.38	0.38	0.21	0.026	49	0.17	1.1	0.21	0.65
P	%	0.33	0.30	0.27	0.18	0.023	54	0.13	1.1	0.17	0.58
S	%	2.7	2.1	2.5	1.6	0.21	62	0.40	7.4	0.75	4.9
Ti	%	0.19	0.18	0.18	0.092	0.012	47	0.068	0.54	0.092	0.30
Ag	mg/kg	0.41	0.30	0.30	0.52	0.067	130	0.10	3.8	0.10	0.65
As	mg/kg	18	12	11	31	4.0	170	2.0	240	5.5	29
Ba	mg/kg	150	110	120	120	16	81	20	730	31	310
Be	mg/kg	1.1	1.0	1.0	0.39	0.051	36	0.50	2.0	0.75	2.0
Bi	mg/kg	0.49	0.44	0.40	0.26	0.033	52	0.20	1.3	0.20	0.80
Cd	mg/kg	2.0	1.4	1.3	2.2	0.29	120	0.50	13	0.80	3.5
Ce	mg/kg	32	30	31	13	1.7	41	9.0	68	15	49
Co	mg/kg	7.0	6.6	6.7	2.5	0.32	35	2.5	16	3.6	9.9
Cr	mg/kg	50	47	47	17	2.2	35	23	100	31	72
Cu	mg/kg	84	66	60	72	9.3	86	19	370	33	180
Hf	mg/kg	0.72	0.68	0.70	0.23	0.030	32	0.40	1.3	0.50	1.1
Hg	mg/kg	2.3	1.1	1.0	4.7	0.60	200	160	26000	310	3800
La	mg/kg	17	15	16	6.9	0.89	41	5.5	36	8.2	26
Li	mg/kg	24	23	24	7.3	0.94	31	9.2	42	14	33
Mn	mg/kg	580	540	540	220	28	38	260	1300	370	900
Mo	mg/kg	2.2	2.0	1.9	1.1	0.15	52	0.70	7.4	1.1	3.7
Nb	mg/kg	4.6	4.2	4.4	2.0	0.26	44	1.4	13	2.4	7.0
Ni	mg/kg	29	27	27	10	1.3	35	13	67	18	41
Pb	mg/kg	190	140	120	250	32	130	51	1800	69	290
Rb	mg/kg	56	54	53	17	2.2	31	24	92	36	84
Sb	mg/kg	3.3	2.8	3.0	2.2	0.29	67	0.60	17	1.2	5.5
Sc	mg/kg	5.5	5.1	5.0	2.1	0.27	39	2.0	13	3.0	8.0
Sn	mg/kg	13	8.4	8.6	15	1.9	120	0.50	95	2.9	25
Sr	mg/kg	130	120	130	50	6.4	39	43	320	76	180
Ta	mg/kg	0.42	0.38	0.40	0.19	0.025	46	0.10	1.3	0.20	0.60
Th	mg/kg	5.1	4.7	5.0	2.0	0.26	40	1.5	12	2.5	7.5
U	mg/kg	1.9	1.7	1.9	0.69	0.090	37	0.40	3.5	1.1	2.8
V	mg/kg	67	59	64	38	4.9	57	20	260	31	100
W	mg/kg	1.5	1.3	1.4	0.73	0.094	50	0.50	4.7	0.75	2.4
Y	mg/kg	9.1	8.6	8.7	3.2	0.41	35	3.2	21	5.8	13
Zn	mg/kg	450	380	320	280	36	62	130	1200	190	900
Zr	mg/kg	21	20	21	7.4	0.96	35	9.1	42	12	32

X – mean; Xg – geometrical mean; Md – median; S – standard deviation; Sx – standard error of mean; CV – coefficient of variance; Min – minimum; Max – maximum; P10 – 10th percentile; P90 – 90th percentile; Data rounded to two digits.

Table 3. Descriptive Statistics of Elemental Concentrations in Household Dust - Countryside (N=60)

	unit	X	Xg	Md	S	Sx	CV	Min	Max	P10	P90
Al	%	1.9	1.8	1.7	0.84	0.12	44	0.65	4.5	1.1	3.4
Ca	%	7.2	6.5	7.0	2.9	0.40	40	1.4	16	3.9	11
Fe	%	1.4	1.3	1.3	0.53	0.074	37	0.59	3.0	0.90	1.9
K	%	0.81	0.78	0.74	0.22	0.031	27	0.44	1.3	0.56	1.2
Mg	%	1.6	1.4	1.7	0.93	0.13	57	0.42	4.5	0.52	2.6
Na	%	0.89	0.75	0.75	1.1	0.15	120	0.41	8.4	0.47	0.98
P	%	0.17	0.15	0.14	0.16	0.022	94	0.084	1.2	0.11	0.19
S	%	-	-	-	-	-	-	-	-	-	-
Ti	%	0.13	0.12	0.12	0.055	0.0077	42	0.060	0.30	0.080	0.20
Ag	mg/kg	1.9	1.2	1.1	2.3	0.32	120	0.25	14	0.25	4.6
As	mg/kg	3.3	3.0	2.5	1.8	0.25	54	2.5	10	2.5	5.0
Ba	mg/kg	410	370	350	210	30	52	140	1200	230	730
Be	mg/kg	-	-	-	-	-	-	-	-	-	-
Bi	mg/kg	-	-	-	-	-	-	-	-	-	-
Cd	mg/kg	3.1	2.6	2.3	2.4	0.33	76	1.1	13	1.5	5.8
Ce	mg/kg										
Co	mg/kg	6.8	6.3	6.0	3.4	0.47	50	3.0	20	4.0	10
Cr	mg/kg	100	90	83	55	7.7	55	45	340	57	150
Cu	mg/kg	180	140	140	160	23	93	59	1100	70	350
Hf	mg/kg	-	-	-	-	-	-	-	-	-	-
Hg	mg/kg	2.1	0.80	0.67	5.1	0.71	240	0.14	27	0.23	3.3
La	mg/kg	20	17	18	15	2.1	75	5.0	97	8.0	29
Li	mg/kg	-	-	-	-	-	-	-	-	-	-
Mn	mg/kg	410	390	360	140	20	34	190	780	260	590
Mo	mg/kg	4.3	3.1	3.0	4.5	0.63	110	1.0	28	1.0	7.0
Nb	mg/kg	4.2	3.9	4.0	2.2	0.30	51	2.0	17	3.0	6.0
Ni	mg/kg	47	43	42	21	3.0	45	19	140	25	75
Pb	mg/kg	160	140	130	110	15	66	55	520	71	300
Rb	mg/kg	-	-	-	-	-	-	-	-	-	-
Sb	mg/kg	8.6	6.6	7.0	8.1	1.1	94	2.5	47	2.5	14
Sc	mg/kg	2.6	2.4	2.0	1.3	0.18	48	1.0	7.0	2.0	4.0
Sn	mg/kg	37	25	21	52	7.3	140	7.0	260	11	51
Sr	mg/kg	110	100	100	45	6.4	41	58	350	75	150
Ta	mg/kg	-	-	-	-	-	-	-	-	-	-
Th	mg/kg	2.8	2.5	2.0	1.4	0.20	51	1.0	6.0	1.0	5.0
U	mg/kg	-	-	-	-	-	-	-	-	-	-
V	mg/kg	29	27	27	11	1.6	39	13	70	18	43
W	mg/kg	-	-	-	-	-	-	-	-	-	-
Y	mg/kg	5.8	5.5	5.0	1.7	0.24	30	3.0	10	4.0	8.0
Zn	mg/kg	660	630	660	210	29	32	340	1200	410	950
Zr	mg/kg	19	17	17	14	2.0	75	7.0	111	11	24

X – mean; Xg – geometrical mean; Md – median; S – standard deviation; Sx – standard error of mean; CV – coefficient of variance; Min – minimum; Max – maximum; P10 – 10th percentile; P90 – 90th percentile; Data rounded to two digits.

Table 4. Descriptive Statistics Elemental Concentrations in Topsoil - Slovenian Towns (N=23)

	unit	Xz	Xg	Md	S	Sx	CV	Min	Max	P10	P90
Al	%	4.7	4.6	5.0	1.2	0.24	25	1.7	6.7	3.4	5.8
Ca	%	7.7	6.7	6.3	4.5	0.93	58	3.0	22	3.5	13
Fe	%	2.9	2.7	2.7	1.5	0.32	51	0.99	8.9	2.0	4.0
K	%	1.0	0.97	0.94	0.30	0.062	29	0.37	1.7	0.77	1.4
Mg	%	2.4	2.2	2.4	1.0	0.21	42	0.53	4.1	1.4	3.7
Na	%	0.52	0.46	0.49	0.26	0.055	51	0.20	1.1	0.23	0.97
P	%	0.20	0.19	0.18	0.078	0.016	39	0.10	0.40	0.11	0.31
S	%	0.11	0.10	0.10	0.041	0.0085	38	0.050	0.20	0.050	0.20
Ti	%	0.27	0.26	0.27	0.094	0.020	35	0.11	0.47	0.17	0.43
Ag	mg/kg	0.56	0.45	0.40	0.40	0.084	73	0.20	1.7	0.20	1.0
As	mg/kg	15	13	12	8.3	1.7	54	3.0	35	8.0	28
Ba	mg/kg	530	460	440	380	79	72	180	2000	330	750
Be	mg/kg	1.7	1.6	2.0	0.69	0.14	40	1.0	3.0	1.0	3.0
Bi	mg/kg	0.62	0.47	0.40	0.67	0.14	110	0.20	3.3	0.30	1.3
Cd	mg/kg	2.0	1.4	1.2	2.2	0.46	110	0.30	10	0.60	3.9
Ce	mg/kg	45	43	48	13	2.6	28	15	74	30	58
Co	mg/kg	11	10	10	2.9	0.60	26	6.4	19	8.4	14
Cr	mg/kg	83	68	60	66	14	80	37	310	41	160
Cu	mg/kg	77	74	74	25	5.2	32	35	140	47	110
Hf	mg/kg	1.2	1.1	1.2	0.48	0.099	41	0.40	2.4	0.70	1.8
Hg	mg/kg	0.72	0.55	0.48	0.71	0.15	98	240	3500	260	1300
La	mg/kg	24	23	24	7.2	1.5	30	8.7	40	17	30
Li	mg/kg	34	33	35	9.4	2.0	27	13	51	23	48
Mn	mg/kg	980	840	720	720	150	74	380	3800	520	1300
Mo	mg/kg	2.4	1.9	1.8	2.0	0.42	84	0.70	9.6	1.2	4.2
Nb	mg/kg	6.6	6.3	6.5	1.9	0.40	29	2.4	11	4.0	9.3
Ni	mg/kg	38	34	30	23	4.7	60	20	110	24	58
Pb	mg/kg	280	230	220	200	41	71	66	820	92	640
Rb	mg/kg	66	64	66	18	3.8	27	25	110	52	86
Sb	mg/kg	2.8	2.6	2.4	1.2	0.25	42	1.2	5.8	1.6	4.4
Sc	mg/kg	8.2	7.9	9.0	2.2	0.47	27	3.0	12	6.0	11
Sn	mg/kg	7.3	6.8	6.9	2.8	0.59	39	3.2	14	3.7	11
Sr	mg/kg	130	120	130	48	9.9	37	71	300	89	170
Ta	mg/kg	0.59	0.56	0.60	0.19	0.040	32	0.20	1.0	0.30	0.80
Th	mg/kg	7.6	7.3	7.8	2.0	0.41	26	2.6	11	5.3	10
U	mg/kg	3.1	2.9	2.6	1.3	0.26	41	1.4	6.6	1.9	4.8
V	mg/kg	74	72	80	19	4.0	26	44	120	50	95
W	mg/kg	1.8	1.6	1.5	1.2	0.24	65	0.50	5.1	1.0	3.5
Y	mg/kg	14	13	13	4.6	0.97	33	6.0	24	10	23
Zn	mg/kg	600	460	390	560	120	92	160	2300	230	1400
Zr	mg/kg	37	34	34	16	3.4	43	16	80	20	59

X – mean; Xg – geometrical mean; Md – median; S – standard deviation; Sx – standard error of mean; CV – coefficient of variance; Min – minimum; Max – maximum; P10 – 10th percentile; P90 – 90th percentile; Data rounded to two digits.

Table 5. Descriptive Statistics of Elemental Concentrations in Attic Dust - Slovenian Towns (N=23)

	unit	X	Xg	Md	S	Sx	CV	Min	Max	P10	P90
Al	%	2.8	2.7	2.7	0.76	0.16	27	1.9	4.7	2.0	4.0
Ca	%	9.8	9.4	9.5	3.0	0.63	31	4.8	17	6.4	12
Fe	%	4.1	2.8	2.4	5.7	1.2	140	1.4	27	1.5	5.4
K	%	0.73	0.69	0.66	0.29	0.060	40	0.46	1.8	0.47	0.93
Mg	%	2.1	1.9	2.1	0.82	0.17	39	0.88	3.8	1.2	3.2
Na	%	0.44	0.38	0.34	0.25	0.052	57	0.18	1.1	0.19	0.79
P	%	0.13	0.11	0.10	0.089	0.019	70	0.053	0.49	0.065	0.20
S	%	5.8	5.1	6.8	2.5	0.53	43	1.1	9.7	2.0	8.6
Ti	%	0.19	0.17	0.15	0.081	0.017	43	0.10	0.35	0.12	0.33
Ag	mg/kg	1.1	0.79	0.80	1.0	0.21	92	0.30	4.0	0.30	2.3
As	mg/kg	38	29	24	36	7.4	94	11	170	13	70
Ba	mg/kg	83	47	42	120	25	150	12	570	15	170
Be	mg/kg	1.0	1.0	1.0	0.21	0.043	20	1.0	2.0	1.0	1.0
Bi	mg/kg	1.2	0.79	0.60	1.5	0.31	130	0.30	6.7	0.40	2.2
Cd	mg/kg	20	4.2	3.1	53	11	260	0.50	240	1.2	73
Ce	mg/kg	24	23	23	8.1	1.7	34	12	48	16	34
Co	mg/kg	12	10	9.1	6.9	1.4	60	4.7	32	6.3	23
Cr	mg/kg	120	70	50	260	55	200	36	1300	40	140
Cu	mg/kg	170	130	110	170	35	99	41	870	72	260
Hf	mg/kg	0.86	0.81	0.80	0.32	0.066	37	0.50	2.0	0.60	1.1
Hg	mg/kg	1.0	0.74	0.73	1.1	0.24	110	210	5700	280	1700
La	mg/kg	12	12	12	4.3	0.90	36	4.8	24	8.0	18
Li	mg/kg	30	27	26	15	3.0	49	18	86	20	41
Mn	mg/kg	1300	600	480	2900	600	200	230	14000	340	1500
Mo	mg/kg	7.9	5.9	5.7	9.9	2.1	120	1.6	51	3.2	11
Nb	mg/kg	4.9	4.5	4.1	2.2	0.46	46	2.6	12	2.9	7.1
Ni	mg/kg	75	56	49	100	21	130	25	520	31	97
Pb	mg/kg	820	600	500	760	160	93	190	3200	250	2000
Rb	mg/kg	40	39	38	8.0	1.7	20	30	61	31	51
Sb	mg/kg	10	8.8	9.0	5.5	1.2	55	3.3	28	4.2	16
Sc	mg/kg	5.2	5.0	5.0	1.7	0.35	32	3.0	9.0	4.0	8.0
Sn	mg/kg	12	6.0	5.7	18	3.7	160	0.80	87	1.5	22
Sr	mg/kg	160	160	150	53	11	32	100	310	120	240
Ta	mg/kg	0.40	0.38	0.40	0.12	0.025	30	0.20	0.60	0.30	0.60
Th	mg/kg	4.0	3.8	3.7	1.3	0.27	33	1.8	6.8	2.7	6.5
U	mg/kg	4.3	3.9	3.8	2.3	0.49	54	2.0	13	2.5	6.3
V	mg/kg	91	86	92	33	6.9	37	39	190	48	120
W	mg/kg	3.2	2.7	3.0	2.1	0.43	64	1.0	9.8	1.4	5.8
Y	mg/kg	9.7	9.2	8.3	3.1	0.66	33	5.6	17	6.8	15
Zn	mg/kg	3500	1600	1300	6000	1300	170	250	28000	450	8800
Zr	mg/kg	27	26	24	9.5	2.0	35	19	60	20	38

X – mean; Xg – geometrical mean; Md – median; S – standard deviation; Sx – standard error of mean; CV – coefficient of variance; Min – minimum; Max – maximum; P10 – 10th percentile; P90 – 90th percentile; Data rounded to two digits.

Table 6. Descriptive Statistics of Elemental Concentrations in Household Dust - Slovenian Towns (N=23)

	unit	X	Xg	Md	S	Sx	CV	Min	Max	P10	P90
Al	%	1.8	1.7	1.6	0.61	0.13	34	0.92	3.4	1.3	2.8
Ca	%	8.1	7.6	8.1	2.8	0.59	35	3.7	14	4.9	12
Fe	%	1.9	1.7	1.5	1.1	0.22	57	0.79	4.7	0.84	3.7
K	%	0.70	0.69	0.69	0.13	0.027	18	0.48	0.97	0.56	0.86
Mg	%	2.0	1.8	1.9	0.87	0.18	43	0.56	3.6	0.83	3.1
Na	%	0.86	0.84	0.79	0.20	0.041	23	0.52	1.2	0.65	1.1
P	%	0.13	0.13	0.13	0.033	0.0068	25	0.085	0.22	0.090	0.17
S	%	-	-	-	-	-	-	-	-	-	-
Ti	%	0.15	0.15	0.14	0.052	0.011	34	0.080	0.26	0.10	0.21
Ag	mg/kg	2.7	2.1	1.8	2.0	0.41	73	0.60	8.1	0.60	4.8
As	mg/kg	6.7	6.1	6.0	3.1	0.64	46	2.5	15	2.5	10
Ba	mg/kg	770	710	730	330	68	43	320	1700	410	1200
Be	mg/kg	-	-	-	-	-	-	-	-	-	-
Bi	mg/kg	-	-	-	-	-	-	-	-	-	-
Cd	mg/kg	4.2	3.6	3.5	2.4	0.50	57	1.5	9.4	2.0	8.2
Ce	mg/kg	-	-	-	-	-	-	-	-	-	-
Co	mg/kg	12	6.5	6.0	27	5.7	230	2.0	140	4.0	9.0
Cr	mg/kg	170	130	110	140	30	84	52	610	68	290
Cu	mg/kg	220	190	170	190	39	83	89	840	110	290
Hf	mg/kg	-	-	-	-	-	-	-	-	-	-
Hg	mg/kg	0.57	0.45	0.39	0.42	0.088	74	0.17	1.6	0.22	1.2
La	mg/kg	29	25	27	17	3.5	58	7.0	71	12	48
Li	mg/kg	-	-	-	-	-	-	-	-	-	-
Mn	mg/kg	450	380	350	360	76	80	160	1700	230	580
Mo	mg/kg	11	5.6	4.0	19	4.0	170	2.0	87	2.0	23
Nb	mg/kg	3.8	3.5	3.0	1.8	0.38	48	2.0	11	2.0	5.0
Ni	mg/kg	76	66	56	51	11	66	38	250	40	140
Pb	mg/kg	270	250	240	110	22	39	140	520	170	480
Rb	mg/kg	-	-	-	-	-	-	-	-	-	-
Sb	mg/kg	13	9.4	9.0	18	3.7	140	2.5	91	5.0	18
Sc	mg/kg	3.3	3.0	3.0	1.5	0.31	45	2.0	7.0	2.0	6.0
Sn	mg/kg	23	21	20	11	2.3	48	11	64	13	30
Sr	mg/kg	120	120	110	38	8.0	32	84	260	88	150
Ta	mg/kg	-	-	-	-	-	-	-	-	-	-
Th	mg/kg	4.1	2.7	3.0	7.3	1.5	180	1.0	37	1.0	5.0
U	mg/kg	-	-	-	-	-	-	-	-	-	-
V	mg/kg	31	28	28	14	2.9	46	16	78	18	44
W	mg/kg	-	-	-	-	-	-	-	-	-	-
Y	mg/kg	6.4	6.1	6.0	2.1	0.44	33	3.0	11	4.0	10
Zn	mg/kg	1300	1200	1300	520	110	39	640	2600	810	1900
Zr	mg/kg	16	15	14	5.1	1.1	33	10	34	11	21

X – mean; Xg – geometrical mean; Md – median; S – standard deviation; Sx – standard error of mean; CV – coefficient of variance; Min – minimum; Max – maximum; P10 – 10th percentile; P90 – 90th percentile; Data rounded to two digits.

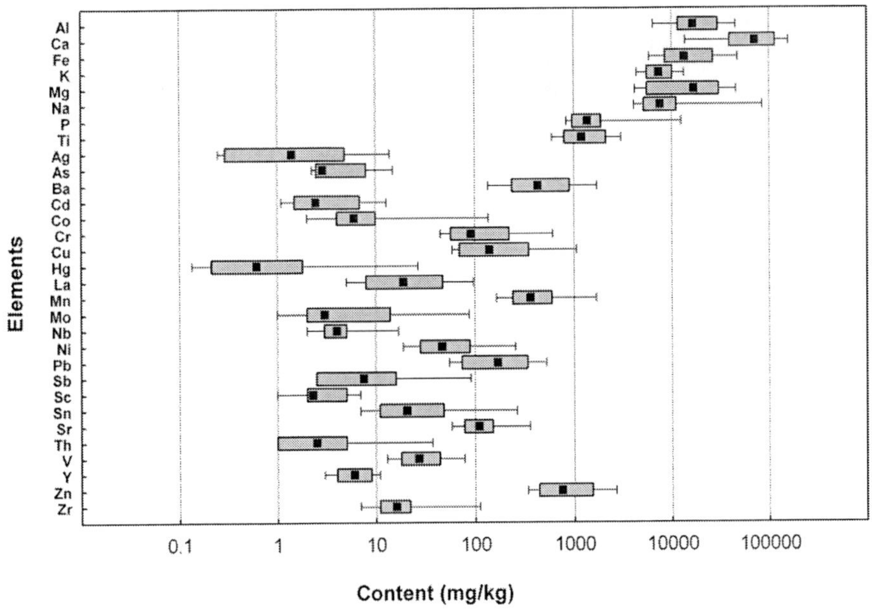

Figure 8. Distribution of chemical elements in household (N=83). Whiskers - range (minimum and maximum value); Box - 10th and 90th percentile values; Square – median value.

4.2. Enrichment Factors (EF)

Separation of chemical elements in the different sampling materials due to their origin (natural or anthropogenic) is possible by using the enrichment factor - (Fergusson and Kim, 1991, Fergusson and Ryan, 1984). The enrichment factors allow us to determine the enrichment of a particular element in the sampling material according to the reference element in the soil. Values between 0 and 3 show the natural origin of chemical elements, but higher levels might be regarded as a result of some other processes (such as an anthropogenic pollution) that are not associated with the bedrock lithology. Reference elements (typical elements of the soil and the earth's crust) have been used by different authors in studies of urban sediments: Al (Fergusson and Ryan, 1984; Ogunsola et al., 1994), Sc (Ogunsola et al., 1994) or Ce (Fergusson and Kim, 1991). However, Al might be used in the fertilizers, and Ce or Sc can be enriched in carbonates (approximately half of the territory f Slovenia is composed of carbonates), so these three elements might not be good representatives for the reference elements. This is why Ti has been used as a reference element in this study, because we believe that Ti is the best representative of the uniform natural (Gaussian) elemental distribution. Its concentrations are probably not influenced by human activities.

Values of the average enrichment factor in household dust and attic dust in the relation to concentrations in topsoil are represented in Figure 9. The lowest values of EF (lower than 1.0) in both types of dusts, household and attic dust, show elements that are most probably the consequence of the bedrock weathering. These are Al, Ba, Be, Ce, Co, La, Li, Mn, Nb, Rb, Sc, Th, Ti and Y in attic dust and Al, As, Co, Fe, Mn, Sc, Th, Ti, V, Y, Zr in household dust. Enrichment factors of other elements, which are between 1.0 and 3.0, indicate the influence of natural regional geological background. In this case, anthropogenic influence is completely subordinate to the natural influence.

The limit value of enrichment factor (3.0) is exceeded at anthropogenically enriched metals. In attic dust this value is exceeded by following elements: Sn (3.5), P (3.7), Mo (3.7), Sb (4.1), Cd (4.3), Zn (4.8), Pb (5.5), Ag (5.5), Ca (7.9), Hg (19) and S (67). A similar pattern of enrichment factor was also observed in household dust, but the order is slightly different: Mo (5.9), Pb (6.1), Cu (7.3), Cd (8.6), Zn (10.5), Sn (6.11), Sb (7.11), Ca (12.5), Hg (8.13), Ag (25.5).

High enrichment of Ca is mainly due to weathering of building materials. Slightly elevated enrichment factors of Cr (2.8) and Ni (3.2) in the household dust, which is probably due to metallurgical activities, are also notable.

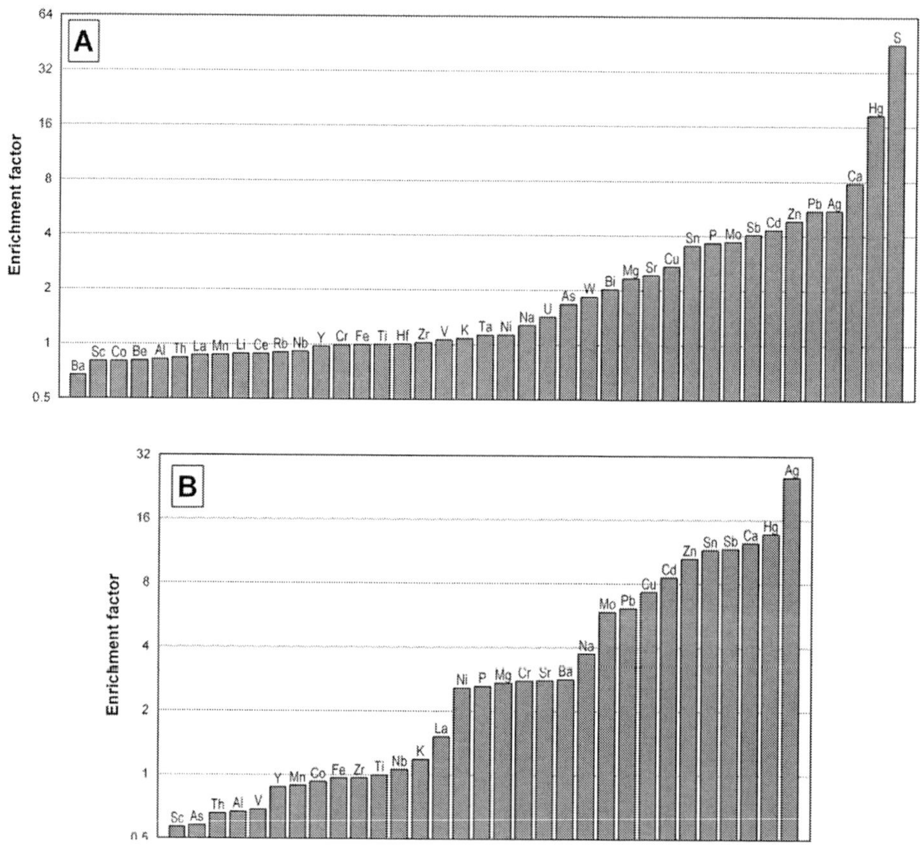

Figure 9. Enrichment factors of chemical elements (A: attic dust regarding to topsoil; B: household dust regarding to topsoil).

4.3. Correlations between Chemical Compositions of Sampling Media

The degree of similarity of the concentrations of particular elements between different sampling materials was assessed with the linear coefficient of correlation "r" (Le Maitre, 1982). Coefficient r was calculated using concentrations of elements between different sampling materials collected at the same sampling point. It was qualitatively assumed that the absolute values of r between 0.25 (limit significance) and 0.6 indicate good association, and those between 0.60 and 1.0 indicate strong association between elements.

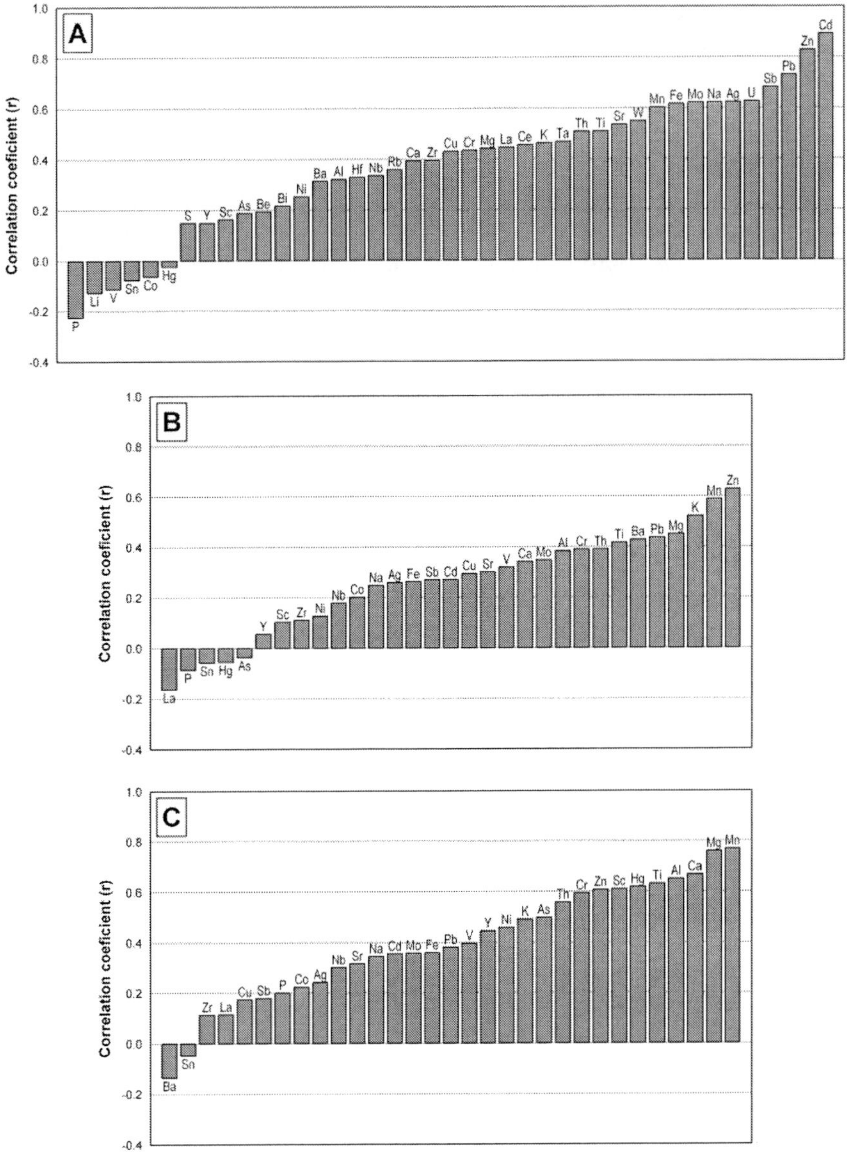

Figure 10. Correlation coefficients between chemical elements of different sampling media (A: attic dust – topsoil; B: household dust – topsoil; C: attic dust – household dust).

Concentrations of chemical elements in topsoil and attic dust (Figure 10a) are pretty well interconnected, except for 10 following elements: As, Be, Co, Hg, Li, P, S, Sc, Sn, V, and Y, which have no significant correlation coefficients. Other elements show either a good or a strong correlation (such as Ag, Cd, Fe, Mn, Mo, Na, Pb, Sb, U, Zn). Slightly weaker correlation coefficients have been found between the elements in topsoil and household dust (Figure 10b). Eight elements (As, Co, Hg, Nb, Ni, Sc, Y, Zr) have no significant correlation.

Other elements are well correlated; only Zn shows a very strong correlation. Correlation coefficients between elemental concentrations of attic and household dust (Figure 10c) show insignificant correlation among following elements: Ba, Co, Cu, La, P, Sb, Sn, Zr. Strong correlations are observed at Al, Ca, Hg, Mg, Mn, Sc, Ti, Zn. Other elements have significant correlation.

4.4. Multivariate Statistics

Multivariate cluster and R-mode factor analyses (FA) were used to reveal associations of chemical elements (Davis, 1986; Facchinelli et al., 2001; Micó et al., 2006; Zhang, 2006; Luo et al., 2007; Templ et al., 2008). All samples of topsoil, attic dust and household dust were used and combined in the case of multivariate statistical analyses. Cluster analysis is a generic name for a set of multivariate statistical or mathematical methods that can be used to find similarities between different multidimensional vector objects in a data set. The aim of cluster analysis is to establish a set of clusters in such way that cases within a cluster are more similar to each other than they are to cases in other clusters. More cluster techniques, which are based on agglomerating more elements according to their composition, giving different results, are known. Hierarchical clustering based on correlation coefficient (r) has been chosen in this research.

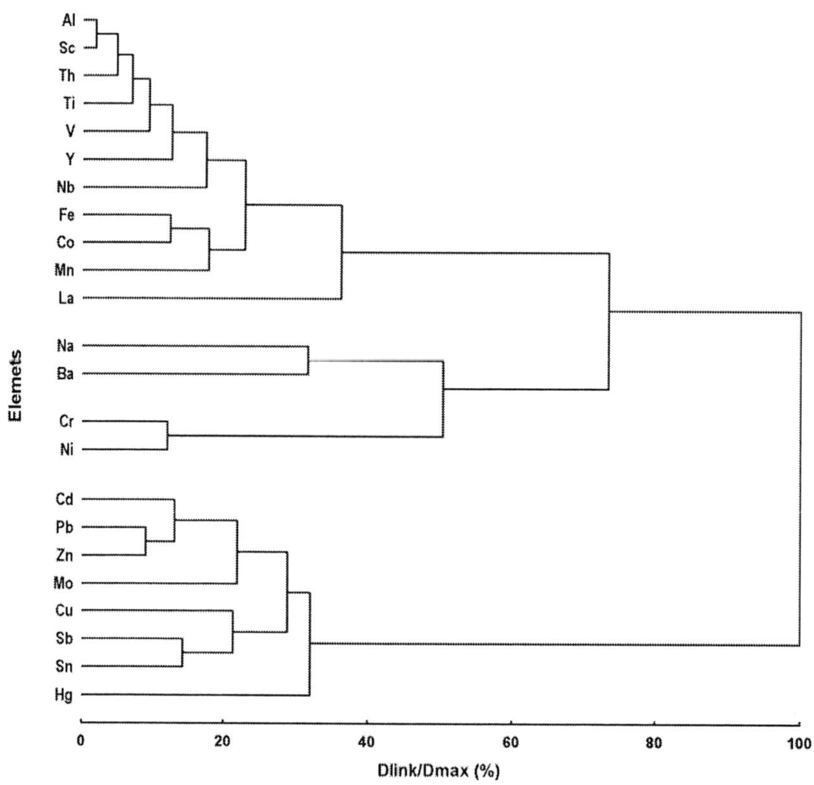

Figure 11. Dendrogram of cluster analysis, single linkage, Euclidean distance (N=249, 23 selected elements).

Table 7. Matrix of Rotated Factor Loadings (n=249, 23 Selected Elements). F1 - F4 – Factor Loadings; Com – Communality in %; Var – Variance in %

	F1	F2	F3	F4	Comm
Ti	**0.90**	-0.26	0.04	0.17	90.6
V	**0.89**	-0.21	0.05	-0.21	89.2
Al	**0.89**	-0.41	-0.01	0.04	95.7
Th	**0.88**	-0.36	-0.10	0.04	92.5
Sc	**0.87**	-0.43	0.02	-0.07	94.6
Fe	**0.85**	-0.13	0.35	-0.06	86.8
Y	**0.83**	-0.32	0.06	-0.07	80.5
Nb	**0.81**	-0.23	0.19	0.22	79.3
Co	**0.71**	-0.21	**0.52**	0.04	81.7
Mn	**0.66**	-0.27	0.42	-0.24	75.0
La	**0.59**	-0.15	0.20	**0.51**	65.8
Pb	-0.16	**0.90**	-0.08	-0.15	86.6
Zn	-0.34	**0.87**	0.03	0.01	86.7
Cd	-0.23	**0.84**	-0.02	-0.01	76.5
Sb	-0.40	**0.81**	0.09	0.13	83.4
Mo	-0.13	**0.76**	0.28	-0.03	67.2
Hg	-0.32	**0.70**	-0.32	-0.17	72.1
Sn	-0.45	**0.67**	0.01	0.27	72.8
Cu	-0.36	**0.67**	0.28	0.15	68.0
Ni	0.16	0.15	**0.90**	0.09	87.0
Cr	0.13	0.16	**0.87**	0.29	87.9
Ba	0.08	-0.12	0.22	**0.82**	74.2
Na	-0.11	0.14	0.03	**0.80**	66.6
Var	**35.5**	**25.6**	**11.1**	**8.7**	**80.9**

The cluster analysis was performed on 23 selected chemical elements (Al, Ba, Cd, Co, Cr, Cu, Fe, Hg, La, Mn, Mo, Na, Nb, Ni, Pb, Sb, Sc, Sn, Th, Ti, V, Y, Zn). Other elements were discarded from multivariate analysis because of their tendency to form their own clusters (or factors in the case of factor analysis). Discarded elements do not show a reasonable connection with other chemical elements. Elemental associations obtained by cluster analysis are: Al-Sc-Th-Ti-V-Y-Nb-Fe-Co-Mn-La, Na-Ba, Cr-Ni and Cd-Pb-Zn-Mo-Cu-Sb-Sn-Hg (Figure 11). Factor analysis (FA) was performed on linearly transformed variables to zero mean and unit standard deviation (Davis, 1986; Reimann et al., 2002; Filzmoser et al., 2005). As a measure of similarity between variables, the product-moment correlation coefficient (r) was applied. For orthogonal rotation, the varimax method was used.

The factor analyses (FA) was performed on the 23 previously mentioned chemical elements. Other elements were discarded from factor analysis in order to satisfy the criteria of the required number of cases, which is based on the number of variables (elements). Elements with a low share of communality or with the tendency to form independent factors were excluded. Factor analysis reduces the dimensionality of the data set, from 23 dimensions (23 elements) to a specific number of the synthetic variables (factors, in our study 4: from F1 to F4). Elements that are loaded in a specific factor tend to express similar geochemical properties / trends. Factor analysis explains 81% of the variability of treated set elements (Table 7). First association explains 36% of total variance and includes 11, mainly lithophile, elements, according to the Goldschmidt geochemical classification: Ti, V, Al, Th, Sc, Fe, Y,

Nb, Co, Mn and La. Second geochemical association explains 26% of total variance and contains eight, mainly chalcophile, elements: Pb, Zn, Cd, Sb, Mo, Hg, Sn and Cu. Third association explains 11% of the total variance and contains three, mainly siderophile, elements: Ni, Cr and Co. The weakest geochemical association identified explains 9% of total variance and contains light litophile elements from the left side of the periodic table: Ba, Na and La.

5. DISCUSSION

The geochemical properties of household and attic dust were compared to topsoil and evaluated in terms of their elemental contents and elemental associations. The proportion of anthropogenically derived toxic metal impact on urban sediments was assessed. As based on comparisons of household dust, attic dust and topsoil with multivariate statistical method (factor and cluster analysis), four patterns of elemental distributions were established. Two natural geochemical associations, Ti-V-Al-Th-Sc-Fe-Y-Nb-Co-Mn-La and Ba-Na-La, are influenced mainly by lithology. The association Pb-Zn-Cd-Sb-Mo-Hg-Sn-Cu represents the anthropogenically introduced group of chemical elements into the environment. The association Ni-Cr-Co is influenced by natural and anthropogenic factors.

5.1. Natural Distribution of Chemical Elements

5.1.1. Association of Ti, V, Al, Th, Sc, Fe, Y, Nb, Co, Mn and La

This association can be attributed to remobilization of dust from soil. Ti, V, Al, Th, Sc, Fe, Y, Nb, Co, Mn and La, which are loaded in the first factor, are representatives of this association. Characteristics of these elements are their abundance in topsoil and depletion in attic and household dust (Figure 12). The same trend is observed in towns and in the countryside (Figs. 12 and 13). The content of elements of this group in household and attic dust in urbanized areas is generally lower than in their content in the countryside.

Because attic dust usually points to the anthropogenic emissions in the atmosphere, we can conclude that this association is naturally distributed and associated with the natural process of weathering. Spatial distribution of factor 1 scores (Figure 13) shows that elements represented by this factor are most abundant in topsoil and depleted in household and attic dust. The exception is the abundance of these elements in attic dust in eastern parts of Slovenia. This is probably because the eastern part of Slovenia is drier, and, moreover, there is extensive farming in the area. This produces dust mobilization in the atmosphere, and this is reflected in the attic dust composition. Contrarily, the "carbonatic realm" in the western part of Slovenia is wetter not suitable for extensive farming, and mainly covered with forest, so the natural process of dusting of topsoil is not so relevant here. This explains the abundance of elements of this factor in soils and the depletion in household and attic dust.

Comparing the different abundances of elements on different geological backgrounds, (Figure 12) we see that this factor dominates in the topsoil on the carbonates, igneous and metamorphic rocks, and on the neogene molasse sediments of the Panonian basin.

This factor is less dominant in the Paleogene flysch area. In attic dust this factor is dominant on Neogene sediments and on igneous and metamorphic rocks. In attic dust, factor 1 is expressed in the areas of igneous and metamorphic rocks and on the areas of Neogene molasse. This factor is not expressed in the household dust. A small exception is the area of metamorphic and igneous rocks.

We can conclude that these elements are connected to the natural process of weathering of bedrocks. These elements are abundant in topsoil on almost the whole territory of Slovenia. Areas of abundance of factor 1 scores and elements, connected to factor 1 in attic dust, indicate that abundance of such elements in attic dust might be connected to the natural process of wind erosion of soil in drier parts of the country.

This geochemical association can denote "soil dusting association," produced by dust mobilization from soil by different means (farming, wind, traffic on dirt roads, etc.). The characteristic for this association is that these elements are generally depleted in attic dust and household dust. Among all four detected geochemical associations, this association shows the most distinctive geochemical trends. The distribution of factor 1 scores shows very similar trends in attic and house dust, and this distribution is very dependant on the regional lithological settings. The highest values of factor 1 scores, which can be connected to the weathering of the igneous and metamorphic rocks of the Central Alps and weathering of the post-orogenic molasse sediments of the Panonian basin, are found on the northeastern part of Slovenia. Fergusson (1992) showed that the dominant source of the majority of the elements in this group (Al, Fe, K, Ti, La, Mn, Sc, Th in V) in ambiental dust is the dust that originates from soil. It is also interesting, that factor 1 scores are abundant in soil on carbonates on the western part of Slovenia, but depleted in the same area in house and attic dust. Moreover, higher factor 1 scores are observed in the countryside than in urbanized areas.

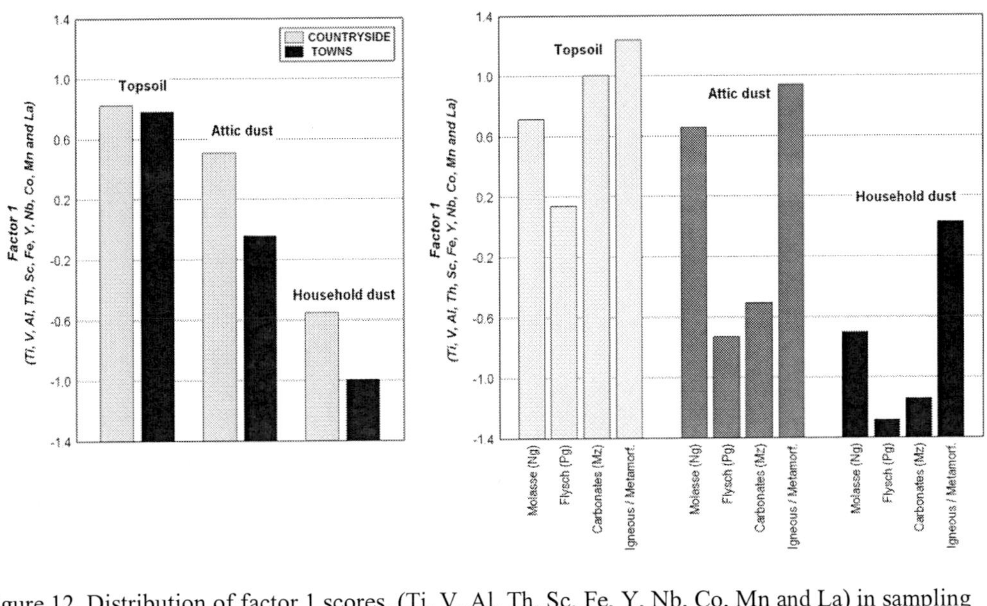

Figure 12. Distribution of factor 1 scores, (Ti, V, Al, Th, Sc, Fe, Y, Nb, Co, Mn and La) in sampling materials according to the urbanization (left) and on areas of different geological settings (right). Bars show averages.

Both facts might indicate that geochemical trends visible in attic and house dust are produced because of higher rate of the soil dusting (farming, wind, drier weather, less forests, etc.) and not because of the particular chemistry of the bedrocks. Still, the discussion of whether the higher content of elements of this group in attic and house dust in eastern Slovenia is the consequence of the particular geological settings or whether it is the consequence of the higher mobilization of soil dust due to extensive farming in this part of the country is still open and can be a topic of further research.

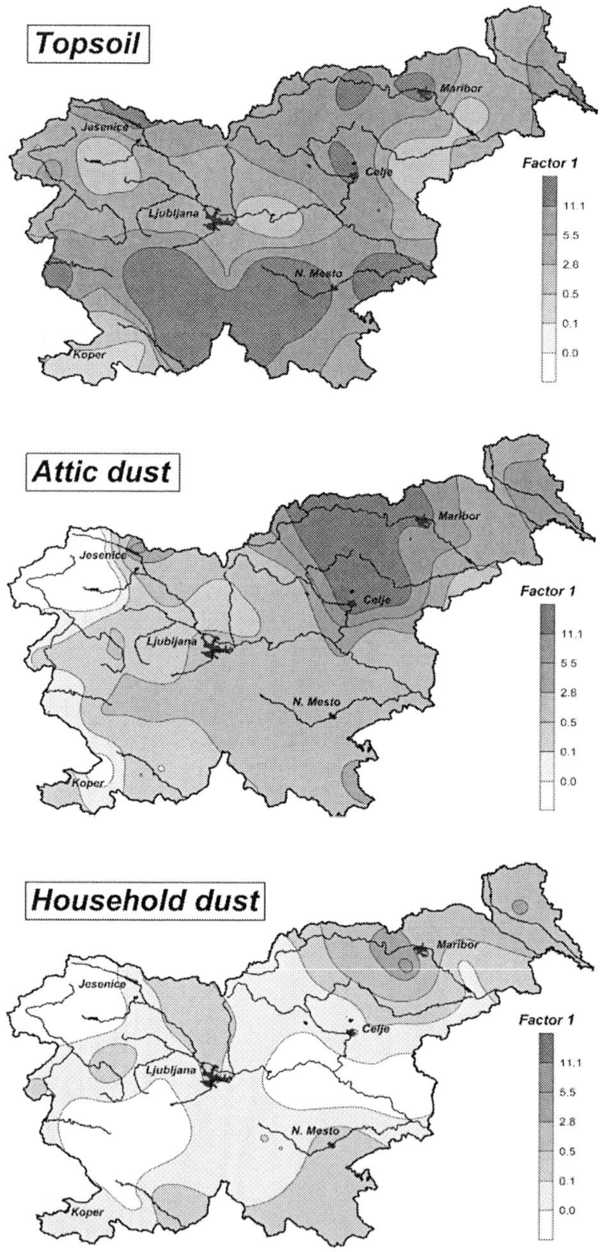

Figure 13. Spatial distribution of Factor 1 scores (Ti, V, Al, Th, Sc, Fe, Y, Nb, Co, Mn and La) in topsoil, attic dust and household dust.

5.1.2. Association of Ba, Na and La

The second identified group of elements is the weakest one, explains about 9% of the variance, and is represented by factor 4 (Table 7). The members of this group are elements of low ionic potential from the left side of the periodic table: Ba, Na and La. These elements form highly soluble salts and are quickly washed away by rain. Characteristic of this group is its abundance in eastern part of Slovenia, on areas of the youngest sedimentary rocks of the Neogene age, formed in the Panonian basin (Figs. 14 and 15). These three elements are characteristically enriched in household dust and depleted in attic dust. This might indicate their quick removal from the atmosphere.

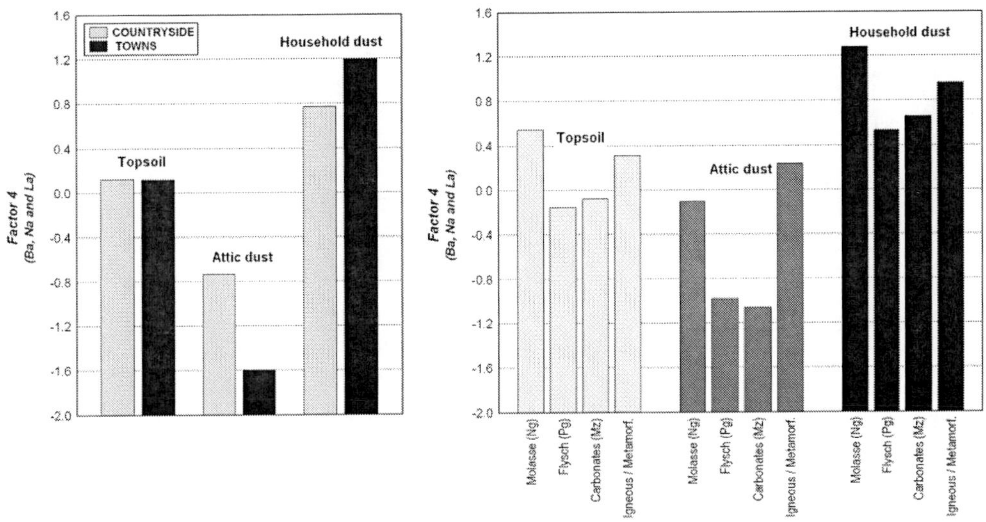

Figure 14. Distribution of factor 4 scores, (Ba, Na and La) in sampling materials according to the rate of urbanization (left) and on areas of different geological settings (right). Bars show averages.

Figure 14 shows the distribution of averages of factor 4 scores on areas of different geological settings. Areas of Neogene molasse sediments of the Panonian basin and areas of igneous and metamorphic rocks show the highest enrichment, while areas of carbonates and Paleogene flysch sediments show the depletion. An interesting thing is that household dust is the most enriched sampled material (Figure 14). However, the reason for this can be only speculated, taking into account the clearly expressed regional trend (Figure 15). The explanation might be that elements from factor 4 are enriched in the quartz, mica and felsic (Na, Albite type) plagioclase minerals, and these minerals are common minerals in household dust (Molhave et al., 2000).

Since the igneous rocks in the area of northern Slovenia are of the felsic type and, moreover, the source for the molasse deposits of the Panonian basin are also felsic magmatic rocks in the Alpine area, which are enriched with aforementioned types of minerals, the abundance of these minerals in household dust is expected.

However, due to their nature the wind can easily mobilize and transport these minerals, even over larger distances (like dust from Sahara across the Atlantic Ocean). The reason for the depletion in attic dust might lie in the fact that Na and Ba bearing minerals might be blown away from the attics, but because erosion does not take place inside the vacuum cleaner bag, these minerals are detected in house dust.

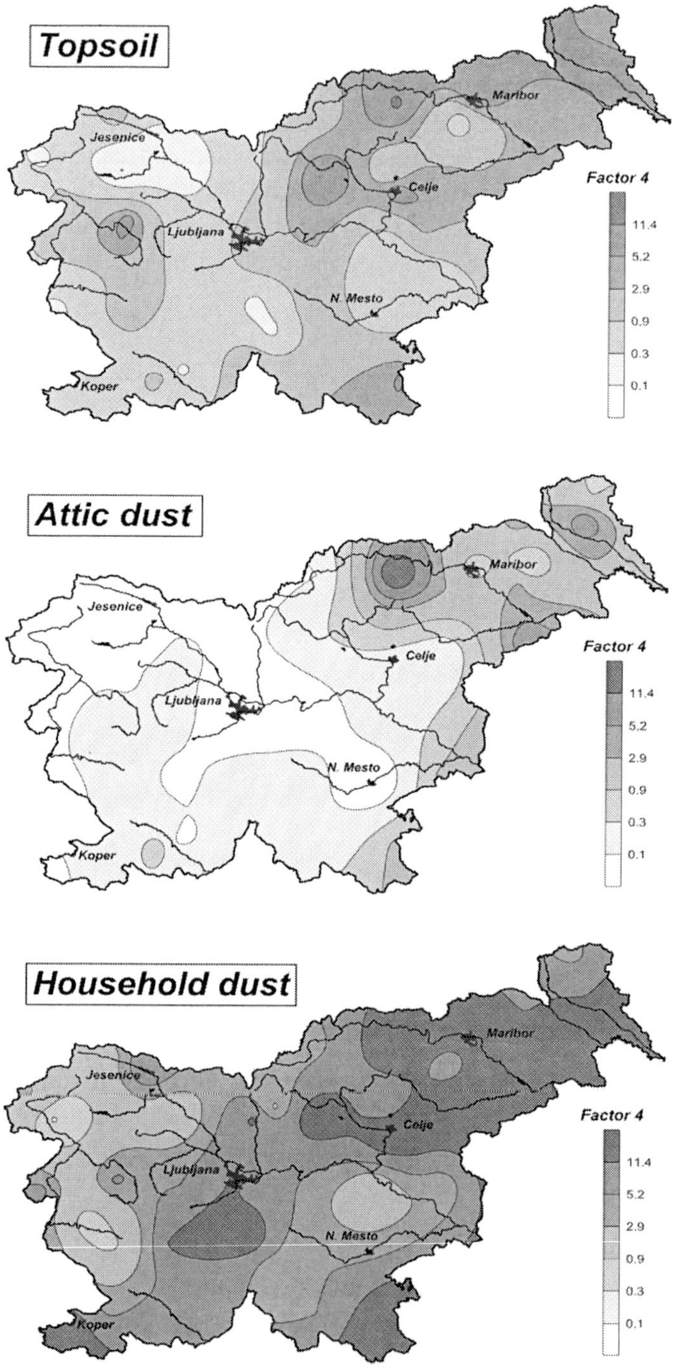

Figure 15. Spatial distribution of Factor 4 scores (Ba, Na and La) in topsoil, attic dust and household dust.

We can conclude that the association of Ba, Na and La can be named as "Neogene molasse natural association," and characteristically, it is most abundant on eastern part of Slovenia in household dust.

5.2. Anthropogenic Distribution of Chemical Elements

5.2.1. Association Pb, Zn, Cd, Sb, Mo, Hg, Sn and Cu

Geochemical association of Pb, Zn, Cd, Sb, Mo, Hg, Sn and Cu is represented by factor 2 and forms the most recognized spatial patterns. This association is also the strongest anthropogenically induced geochemical association in the environment in Slovenia. High factor 2 scores are found in house and attic dust. Contrarily, low factor 2 scores are found in soil. Average contents of the aforementioned elements in the countryside generally increase according to the following pattern: topsoil → attic dust → house dust. In the urbanized areas we can observe a similar pattern. Lowest factor 2 scores are observed in topsoil, and the highest in attic dust (Figure 16). Characteristic of this group of elements is that they are abundant in towns compared to the countryside (Figure 16, left). Moreover, enrichment of elements of this group in urbanized areas is several times higher than geochemical background values. Deviations from this pattern (enrichments for a factor larger than 100) can be attributed to the strong anthropogenic influences, such as metal smelters and ironworks.

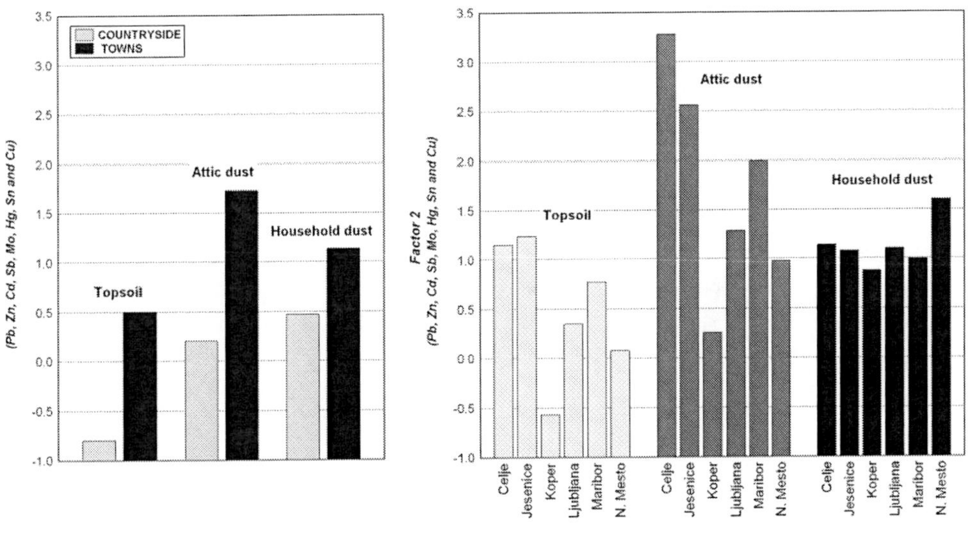

Figure 16. Distribution of factor 2 scores (Pb, Zn, Cd, Sb, Mo, Hg, Sn and Cu) in sampling materials according to the rate of urbanization (left) and between different Slovenian towns (right). Bars show averages.

Representative elements for this factor are commonly associated with anthropogenic pollution, metal mines, smelters and population. Figure 17 shows the spatial distribution of factor 2 scores. The pattern shows that factor 2 scores are most abundant in attic dust in the wider area of mining, smelting and ironworking centers (northern part of Slovenia: Celje, Jesenice and Mežica towns) and in other densely populated areas. Since attic dust reflects the last 100-year history of atmospheric emissions, we can conclude that influences of heavy industry, such as ironworks and smelters, can be detected regionally. This anomaly cannot also be detected in household dust, because household dust represents the geochemical footprint of the area over a shorter time period. In topsoil, elements of this group are abundant only in largely populated areas, especially around industrial town of Jesenice, where large ironworks have existed for more than 100 years.

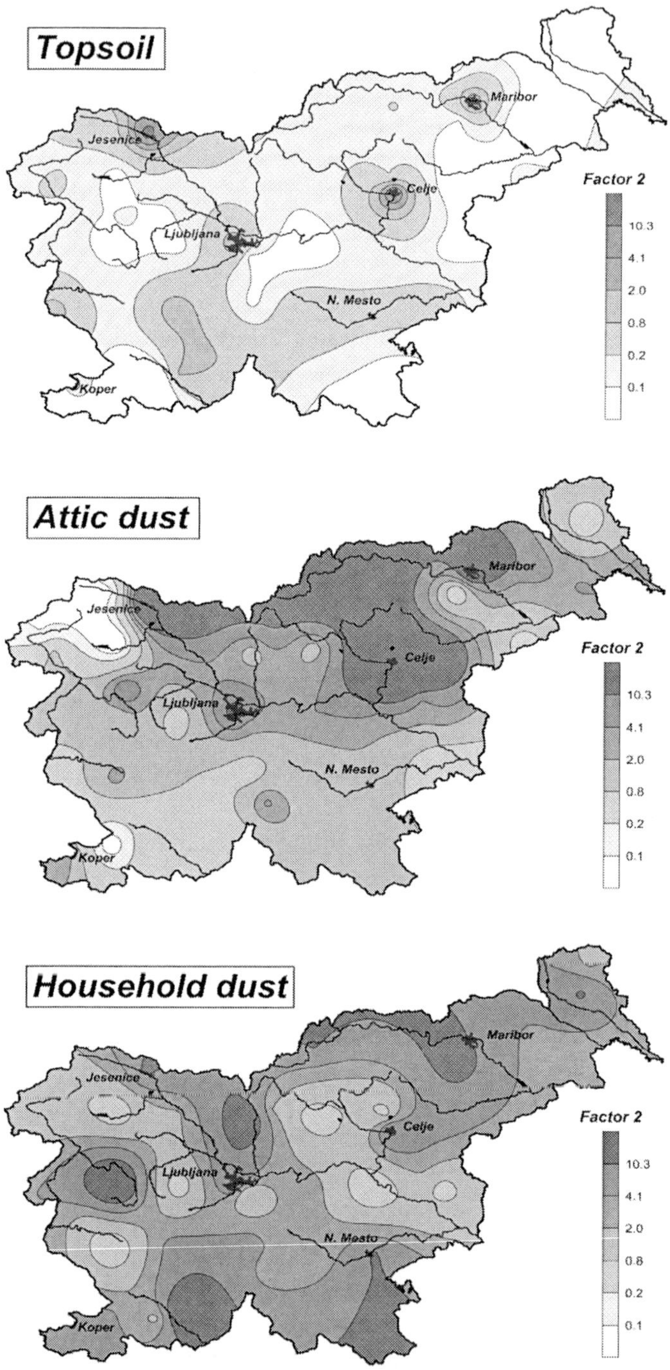

Figure 17. Spatial distribution of Factor 2 scores (Pb, Zn, Cd, Sb, Mo, Hg, Sn and Cu) in topsoil, attic dust and household dust.

Spatial distributions of particular elements show (shown on the distribution of the particular elements in sampling media in the following chapters) a similar pattern: small values in topsoil, with the exception of the areas around the known sources of pollution (like towns, smelters, etc.); high values in attic dust in the larger areas or aureoles around sources of pollution; and quite variable spatial distribution in household dust, without many visible trends connected to the locations of main pollution, towns or geology.

This means that attic dust is a very sensitive medium and atmospheric pollution is the reason for its contamination. Chemical composition of soil also reflects the same pattern, but not at such distances as in attic dust. Spatial concentrations in household dust do not show any specific trends from which to draw solid conclusions.

Observing the factor 2 scores in household dust (Figures 16 and 17), which have very uniform distribution in all towns, and which are higher than in soil, the conclusion can be made that elements loaded in factor 2 might be connected to population influence. However, comparison of average elemental concentrations between different towns and between different sampling materials shows that not all towns are evenly contaminated with all elements.

It is true that attic dust is the material most enriched metals connected to factor 2, but enrichment in attic dust in some towns can reach extreme values, as is the case for Celje and Jesenice (Figure 16 right), which are known for their long history of metallurgy and metal smelting. A long-lasting ironworking tradition exists in Jesenice and Celje. Moreover, the Celje area also has a 100-year-long lasting zinc smelting tradition; it is clear that heavy industry can drastically influence the composition of attic dust. Most polluted towns included in this research are not the largest towns (Ljubljana and Maribor), but towns with known ironworking and smelting tradition. One thing is also evident from the figures, and this is the rate of contamination, which is larger in attic dust than in household dust. Soils have the smallest concentrations, but this can easily be elevated in towns of known heavy industry, as evident from the case of Celje and Jesenice, where the largest values of factor 2 scores in soil are observed (Figure 16 right).

We can conclude that heavy industry can drastically impact the chemical composition of attic dust and soil, but this influence is not visible for household dust. On the other hand, everyday population activities in urbanized areas also produce enrichment in household dust.

This geochemical association can be thus titled the "anthropogenic pollution factor," produced by past metal mining and smelting activities, and pollution caused by urbanization in general (power producing, heating, traffic, etc.). This association is most abundant in attic dust around populated areas and especially around areas of known past and present smelting activities.

5.3. Association of Ni, Cr and Co

A third group of elements represent 11% of the total variance and consists of Ni, Cr and, partially, Co, according to the table of factor 3 loadings (Table 7). The existence of this association can be attributed to the natural and anthropogenic factors.

Factor 3 scores in the countryside are the highest in household dust and lowest in the topsoil. In the urbanized areas, factor 3 scores are the highest in topsoil and house dust and lowest in attic dust (Figure 18). The relative depletion of Ni, Co and Cr in attic dust in the countryside shows that the dust in the atmosphere is depleted of elements in this group. This is probably because minerals containing Ni, Co or Cr have larger specific weight, so particles containing these elements tend to fall down from the atmosphere very quickly.

As urbanized areas are at the same time the areas of past and present ironworking activities and usually aggregated around the locations of past and present industry, the same pattern is not observed in the urbanized areas. Similar effects are also visible from Figure 20, where anomalies around towns are much bigger in household dust and soil compared to attic dust. The small anomalies in attic dust on the other hand clearly point at the sources of these elements in the atmosphere. The comparison between factor 3 scores in Slovenian towns (Figure 18) points out that metallurgy is the dominant source of these metals in the urban environment and can affect all three sampling media, as evident in the case of Jesenice.

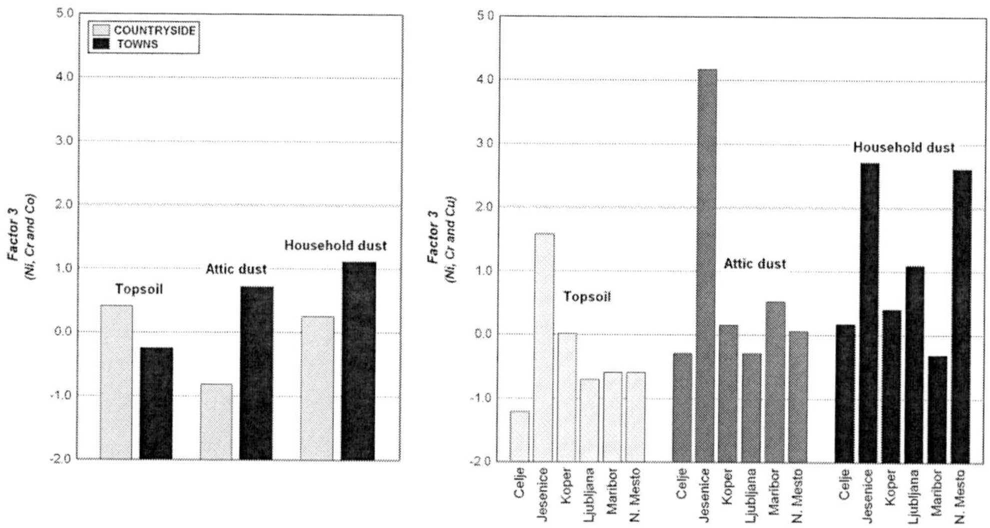

Figure 18. Distribution of factor 3 scores (Ni, Cr and Co) in sampling materials according to the rate of urbanization (left) and between different Slovenian towns (right). Bars show averages.

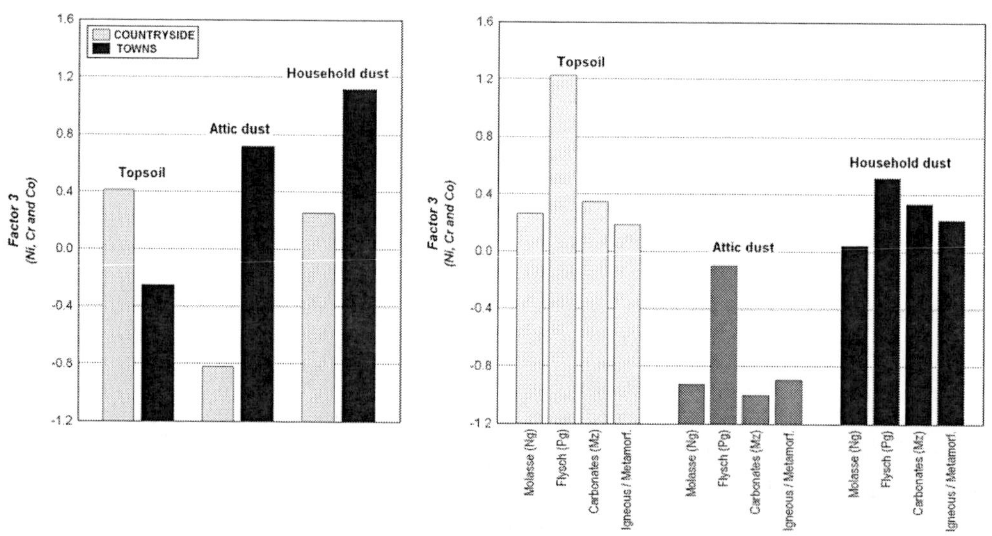

Figure 19. Distribution of factor 3 scores (Ni, Cr and Co) in sampling materials according to the rate of urbanization (left) and according to the different geological settings (right). Bars show averages.

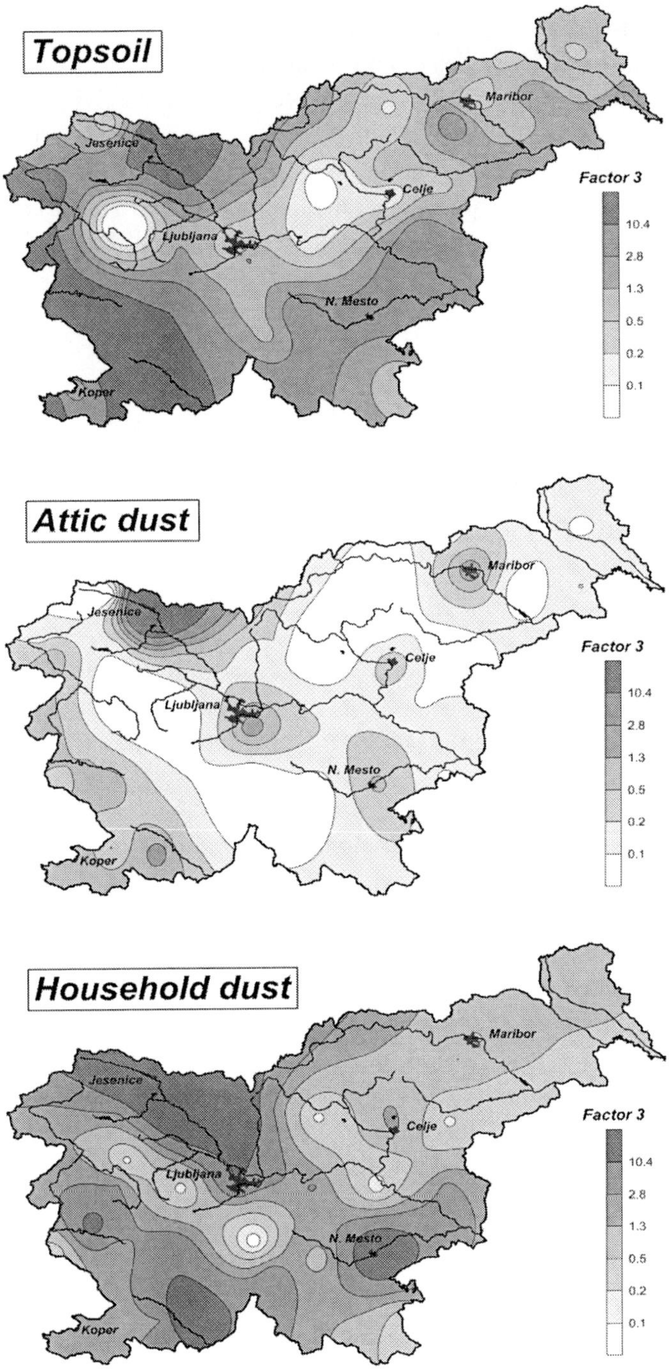

Figure 20. Spatial distribution of Factor 3 scores (Ni, Cr and Co) in topsoil, attic dust and household dust.

Contrarily, the existence of this association can also be attributed to the natural geological conditions, because factor 3 scores are the highest on the Paleogene flysch areas (Figure 19). The spatial distribution of factor 3 scores (Figure 20) in soil also suggest two sources in the environment: the ironworking in Jesenice and Paleogene flysch sediments in the southwestern part of Slovenia. Natural sources of these 3 elements are clearly suggested by the shapes and the locations of ellipsoidal anomalies in topsoil and house dust, corresponding to the areas of paleogene flysch outcropping in southwestern Slovenia.

The spatial distribution of anomalies (Figure 20) also shows that anomalies in topsoil and house dust are very similar to each other, as oppositely as the spatial distribution of factor 3 scores in attic dust. This might indicate that the major source of Ni, Cr and Co in household dust is intake of soil in the apartments and not the sedimentation of airborne particles.

The anomaly has also been detected in attic dust in Maribor. Because attic dust is a geochemical medium very sensitive to the atmospheric emissions, we can conclude that a small anthropogenic anomaly also exists in Maribor, probably because of the car making industry.

To conclude, the composition of the lithology influences most the content of the Ni, Co and Cr in topsoil and attic dust and does not influence the composition of house dust. Contrarily, the ironworking activities influence all three sampling media. Urbanization increases the concentration of Ni, Co and Cr in the attic and house dust. Taking into account all aforementioned interpretations, this factor can be titled the "ironworking and Paleogene flysch factor;" and it is most abundant in attic dust and soil in areas around ironworks and on areas of Paleogene flysch rocks in topsoil.

5.4. Distribution of Particular Metals in Sampled Media

In the following part the distributions of the selected toxic metals will be presented and compared according to the rate of urbanization, geology and towns. Spatial distributions are also presented.

5.4.1. Cadmium (Cd)

Cadmium is a relatively rare, soft, bluish-white, transitional metal in group 12, together with Hg and Zn. Cadmium occurs as a minor component in most zinc ores and therefore is a byproduct of zinc production. The use of Cd is generally decreasing due to competing technologies and its toxicity, with the exception of its use in Ni-Cd batteries and Cd-Te solar panels. Due to color metallurgy, ironworks, and steelworks, but also processes of burning of fossil fuels, Cd goes into the atmosphere. Its concentration in soil increases more and more and already in many places exceeds natural limits of 1 mg/kg (Greenwood, 1984).

Cadmium is decidedly toxic for organisms. The human body contains about 0.4 mg/kg (smokers 0.8 mg/kg). The daily intake with food is usually about 0.03 mg, but the limit of toleration is about 0.07 mg. (Wiberg at al., 2001). Clarke value of Cd in soil is 0.35 mg/kg (Bowen, 1979) and average content in European topsoil is 0.12 mg/kg (Salminen, 2005).The world average content of Cd in household dust is 7 mg/kg (Fergusson and Kim, 1991).

In Slovenian rural areas, the average content of Cd in soil is 0.50 mg/kg, whereas its average concentrations are higher in attic dust (1.4 mg/kg) and household dust (2.6 mg/kg) by several times. In the urban areas, the median values of the studied material significantly exceeded the rural values. It has been found that the median value of Cd in urbanized areas is 1.4 mg/kg in soil, 4.2 mg/kg in attic dust and 3.6 mg/kg in household dust (Figure 21).

According to the obtained data, Cd is enriched in urbanized areas in all three mediums (Figure 21). The highest enrichment is detected in the Celje area with its well known 100-

year-lasting Zn smelting tradition. The average values in soil (5.4 mg/kg), household (6.4 mg/kg) and attic dust (76 mg/kg) exceed the average values of another urban zone in Slovenia by several times and can be regarded as extreme compared to the other Slovenian towns. The highest measured value of Cd in the research was in Celje in attic dust - 240 mg/kg. Spatial distributions of Cd in attic dust (Figure 22) show clear anomalies around urban areas of Celje, Maribor and Jesenice.

The aureole around Celje is also extended towards the second largest metal mining enterprise in Mežica area, where Pb-Zn smelter existed in the past. Excluding the most contaminated area of Celje, the highest values of Cd have been found in attic dust 12.9 mg/kg and 13.3 mg/kg in household dust. A similar pattern is found also in the spatial distribution of Cd in topsoil, but this pattern is not as clearly expressed.

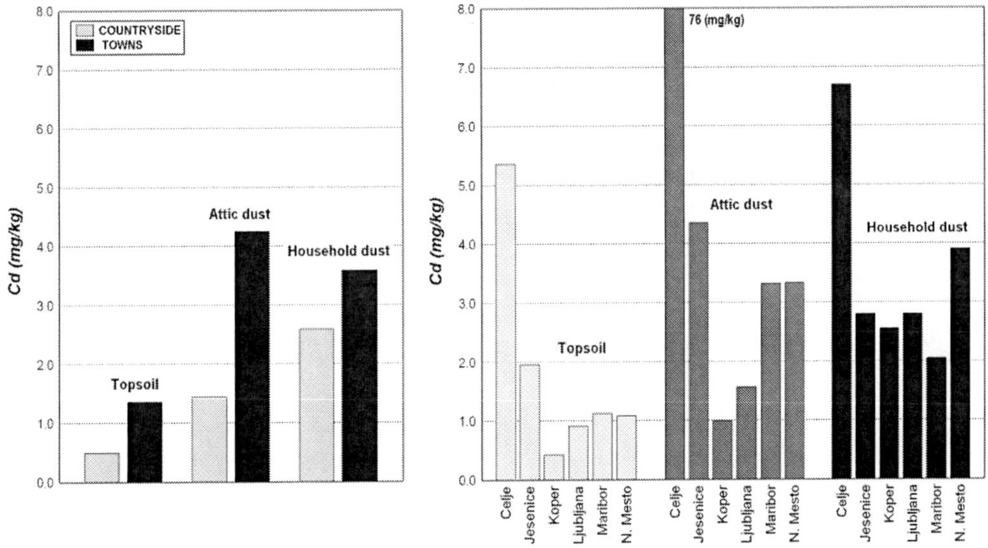

Figure 21. Distribution of cadmium in sampling materials according to the urbanization (left) and in different Slovenian towns (right). Bars show averages.

The presented results are well matched to our previous studies (Šajn, 2005; Žibret, 2008; Žibret and Šajn, 2008a; 2008b; Žibret and Rokavec, 2010).

5.4.2. Lead (Pb)

Elemental lead was known to the most ancient civilized people for daily use. Correspondingly, its toxicity has also been known for a long time. Both elemental and chemically bound Pb are toxic to organisms; human tissue contains about 0.5 mg per kg. However, owing to the relatively weak resorption, acute Pb poisoning occurs only when very large doses have been taken (Wiberg at al., 2001).

Anthropogenic emission of Pb is much higher than natural, whereas the biggest part is from internal combustion engines, roasting of ore that contains Pb, uses of different Pb materials, removal of waste materials, making of colors, uses of some insecticide and burning of fossil fuels. Due to the past use of Pb as a fuel additive, its concentration in the last 80 to 100 years has increased in the environment (Hill, 1992).

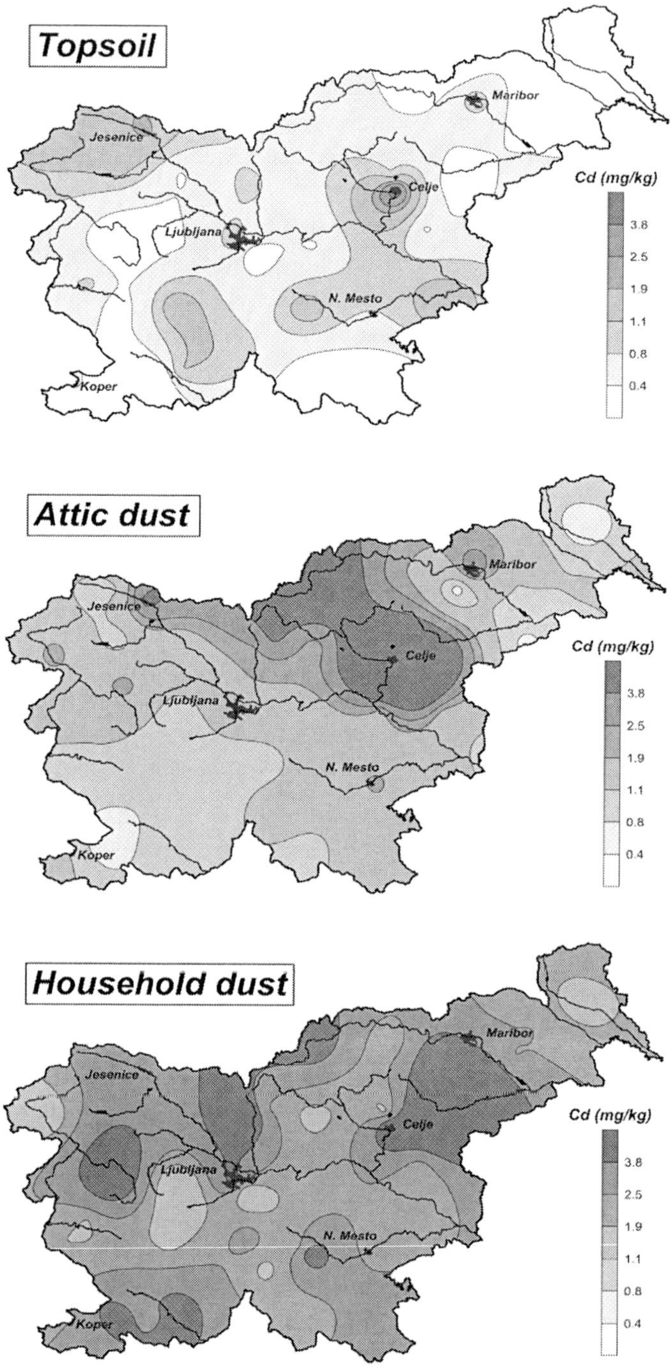

Figure 22. Spatial distribution of Cd in topsoil, attic dust and household dust.

Lead is a pretty well-known toxic metal. This is why lead contamination has quite often been reported in many recent studies, such as with the effect of mining and smelting (Galvin et al.,1993), degradation of lead ore and mining waste dumps (Geater et al., 2000), leaded gasoline combustion (von Storch et al., 2003), the manufacturing of lead batteries and battery dumps (Ibiebele, 1994), the use of lead-containing paint (Bates et al., 1995), the use of lead water pipes (Quinn and Sherlock, 1990) and others.

Clarke of Pb in world soil is 35 mg/kg (Bowen, 1979) and in the European topsoil it is 33 mg/kg (Salminen, 2005). The world average content of Pb in household dust is 7 mg/kg (Fergusson and Kim, 1991). In Slovenian rural areas, the average content of Pb in soil is 40 mg/kg. The average content in household dust and attic dust (140 mg/kg) exceed the average concentration in soil. In the urban areas, the median values of studied materials exceed the rural values significantly. It has been found that the median value of Pb in urban soil is 230 mg/kg. Similar concentrations have been found in household dust (260 mg/kg) and concentrations more than twice as high in attic dust (590 mg/kg, Figure 23).

The spatial distributions of Pb in attic dust and household dust are provided in Figure 24. It is noticeable that concentrations in rural and urban areas are different, especially in the northern part of Slovenia where three important sources of pollution can be isolated: the Pb-Zn ore mining and smelting in Mežica, the Zn smelter plant in Celje, and the ironworks in Jesenice. Correspondingly, it has been found that the maximum values are in attic dust around the known sources of Pb pollution. In the Mežica area the maximum Pb concentration is 1800 mg/kg. It can be observed that atmosphere is very important pathway for transportation of toxic elements, where the small dust particles may be transported far away from the sources of pollution. Regional trends of Pb distribution in household dust are most probably covered by different domestic activities. In the towns (Figure 23), the highest average values have been found in attic dust: Celje (1430 mg/kg) and Jesenice (1330 mg/kg). However, the record values of Pb have been found in Celje 3200 mg/kg and Jesenice 2160 mg/kg in attic dust. On the same diagram it can be observed that the highest average values in soil are also in these two towns. It is very interesting, however, that the average values in household dust are more or less similar in all studied urban zones. Thus we can conclude that the past environmentally unfriendly Zn and iron production had the significant impact on the attic dust, and, partially, also on the soil. Luckily, however, this influence can not be detected in house dust any more.

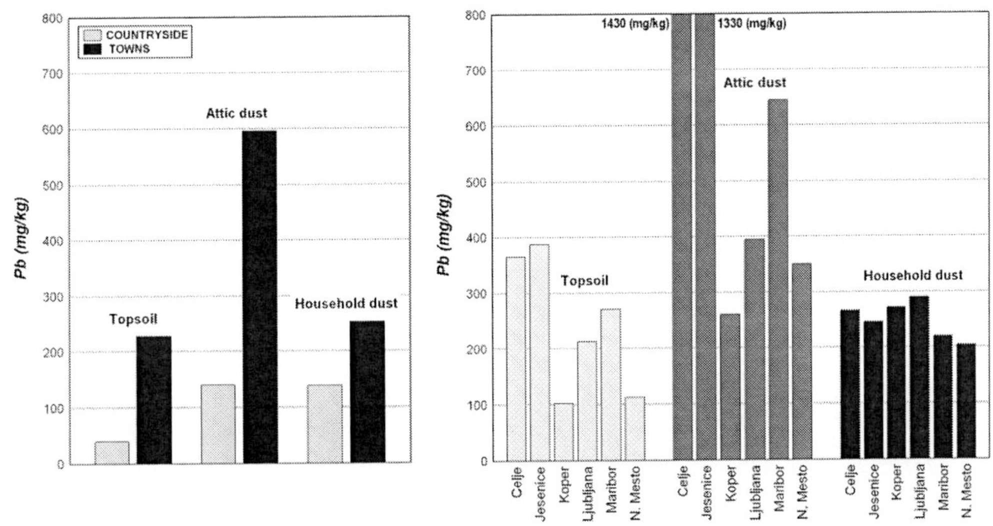

Figure 23. Distribution of lead in sampling materials according to the urbanization (left) and between different Slovenian towns (right). Bars show average values.

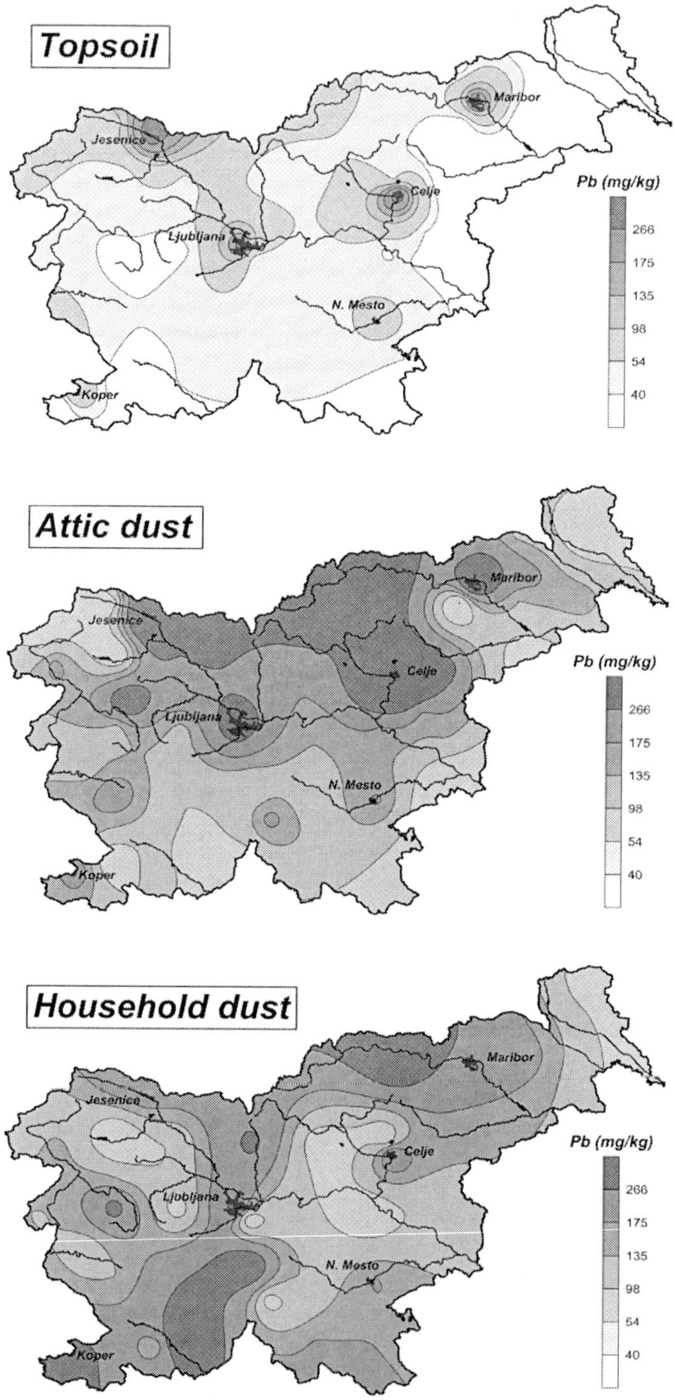

Figure 24. Spatial distribution of Pb in topsoil, attic dust and household dust.

5.4.3. Mercury (Hg)

Mercury and its compounds are non-essential for organisms, and, in fact, they are highly toxic. About half of the mercury in the environment comes from natural process, such as volcanism and weathering of rocks. The other half is of anthropogenic origin, released by mining and purification of Hg, combustion of fossil fuels, etc.

Clarke value of Hg in soil is 0.060 mg/t (Bowen, 1979), and in European topsoil it is 0.061 mg/kg (Salminen, 2005). In Slovenian rural areas, the average content of Hg in soil is 0.6 mg/kg, corresponding to aforementioned literatures. Same as for all other anthropogenic elements, the highest median values have been found in household dust (1.1 mg/kg) and attic dust (0.8 mg/kg). It is interesting to note that the estimated median values in urban areas, except in urban soil (0.55 mg/kg), are much lower than the estimated averages for the entire country. In the urban zones it has been found that the average value of Hg in household dust is 0.74 mg/kg and 0.45 mg/kg in attic dust (Figure 25).

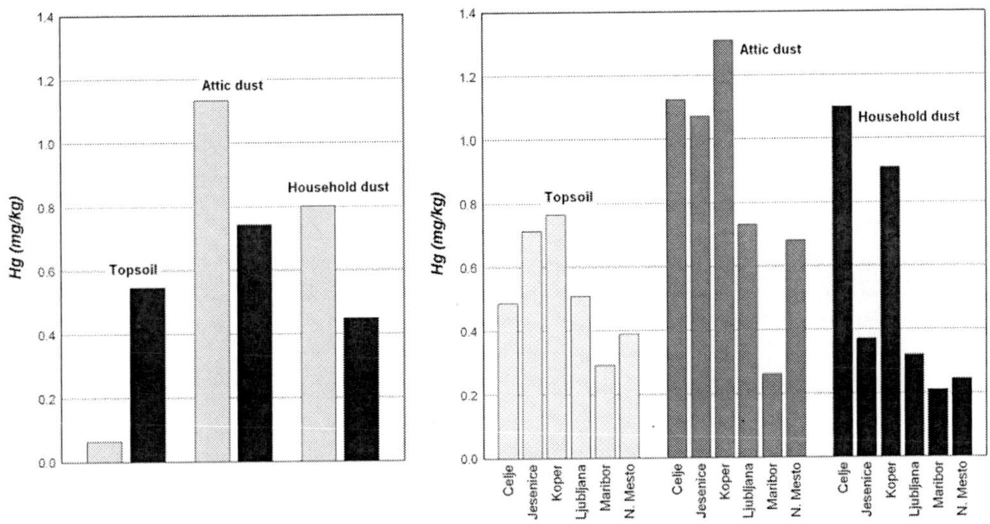

Figure 25. Distribution of mercury in sampling materials according to the urbanization (left) and in different Slovenian towns (right).

Among the analyzed heavy metals (such as Cd, Cu, Pb, Zn and Hg), the distribution of mercury is showing the most pronounced geochemical trends in rural areas. Distribution of Hg in household dust and attic dust is very similar.

The same distribution can be also observed comparing its distribution in soil (Figure 26). The strong anomaly in the distribution of mercury has been found in southwestern Slovenia in sampled material (household dust, attic dust and soil). In urban areas (Figure 25), the same anomaly can be confirmed by high values of Hg in attic dust (two maximum values are 30.1 mg/kg and 29.5 mg/kg) and household dust (26.9 mg/kg and 14.7 mg/kg).

The distribution pattern depends strongly on the high impact of ore deposits, mines and Hg smelter in Idrija. An important source of Hg is natural weathering from soil or rocks, because of its mobility in the natural environment. However, it is assumed that the natural impact is less significant than the anthropogenic because more than 500 years of mining and smelting activities in Europe's second largest mine left an enormous impact on the environment. With the use of high explosives, where some parts were made from Hg in the first world war, the content of Hg increased in SW Slovenia, especially in the Soča Valley (Pirc and Budkovič, 1996). The high content of mercury at some individual samples of household dust originate with domestic activities, such as the breaking of thermometers.

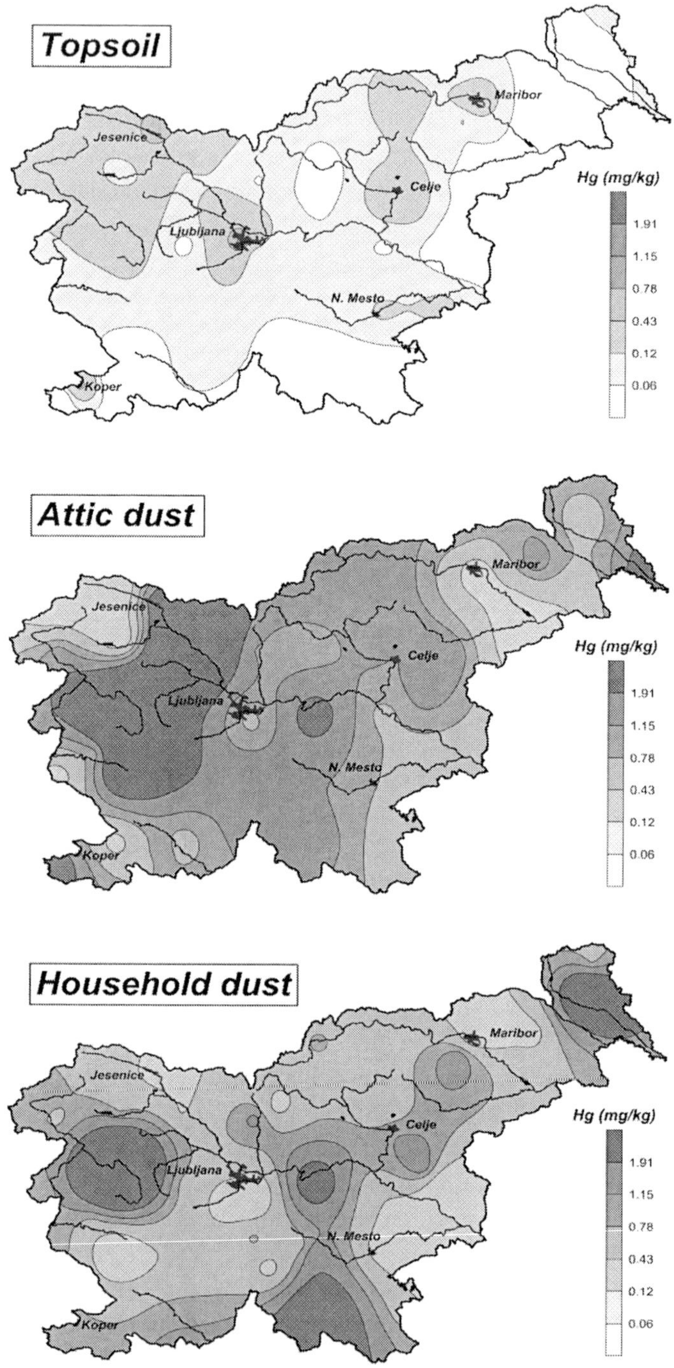

Figure 26. Spatial distribution of Hg in topsoil, attic dust and household dust.

The highest values are found in Koper and Celje (Figure 25) urban zones in all surveyed material. The high contents of mercury in Koper are probably due to sediment transportation of the river Soča. Basically, the whole Trieste bay is enriched with Hg (Ogorelec et al., 1989; Faganeli et al., 1991). However, the highest values observed in urban zones are very low compared to values from the surroundings of Idrija (Gosar at al., 2002; 2006).

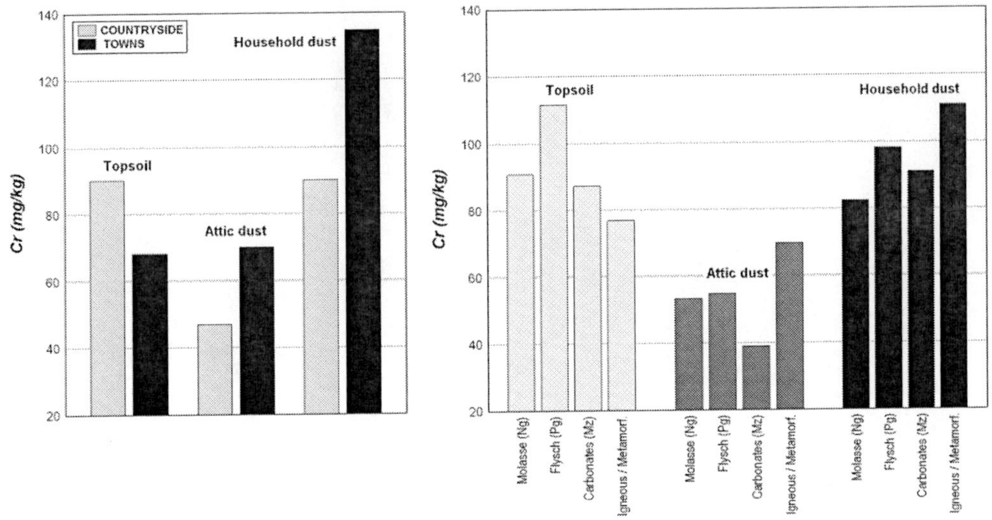

Figure 27. Distribution of Cr in sampling materials according to the urbanization (left) and according to the different geological settings (right).

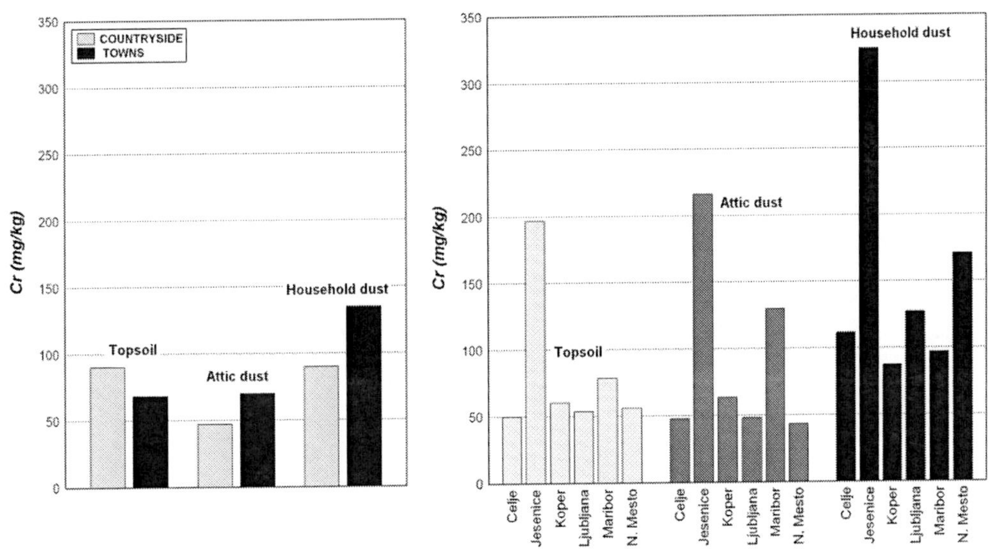

Figure 28. Distribution of Cr in sampling materials according to the urbanization (left) and in different Slovenian towns (right).

5.4.4. Chromium (Cr)

Chromium is a steel grey, lustrous, hard metal in 12th group of the periodic table with atomic number 24. Cr metal and Cr (III) compounds are usually not considered as health hazards, but Cr (VI) compounds can be toxic if orally ingested or inhaled and are irritating to eyes, skin and mucous membranes (Barceloux, 1999). Chromium was regarded with great interest because of its high corrosion resistance and hardness. A major development was the discovery that steel could be made highly resistant to corrosion and discoloration by adding chromium.

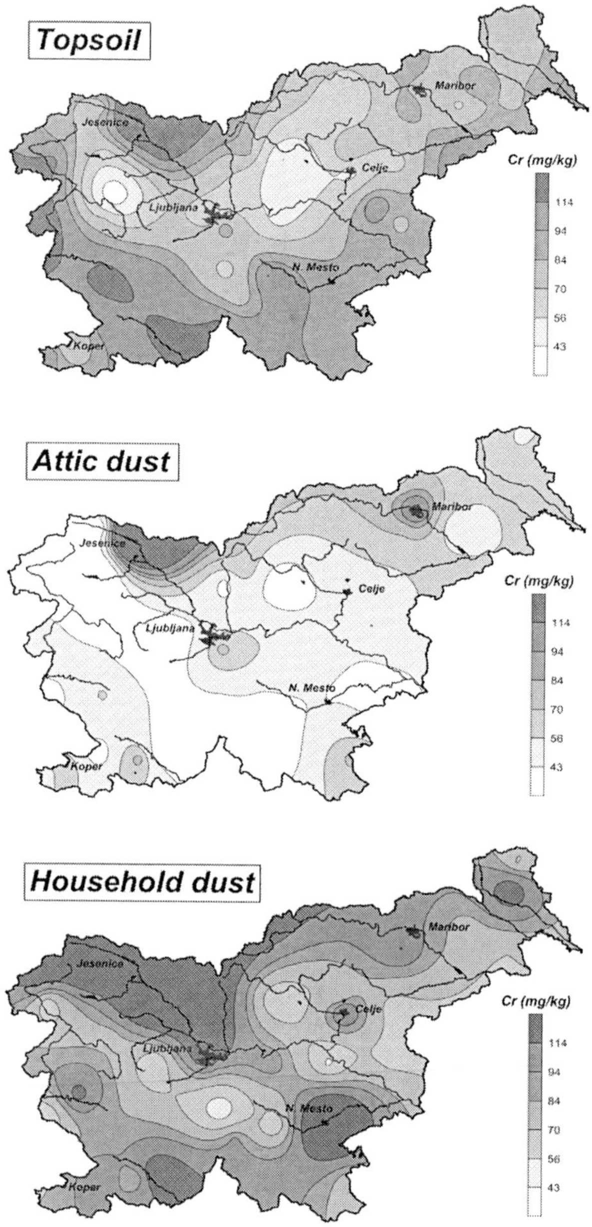

Figure 29. Spatial distribution of Cr in topsoil, attic dust and household dust.

Such steel is named "stainless steel." This application, along with Cr plating, is currently the highest-volume use of the metal. As Cr compounds were used in dyes and paints and the tanning of leather, they are often found in soil and groundwater at abandoned industrial sites (Kotaś, and Stasicka, 2000).

Clarke value of Cr in world soil is 100 mg/kg (Emsley, 2001), and in European topsoils it is95 mg/kg (Salminen, 2005). The world average content of Cr in household dust is 60 mg/kg (Fergusson, 1992). In Slovenian rural areas, the average amount of Cr in topsoil is 90 mg/kg, in household dust (90 mg/kg) and in attic dust (47 mg/kg). A similar distribution pattern of Cr has been found in the sampled media of urban centers.

In the urban areas, the average concentrations of Cr in the attic and house dust exceed the rural values. Average Cr content in urbanized areas is about 130 mg/kg in household dust, and 70 mg/kg in attic dust (Figure 27). Distribution of Cr is influenced by anthropogenic and natural factors (Figures 27, 28 and 29). High concentrations of Cr in soil depend on geological background. In southwestern Slovenia, some chromium is released by the weathering of flysch layers, which are naturally enriched with Cr, and its concentration in soil increases. Despite a strong anthropogenic influence on the distribution of chromium, a connection to the lithological background has been found. According to the geological background, the highest values of Cr in attic dust and household dust have been observed in the igneous and metamorphic rocks (northern Slovenia), and flysch (southwestern Slovenia).

The aforementioned distributions correspond to the high concentrations of Cr in overbank sediments, which had been found as natural in origin (Bidovec et al., 1999). However the areas of igneous and metamorphic rocks also correspond with the areas of ironworking and car producing factories in Jesenice and Maribor, so the natural enrichment of Cr in the northern part of Slovenia is questionable.

In the Slovenian towns, the natural distribution of Cr is almost completely covered by the influence of metallurgic activities or domestic activities. Chromium is typical element of houses. The high content of Cr in household dust is the result of weathering a coating of metallic objects (chrome-plated metal parts and metal sheet metal).

Higher values of Cr are observed in attic dust and household dust in northwestern Slovenia, as a consequence of metallurgic activities in Jesenice (Figure 29). The highest values have been found in soil (310 mg/kg), household dust (610 mg/kg) and attic dust (1320 mg/kg). Similar concentrations have been found in attic dust in Celje (possibly because of Štore ironworks) and around the ironworks Štore and Ravne (Šajn, 2005; 2006), especially with respect to the concentration of Cr in attic dust and soil. The anomaly around Maribor town in attic dust (Figures 28 and 29) can possibly be attributed to the production of different machinery.

5.5. Influence of the Urbanization to the Concentrations of Elements in Topsoil, Attic Dust and Household Dust

Taking into account that samples from natural and urbanized areas are scattered more or less evenly on the whole territory of Slovenia, a comparison between concentrations of elements on natural (N=60) and urbanized areas (N=23) was made using median values. We compared median values (Md) and calculated enrichment ratios (ER) in urbanized areas according to the following equation:

$$ER = Md(C_{town}) / Md(C_{natural\ area}),$$

where $Md(C_{town})$ represents the median value for the specific element in the urbanized areas dataset and $Md(C_{natural\ area})$ represents the same value in the dataset of natural areas. Figure 30 shows the values of the enrichment ratios in towns for different elements and different sampling materials.

Average values of enrichment for all elements are the biggest for topsoil (1.8), slightly smaller for attic dust (1.5) and the smallest for household dust (1.3). We can conclude that the impact of the urbanization is the highest for topsoil. The highest enrichments found for soil are Sn, Sb, Ag, Ta, Cr, Be. In attic dust it is interesting that high enrichments are found among metals connected to smelting industry like Pb, Sb, Mo or Cd (Figure 30). However beside soil and attic dust, household dust is of main interest here, because this is a material everybody is exposed to on a daily basis. The average enrichment of elements in the household dust in urbanized areas is not high, only 1.3, but the problem is that the elements household dust is enriched with are mainly toxic metals, such as Pb and As, which might be a reason for concern. Traffic, heating, using of natural gas for cooking and power generation are suspects for the elevated concentrations of the aforementioned elements in household dust in towns.

The comparison of differences in elemental concentrations between soil/attic dust and between soil/household dust (Figure 31) shows us the scale of anthropogenic emissions of toxic metals into the air from traffic, industry and fossil fuel combustion for heating and other purposes, which is especially valid for attic dust and, to a lesser extent, for house dust. The results show that higher enrichments are found in natural areas than in towns. Moreover, household dust is more enriched material than attic dust. This can be attributed to the fact that soils are less polluted (or not polluted at all) in natural areas, but in towns, due to past industrial activities, soil is already polluted to some extent.

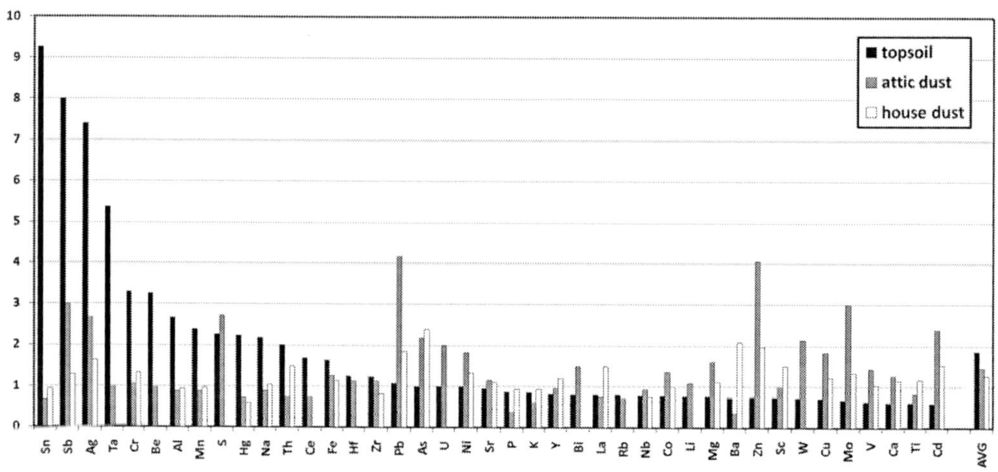

Figure 30. Enrichment ratios (ER) of elements in different sampling materials between natural areas (N=60) and urbanized areas – Slovenian towns (N=23).

On the other hand, everyday dwelling activities (use of metals in houses, gas combustion, paints, plastics etc) have a negative impact on household dust composition more or less evenly (Figure 31), regardless the observation is made in towns or in natural areas. This is the reason for the higher enrichment in household dust in natural areas. Thus it can be concluded from the data in Figure 30 that in houses and apartments the population has the highest impact to the concentrations of Ag, Hg, Ca, Sn, Sb, Zn, Cd, Cu, Mo, Pb and Mg. Ag, Cd, Sb, Zn, Cd, Cu and Mo enrichment can be attributed to the use of metal and electronics in everyday activities. The cause of Hg enrichment can be attributed to natural gas combustion for the cooking or heating purposes and the enrichment of Ca and partially Mg to the use and decoration of indoor construction materials, like white wall paint, mortar, etc.

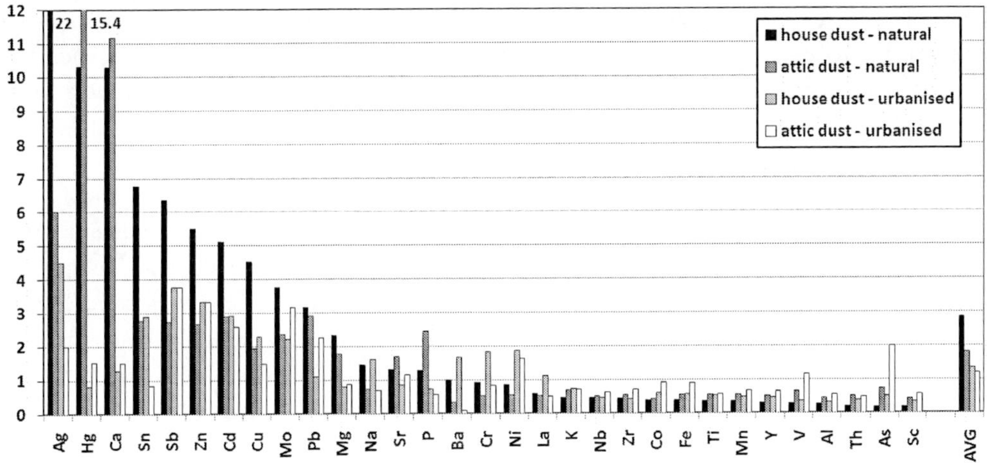

Figure 31. Enrichment ratios (ER) of elements between concentrations in topsoil - attic and house dust for natural areas and towns.

CONCLUSION

The goal of the research work was the introduction of studies of household and attic dust in Slovenia and to assess the abundances and areal distribution of chemical elements in anthropogenic urban sediments in Slovenia - its countryside and larger towns- and to define them with respect to lithological settings (geology). The main conclusions of the research work are that ironworking and smelting activities influence household and attic dust the most. Contrarily, the geological settings influence the composition of soil the most. Population or urbanization influence the soil composition, and this is also valid for attic dust, mostly for toxic metals, connected to metallurgy and smelting. Household dust is the least susceptible to the influential factors discussed in this paper.

In the work, distributions of 41 chemical elements (Al, Ca, Fe, K, Mg, Na, P, S, Ti, Ag, As, Au, Ba, Be, Bi, Cd, Ce Co, Cr, Cu, Hf, La, Li, Mn, Mo, Nb, Ni, Pb, Rb, Sb, Sc, Sn, Sr, Ta, Th, U, V, W, Y, Zn and Zr) were assessed. Dominant geochemical associations have been recognized and presented together with their areal distributions in urban sediments across the Slovenian countryside. The geochemical properties of household and attic dust were compared to topsoil and evaluated in terms of their elemental contents and elemental associations. The proportion of anthropogenically derived toxic metal's impact on urban sediments was assessed. As based on comparisons of household dust, attic dust and topsoil with multivariate statistical method (factor and cluster analysis), four patterns of elemental distributions were established

The first group of chemical elements comprises Ti, V, Al, Th, Sc, Fe, Y, Nb, Co, Mn and La. The Slovenian averages of these elements in household and attic dust are much lower than in topsoil. In the Slovenian countryside these elements have a similar distribution in household and attic dust. The highest elemental values occur in northeastern Slovenia. With respect to lithology, the highest medians of mentioned elements were found in household and attic dust in the area of igneous and metamorphic rocks (Pohorje) and post-orogenic deposits of the Pannonian Sea (molasse) in NE Slovenia. The lowest values were found in areas of outcropping flysch rocks. The contents of these elements in attic and household dust in Slovenian towns oscillate around the Slovenian averages. The highest values of mentioned elements occur in the area of the Maribor town, which confirms the hypothesis of an association of these elements with the weathering of igneous rocks. The principal source of these elements is most likely the weathering of crust or soil.

The second group unites Ba, Na and La. Typical for them is the highest contents in house dust, and very small deviations of contents within sampled materials. The contents might be dependent upon the amount of plagioclases in samples. In the countryside, Na, Ba and Sr are almost equally distributed among house and attic dust. The highest contents were found in northeastern Slovenia. High contents of these elements in house dust might be connected to the use of salt for cooking. Ba, Na and La, especially in attic dust, indicate a well recognized regional geochemical trend. Significant differences within the settlements are attributed to household activities. With respect to the lithology, the highest averages of elements of this group were found in house and attic dust in regions of igneous and metamorphic rocks (Pohorje) and their lowest contents on carbonate rocks. In towns a similar association was detected consisting of Ba, Na and La. Their distributions between towns show high values in the Maribor area which is attributed mainly to geogenic causes. With regard to the other towns included in this research, high deviations in concentrations of elements of this group in attic and house dust is recognized.

The third group associating Pb, Zn, Cd, Sb, Mo, Hg, Sn and Cu represents the chemical elements anthropogenically introduced into the living environment. The Slovenian averages of these elements in attic and household dust are much higher than their Slovenian averages for topsoil. High values of Cd, Pb and Zn in the N part of Slovenia are mainly ascribed to the influence of Mežica mines and smelters and of the Celje zinc smelters. High Hg content in western Slovenia in all considered sample media is a consequence of centuries of operation of the Idrija mercury mine and smelters and of military activities during the first world war. In Slovenian towns, the contents of mentioned chemical elements display similar tendencies of accumulation, with the difference, however, being that the estimated means of these elements are in all analyzed materials much higher than the Slovenian averages.

Interesting exceptions are the mercury means in house and attic dust of towns that are lower than the Slovenian averages. However, the Slovenian average for Hg might be tilted because in the western part of the country a large natural and anthropogenic Hg anomaly exists and is connected to Idrija Hg mining. Very high contents of mentioned elements in all sampled materials are characteristic for the town of Celje, which should be explained by operation of the Celje zinc smelters. High mercury in Koper might be the result of deposition of highly contaminated (with Hg) alluvial sediments of the Soča River to the Gulf of Trieste.

The fourth group comprises Ni, Cr and Co. Their highest contents among all sampled materials were found in house dust. Ni and Cr, and Co are the chemical elements most typical for households. The distributions of mentioned elements in the countryside of Slovenia are influenced by natural and anthropogenic factors. The natural distribution is confirmed by their high contents in the flysch areas of SE Slovenia. The anthropogenic influence is reflected in high values in NW part of the country, as a result of metallurgy in Jesenice. Their contents vary much within the settlements. This fact is attributed to the household activities.

An alarming fact is the very high contents of Cd, Cu, Pb, Zn, Sn and Hg that are found in house dust. They exceed their corresponding values in natural sediments by more than twenty times. The urban sediments, especially the house dust, are substances to which humans are incessantly exposed. Dust particles containing toxic metals enter human organisms through swallowing or inhaling, and they accumulate in it.

Several authors established significant associations between the toxic metal contents in house dust, and concentrations in body liquids (blood, urine). Thereof direct hazards to the population in Slovenia may be derived. High contents of toxic metals in urban dust are potentially dangerous, especially to children. Small infants are the most endangered group because of their higher intake of dust than adults and owing to their higher sensitivity to the influence of toxic metals.

ACKNOWLEDGMENTS

The authors would like to thank the Research Agency of the Republic of Slovenia for funding the research (research programs: Geochemistry and underground waters and Sedimentology and mineral resources), all the people who participated in the research and allowed the sampling of attic and house dust in their houses, and finally, NOVA Science Publishing house for language editing.

REFERENCES

Agarval, S. K. *Heavy metal pollution.* APH Publishing Corporation: New Delhi, 2009; pp 267.

Alijagić, J.; Šajn, R. Distribution of chemical elements in an old metallurgical area, Zenica (Bosnia and Herzegovina). *Geoderma*. 2011, 162/1-2, 71-85.

Alijagić, J.; Šajn, R. Influence of ironworks on distribution of chemical elements in Bosnia and Herzegovina and Slovenia. *Geologija*. 2006, 49/1, 123-132.

Analytical Methods Committee. Simple fitness-for-purpose control chart based on duplicate results obtained from routine test materials. *AMC Technical Brief, Royal Society of Chemistry*. 2002, 9, pp 2.

Avila, A.; Rodrigo, A. Trace metal fluxes in bulk deposition, through fall and stemflow at two evergreen oak stands in NE Spain subject to different exposure to the industrial environment. *Atmospheric Environment.* 2004, 38, 171-180.

Bačeva, K.; Stafilov, T.; Šajn, R.; Tanaselina, C.; Ilić Popov, S. Distribution of chemical elements in attic dust in the vicinity of a ferronickel smelter plant. *Fresenius environmental bulletin.* 2011, 20/9, 2306-2314.

Balabanova, B.; Stafilov, T.; Bačeva, K.; Šajn, R. Biomonitoring of atmospheric pollution with heavy metals in the copper mine vicinity located near Radoviš, Republic of Macedonia. *Journal of Environmental Science and Health, Part A - Toxic / Hazardous Substances and Environmental Engineering.* 2010, 45/12, 1504-1518.

Balabanova, B.; Stafilov, T.; Šajn, R.; Bačeva, K. Distribution of chemical elements in attic dust as reflection of their geogenic and anthropogenic sources in the vicinity of the copper mine and flotation plant. *Archives of Environmental Contamination and Toxicology*. 2011, 61/2, 173-184.

Banerjee, A. Heavy metal levels and solid phase speciation in street dusts of Delhi, India. *Environmental Pollution.* 2003, 123, 95-105.

Barceloux, D. G. Chromium. *Clinical Toxicology.* 1999, 37/2, 173–194.

Bates, M.; Malcolm, M.; Wyatt, R.; Garrett, N.; Galloway, Y.; Speir, T.; Read, D. Lead in children from older housing areas in the Wellington region. *The New Zealand medical journal.* 1995, 108, 400-404.

Beelen, R.; Hoek, G.; Pebesma, E.; Vienneaud, D.; de Hoogh, K.; Briggs, D.J. Mapping of background air pollution at a fine spatial scale across the European Union. *Science of the Total Environment.* 2009, 407, 1852-1867.

Bidovec, M.; Šajn, R.; Gosar, M. Uporaba recentnih poplavnih sedimentov v geokemičnem kartiranju Slovenije = The use of recent overbank sediments in geochemical mapping of Slovenia. *Geologija.* 1999, 41, 275-317.

Boutron, C.; Vandal, G.; Fitzgerald, W.; Ferrari, C. A forty year record of mercury in central Greenland snow. *Geophysical Research Letters.* 1998, 25, 3315-3318.

Bowen, H.J. *Environmental chemistry of the elements.* Academic Press: London, 1979; pp 318.

Budkovič, T.; Šajn, R.; Gosar, M. Vpliv delujočih in opuščenih rudnikov kovin in topilniških obratov na okolje v Sloveniji = Influence of abandoned metal mines and smelters to the environment of Slovenia. *Geologija.* 2003, 46/1, 135-140.

Buser, S.; Draksler, V. *Geološka karta Slovenije 1:500.000* = Geological map of Slovenia 1:500.000. Geodetski zavod Slovenije: Ljubljana, 1989; pp 1.

Buser, S.; Ramovš, A.; Drovenik, M.; Pleničar, M. *Geološka zgradba Slovenije = Geological composition of Slovenia.* In: Enciklopedija Slovenije = Encyclopedia of Slovenia, 3rd book. Mladinska knjiga: Ljubljana, 1989; 195-203 pp.

Candelone, J.; Hong, S.; Pellone, C.; Boutron, C. Post-industrial revolution changes in large-scale atmospheric pollution of the northern hemisphere by heavy metals as documented in central Greenland snow and ice. *Journal of Geophysical Research.* 1995, 100, 16605-16616.

Cizdziel, J.V.; Hodge, V.F.; Faller S. Resolving Nevada Test Site and global fallout plutonium in attic dust and soils using 137Cs/239+240 Pu activity ratios. *Health Physics.* 1999, 77/1, 67-75.

Cizdziel, J.V.; Hodge, V.F.; Faller, V.F. Plutonium anomalies in attic dust and soils at locations surrounding the Nevada Test Site. *Chemosphere.* 1998, 37/6, 1157-1168.

Cizdziel, J.V.; V.F. Hodge. Attics as archives for house infiltrating pollutants: trace elements and pesticides in attic dust and soil from southern Nevada and Utah. *Microchemical Journal.* 2000, 64, 85-92.

Culbard, E.; Thornton, I.; Brooks, K.; Watt, J.; Thomson, M.; Moorcroft, S. Metal contamination of dusts and soils in urban and rural households in the United Kingdom: National reconnaissance survey of the united kindom to determine metal concentration. In: Hemphill, D.D. Eds.: *Trace substances in environmental health XVII.* University of Missouri: Columbia, 1983, pp 236-241.

Davies, D.J.; Thornton, I.; Watt, J.M.; Culbard, E.B.; Harvey, P.G.; Delves, H.T.; Sherlock, J.C.; Smart, G.A.; Thomas, J.F.; Quinn, M.J. Lead intake and blood lead in two-year-old UK urban children. *Science of the Total Environment.* 1990, 90, 13-29.

Davis, J.C. *Statistic and data analysis in geology.* Willey and Sons: New York; 1986, pp 651.

Drev, J.; Lukan, B.; Gros, I. *Meritve zaprašenosti in koncentracije azbestnih vlaken v DO Salonit Anhovo TOZD Cevi* = Measurements of the dust and concentrations of the azbestous fibres in the tubes production sector in the Salonit Anhovo company. Zavod za zdravstveno varstvo Maribor: Maribor; 1986, pp 16.

Drev, J.; Roter, M. *Meritve imisije v okolici tovarne Impol v Slovenski Bistrici* = Measurements of the imissions in the surroundings of the Impol factory in Slovenska Bistrica. Zavod za zdravstveno varstvo Maribor: Maribor; 1992, pp 7.

Emsley, J. Chromium. Nature's Building Blocks: An A-Z Guide to the Elements. Oxford University Press: Oxford; 2001, pp. 495-498.

Facchinelli, A.; Sacchi, E.; Mallen, L. Multivariate statistical and GIS-based approach to identify heavy metal sources in soils. *Environmental Pollution*. 2001, 114, 313-324.

Faganeli, J.; Planinc, R.; Pezdič, J.; Smodiš, B.; Stegnar, P.; Ogorelec, B. Marine geology of the Gulf of Trieste (northern Adriatic), Geochemical aspect. *Marine Geology*. 1991, 99, 93-108.

Fergusson, J.E. Dust in the environment, elemental composition and sources. In: Dunnette, D.A.; O'Brien, R.J. Eds.: The science of global change, the impact of human activities on the environment. Portland State University, American Chemical Society: Washington D.C.; 1992, pp 116-133.

Fergusson, J.E.; Kim, N.D. Trace elements in street and house dusts: Sources and speciation. *The science of the total environment.* 1991, 100, 125-150.

Fergusson, J.E.; Ryan, D.E. The elemental composition of street dust from large and small urban areas related to city type, source and particle size. *The science of the total environment.* 1984, 34, 101-116.

Filipović, I.; Lipanović S. Opća i anorganska kemija = General and anorganic chemistry. Školska knjiga: Zagreb; 1995, pp 1145.

Filzmoser, P.; Garrett, R.G.; Reimann, C. Multivariate outlier detection in exploration geochemistry. *Computers and Geosciences*. 2005, 31, 579–587.

Finlayson-Pitts, B.; Pitts, J.N. Chemistry of the Upper and Lower Atmosphere. Academic Press: San Diego; 2000, pp 969.

Gabor, B. Problem emisij asfaltne baze v Lipovcih = The problematics of the emissions of the asphalt plant in Lipovci, diploma thesis. Fakulteta za strojništvo, University of Maribor: Maribor; 1994, pp 94.

Galvin, J.; Stephenson, J.; Wlodarczyk, J.; Loughran, R.; Waller, G. Living near a lead smelter: an environmental health risk assessment in Boolaroo and Argenton, New South Wales. *Australian Journal of Public Health*. 1993, 17, 373-378.

Gauderman, W.J.; Avol, E.; Gilliland, F.; Vora, H.; Thomas, D.; Berhane, K.; McConnell, R.; Kuenzli, N.; Lurmann, F.; Rappaport, E.; Margolis, H.; Bates, D.; Peters, J. The effect of air pollution on lungs from 10 to 18 Years of age. *The New England Journal of Medicine*. 2004, 351, 1057-1067.

Geater, A.F.; Duerrawee, M.; Chompikul, J.; Chairatanamanokorn, S.; Pongsuwan, N.; Chongsuvivatwong, V.; McNeil, D. Blood lead levels among schoolchildren living in the Pattani River: two contamination scenarios? *Journal of Occupational and Environmental Medicine*. 2000, 2, 11–16.

Godish, T. *Air quality*, 4th edition. CRC Press LLC: Boca Raton; 2004, pp 480.

Gorlach, U.; Boutron, C. Variations in heavy metals concentrations in Antarctic snows from 1940 to 1980. *Journal of Atmospheric Chemistry*. 1992, 14, 205-222.

Gosar, M.; Šajn, R.; Biester, H. Zvrsti živega srebra v tleh in podstrešnem prahu na Idrijskem = Mercury speciation in soils and attic dust in the Idrija area. *Geologija*. 2002, 45-2, 373-378.

Gosar, M.; Žibret, G. Mercury contents in the vertical profiles through alluvial sediments as a reflection of mining in Idrija (Slovenia). *Journal of Geochemical Exploration*. 2011, 110/2, 81-91.

Gosar, M; Šajn, R.; Biester, H. Binding of mercury in soils and attic dust in the ,Idrija mercury mine area (Slovenia). *The Science of the Total Environment*. 2006, 369/1-3, 150-162.

Govindaraju, K. 1989: Geostandards newsletter. -International working group "Analytical standard of minerals, ores and rocks", Vol. 13, Special Issue, 133 p. Vandoeuvre-les-Nancy Cedax.

Greenwood, N.N. *Chemistry of the Elements*. Butterworth-Heinemann: Oxford; 1984, pp 1542.

Gspan, P.; Hrašovec, B. *Prah v proizvodnji* = Dust in production. Zavod Republike Slovenije za varnost pri delu: Ljubljana; 1993, pp 160.

Hill, S.J. Lead. In: Stoeppler, M. Eds.: *Hazardous metal in the environment*. Elsevier: Amsterdam; 1992, pp 231-255.

HIRS. *Klimatografija Slovenije, padavine 1951-1980, drugi zvezek* = Climatography of Slovenia, precipitation 1951-1980, second book. Hidrometeorogical Institute of Slovenia: Ljubljana; 1989b, pp 393.

HIRS. *Klimatografija Slovenije, temperatura zraka 1951-1980, prvi zvezek* = Climatography of Slovenia, air temperatures 1951-1980, first book. Hidrometeorogical Institute of Slovenia: Ljubljana; 1989a, pp 331.

Hollander, M.; Wolfe, D.A. *Nonparametric Statistical Methods, 2nd Edition*. John Wiley and Sons: New York, Chichster, Weinheim, Brisbane, Singapore, Toronto; 1999, pp 787.

Hong, S.; Candelone, J.; Patterson, C.; Boutron, C. History of ancient copper smelting pollution during Roman and medieval times recorded in Greenland ice. *Science*. 1996, 272, 246–249.

Hou, H.; Takamatsu, T.; Koskiwa, M.K.; Hosomi, M. Trace metal sin bulk precipitation and through fall in a suburban arca of Japan. *Atmospheric Environment*. 2005, 39, 3583–3595.

Hršak, J.; Škrbec, A.; Balagovič, I.; Šega, K. Thallium content in Zagreb air. *Bulletin of Environmental Contamination and Toxicology*. 2003, 71, 131-134.

Hur, S.; Cunde, X.; Hong, S.; Barbante, C.; Gabrielli, P.; Lee, K.; Boutron, C.; Ming, Y. Seasonal patterns of heavy metal deposition to the snow on Lambert Glacier basin, East Antarctica. *Atmospheric Environment*. 2007, 41, 8567-8578.

Ibiebele, D.D. Air and blood lead levels in a battery factory. *Science of the Total Environment*. 1994, 152, 269-273.

Imperato, M.; Adamo, P.; Naimo, D.; Arienzo, M.; Stanzione, D.; Violante, P. Spatial distribution of heavy metals in urban soils of Naples city (Italy). *Environmental Pollution*. 2003, 124, 247–256.

Jemec, M.; Šajn, R. Geokemične raziskave tal in podstrešnega prahu na območju Litije = Geochemical research of soil and attic dust in Litija area, Slovenia. *Geologija*. 2007, 50/2, 497-505.

Kotaś, J.; Stasicka, Z. Chromium occurrence in the environment and methods of its speciation. Environmental Pollution. 2000, 107/3, 263-283.

Lado, L.R.; Hengl, T.; Reuter, H.I. Heavy metals in European soils: A geostatistical analysis of the FOREGS Geochemical database. *Geoderma*. 2008, 148, 189-199.

Le Maitre, R.W. *Numerical Petrology, Statistical interpretation of geochemical data.* Elsevier: Amsterdam; 1982, pp 281.

Lioy P.J. Assessing total human exposure to contaminants. *Environmental Science and Technology*. 1990, 24, 938-945.

Lukan, B.; Pohorec, I.; Medved, M. *Emisije prašnih delcev iz tovarne Belt črnomelj (strokovno poročilo)* = Emissions of dust particles from the Belt factory in Črnomelj (Expert report). Zavod za zdravstveno varstvo Maribor: Maribor; 1990, pp 18.

Lukan, B.; Roter, M. *Emisije prašnih delcev iz rotacijske peči stare cementarne v DO Salonit Anhovo (strokovno poročilo)* = Emissions of dust particles from the rotary furnace in Salonit Anhovo (expert report). Zavod za zdravstveno varstvo Maribor: Maribor; 1990, pp 6.

Luo, W.; Wang, T.; Lu, Y.; Giesy, J.P.; Shi, Y.; Zheng, Y.; Xing, Y.; Wu, G. Landscape ecology of the Guanting reservoir, Beijing, China: multivariate and geostatistical analyses of metals in soils. *Environmental Pollution*. 2007, 146, 567-576.

Micó, C.; Recatalá, L.; Peris, M.; Sánchez, J. Assessing heavy metal sources in agricultural soils of an European Mediterranean area by multivariate analysis. *Chemosphere*. 2006, 65, 863-872.

Mielke, H.W; Adams, J.L.; Reagan, P.L.; Mielke Jr., P.W. Soil-dust lead and childhood lead exposure as a function of city size and community traffic flow: The case for lead abatement in Minnesota. In: Davies, B.E.; Wixson, B.G. Eds.: Lead in soil: issues and guidelines. *Environmental Geochemistry and Health Supplement.* 1989, 9, 253-271.

Molhave, L.; Schneider, T.; Kjvrgaard, S.K.; Larsen, L.; Norn, S.; Jorgensen, O. House dust in seven Danish offices. *Atmospheric Environment*. 2000, 34, 4767-4779.

Ogorelec, B.; Mišič, M.; Faganeli, J.; Stegnar, P.; Vrišer, B.; Vukovič, A. Recentni sedimenti Koprskega zaliva. *Geologija*. 1988, 30, 87-121.

Ogunsola, O.J.; Oluwole, A.F.; Asubiojo, O.I.; Olaniyi, H.B.; Akeredolu, F.A.; Akanle, O.A.; Spyrou, N.M. Traffic pollution: Preliminary elemental characterisation of roadside dust in Lagos, Nigeria. *The science of the total environment*. 1994, 146/147, 175-161.

Ozaki, H.; Watanabe, I.; Kuno, K. As, Sb and Hg distribution and pollution sources in the roadside soil and dust around Kamikochi, Chubu Sangaku National Park, Japan. *Geochemical Journal*. 2004, 38, 473-484.

Pacyna, E.G.; Pacyna, J.M.; Fudala, J.; Strzelecka-Jastrzab, E.; Hlawiczka, S. Panasiuk, D.; Nitter, S.; Pregger, T.; Pfeiffer, H.; Friedrich, R. Current and future emissions of selected heavy metals to the atmosphere from anthropogenic sources in Europe. *Atmospheric Environment*. 2007, 41, 8557-8566.

Peh, Z.; Šajn, R.; Halamić, J.; Galović, L. Multiple discriminant analysis of the Drava River alluvial plain sediments. *Environmental Geology*. 2008, 55/7, 1519-1535.

Pirc, S.; Budkovič, T. *Remains of world war I geochemical pollution in the landscape*. In: Richardson, M. Eds.: Environmental xenobiotics. Taylor and Francis: London. 1996, pp 375-418.

Quinn, M.J.; Sherlock, J.C. The correspondence between UK ‘‘action levels’’ for lead in blood and in water. *Food Additives and Contaminants*. 1990, 7, 387-424.

Rasmussen, P.E.; Subramanian, K.S.; Jessiman, B.J. A multi-element profile of housedust in relation to exterior dust and soils in the city of Ottawa, Canada. *Science of the Total Environment*. 2001, 267, 125-140.

Reimann, C.; Filzmoser, P.; Garrett, R.G. Factor analysis applied to regional geochemical data: problems and possibilities. *Applied Geochemistry*. 2002, 17/3, 185-206.

Rose, A. W.; Hawkes, H.E.; Webb, J.S. *Geochemistry in mineral exploration, II ed*. - Academic Press: London; 1979, pp 657.

Salminen, R.; Batista, M.J.; Bidovec, M.; Demetriades, A.; De Vivo, B.; De Vos, W.; Duris, M.; Gilucis, A.; Gregorauskiene, V.; Halamic, J.; Heitzmann, P.; Jordan, G.; Klaver, G.; Klein, P.; Lis, J.; Locutura, J.; Marsina, K.; Mazreku, A.; O'Connor, P.J.; Olsson, S.Å.; Ottesen, R.T.; Petersell, V.; Plant, J.A.; Reeder, S.; Salpeteur, I.; Sandström, H.; Siewers, U.; Steenfelt, A.; Tarvainen, T.; *Geochemical Atlas of Europe, Part 1, Background Information, Methodology and Maps*. Geological Survey of Finland: Espoo; 2005, pp 526.

Skumavec, D.; Šabić, D. Pokrovnost tal v Sloveniji 1993-2001 = Land cover in Slovenia 1993-2001. Statistical Office of the Republic of Slovenia: Ljubljana. 2005, 851, pp 66.

Snedecor, G.W.; Cochran, W.G. *Statistical methods*. The Iowa State University Press: Ames; 1967, pp 575.

SURS. Statistične informacije – nacionalni računi = Statistical rapid reports = National Accounts. Statistical Office of the Republic of Slovenia: Ljubljana. 2011, 15, pp. 19.

SURS. Statistične novice – E-objava = Statistical news – E-publication. Statistical Office of the Republic of Slovenia: Ljubljana. 2009, 12. february.

Šajn R. Distribution of chemical elements in attic dust and soil as reflection of lithology and anthropogenic influence in Slovenia. *Journal de Physique*. 2003, 107, 1173-1176.

Šajn, R. Factor analysis of soil and attic-dust to separate mining and metallurgy influence, Meža Valley, Slovenia. *Mathematical geology*. 2006, 38/6, 735-747.

Šajn, R. *Geokemične lastnosti urbanih sedimentov na ozemlju Slovenije* = Geochemical properties of urban sediments on the territory of Slovenia. Geological survey of Slovenia: Ljubljana; 1999, pp 136.

Šajn, R. Using attic dust and soil for the separation of anthropogenic and geogenic elemental distributions in an old metallurgic area (Celje, Slovenia). *Geochemistry: Exploration Environment Analysis*; 2005, 5/1, 59-67.

Šajn, R. Vpliv litologije in antropogenih dejavnosti na porazdelitev kemičnih prvin v bivalnem prahu Slovenije = Influence of lithology and antropogenic activity on distribution of chemical elements in dweling dust, Slovenia. *Geologija*. 2000, 43/1, 85-101.

Šajn, R.; Bidovec, M.; Gosar, M.; Pirc, S. Geochemical soil survey at Jesenice area, Slovenia. *Geologija*. 1998, 41, 319-338.

Šajn, R.; Gosar, M. Onesnaženost tal v okolici Litije kot posledica rudarskih in metalurških dejavnosti ter naravnih danosti = Soil pollution in surroundings of Litija as a reflection of mining, metallurgy and natural conditions. *Geologija*. 2007, 50/1, 131-145.

Šajn, R.; Gosar, M.; Bidovec, M. Geokemične lastnosti tal, poplavnega sedimenta ter stanovanjskega in podstrešnega prahu na območju Mežice = Geochemical properties of soil, overbank sediment, household and attic dust in Mežica area (Slovenia). *Geologija*. 2000, 43/2, 235-245.

Šajn, R.; Halamić, J.; Peh, Z.; Galović, L.; Alijagić, J. Assessment of the natural and anthropogenic sources of chemical elements in alluvial soils from the Drava River using multivariate statistical methods. *Journal of geochemical exploration*. 2011, 110/3, 278-289.

Tasdemir, Y.; Kural, C. Atmospheric dry deposition fluxes of trace elements measured in Bursa, Turkey. *Environmental Pollution*. 2005, 138, 462-472.

Templ, M.; Filzmoser, P.; Reimann, C. Cluster analysis applied to regional geochemical data: Problems and possibilities. *Applied Geochemistry*. 2008, 23, 2198-2213.

Theocharopoulos, S.P.; Wagner, G.; Sprengart, J.; Mohr, M.E.; Desaules, A.; Muntau, H.; Christou, M.; Quevauviller, P. European soil sampling guidelines for soil pollution studies. *The Science of the Total Environment*. 2001, 264, 51-62.

Vallero, D. Fundamentals of air pollution, 4th edition. Academic Press: Amsterdam, Boston, Heidelberg, London, New York, Oxford, Paris, San Diego, San Francisco, Singapore, Sydney, Tokyo; 2008, pp 968.

Van de Velde, K.; Vallelonga, P.; Candelone, J.; Rosman, K.; Gaspari, V.; Cozzi, G.; Barbante, C.; Udisti, R.; Cescon, P.; Boutron, C. Pb isotope record over one century in snow from Victoria Land, Antarctica. *Earth and Planetary Science Letters*, 2005, 232, 95-108.

von Storch, H.; Costa-Cabral, M.; Hagner C.; Feser, F.; Pacyna, J.; Pacyna, E.; Kolbc, S. Four decades of gasoline lead emissions and control policies in Europe: a retrospective assessment. *Science of The Total Environment Volume*. 2003, 311/1-3, 151-176.

Vrabec, M.; Šmuc, A.; Pleničar, M.; Buser, S. *Geološki razvoj Slovenije - Povzetek* = Geological evolution of Slovenia - An Overview. In.: Pleničar, M.; Ogorelec, B.; Novak, M. Eds.: Geologija Slovenije = Geology of Slovenia. Geološki zavod Slovenije: Ljubljana; 2009, pp 23-40.

Vreča, P.; Pirc, S.; Šajn, R. Natural and anthropogenic influences on geochemistry of soils in terrains of barren and mineralized carbonate rocks in the Pb-Zn mining district of Mežica, Slovenia. *Journal of geochemical exploration*. 2001, 74, 99-108.

Wiberg, E.; Wiberg, N.; Holleman A.F. *Holleman-Wiberg Inorganic Chemistry*. Academic Press: San Diego, London, Boston, New York, Sydney, Tokyo, Toronto; 2001, pp 1884.

Wilson, B.; Pyatt, F.Heavy Metal Bioaccumulation by the Important Food Plant, Olea europaea L., in an Ancient Metalliferous Polluted Area of Cyprus. *Bulletin of Environmental Contamination and Toxicology*. 2007, 78/5, 390-394.

Zhang, C. Using multivariate analyses and GIS to identify pollutants and their spatial patterns in urban soils in Galway, Ireland. *Environmental Pollution*. 2006, 142, 501-511.

Zupančič, B. *Klima* = Climate. In: Enciklopedija Slovenije, 5. zvezek = Encyclopedia of Slovenia, 5th book. Mladinska knjiga: Ljubljana; 1989, pp 96-100,

Žibret, G. Determination of historical emission of heavy metals into the atmosphere: Celje case study. *Environmental geology*. 2008, 56/1, pp 189-196.

Žibret, G.; Gosar, M. Calculation of the mercury accumulation in the Idrijca River alluvial plain sediments. *Science of the total environment*. 2006, 368/1, pp 291-297.

Žibret, G.; Rokavec, D. Household dust and street sediment as an indicator of recent heavy metals in atmospheric emissions: a case study on a previously heavily contaminated area. *Environmental earth sciences*. 2010, 61/3, pp 443-453.

Žibret, G.; Šajn, R. *Impacts of the mining and smelting activities to the environment - Slovenian case studies*. In: Sánchez, M.L. Eds.: Causes and effects of heavy metal pollution. Nova Science Publishers: New York; 2008a, 1-80.

Žibret, G.; Šajn, R. Modeling of atmospheric dispersion of heavy metals in the Celje area, Slovenia. *Journal of geochemical exploration*. 2008b, 97, 29-41.

In: Dust: Sources, Environmental Concerns and Control ISBN: 978-1-61942-547-7
Editors: Laurent B. Wouters and Michel Pauwels

Chapter 2

DISTRIBUTION OF HEAVY METALS IN ATTIC AND DEPOSITED DUST IN THE VICINITY OF COPPER ORE PROCESSING AND FERRONICKEL SMELTER PLANTS IN THE REPUBLIC OF MACEDONIA

***Trajče Stafilov*[1,*], *Robert Šajn*[2], *Biljana Balabanova*[3] and *Katerina Bačeva*[1]**

[1]Institute of Chemistry, Faculty of Science,
Sts. Cyril and Methodius University, Skopje, Macedonia
[2]Geological Survey of Slovenia, Ljubljana, Slovenia
[3]Faculty of Agriculture, Goce Delčev University, Štip, Macedonia

ABSTRACT

A comprehensive monitoring was applied to assess the environmental pollution in the Republic of Macedonia. Large amounts of fine dust are generated during blasts and excavations of mining minerals and from highly technological industrial processes, whereas they are distributed in the air by the winds. For that issue characterization of heavy metals content in attic dust and total deposited matter was performed. The significant emission sources that contribute to atmospheric pollution with heavy metals on the territory of the Republic of Macedonia appear to be all mines and smelter plants. In one of the region three collection stations were placed for monthly monitoring of heavy metals content in total deposited dust is near a copper mine in the Radoviš region and the second location is in the vicinity of the smelter plant known for its ferronickel industrial activity in the Kavadarci region. In both locations two most affected villages and the nearest town was monitor within one year. Characterisation of contents showed higher content of heavy metals in deposited dust, which is due to distribution of fine particles from mine and smelter plant respectively, in air and their deposition in settlements in the close vicinity.

Attic dust samples were also used in order to determine and monitor long-term air pollution with heavy metals around the copper mine and the ferronickel smelter plant.

* E-mail: trajcest@pmf.ukim.mk.

The total contents of about 20-40 elements were analysed by inductively coupled plasma - mass spectrometry (ICP-MS), inductively coupled plasma - atomic emission spectrometry (ICP-AES) and electrothermal atomic absorption spectrometry (ETAAS). Basic statistical methods and multivariate exploratory (factor and cluster analysis) techniques were applied for data processing. The results of the study of spatial distribution of different trace elements over these regions show some areas with critically high content of some elements. It was found that high contents of those anthropogenic elements are deposited in the close vicinity of the mine and the smelter plant.

Keywords: Air pollution, monitoring, heavy metals, attic dust, total deposited matter, copper mine, ferronickel smelter plant, Republic of Macedonia

1. Introduction

Heavy metals is the term applied to a large group of trace elements which are both industrially and biologically important. 'Heavy metals' is the most widely recognized and used term for a large group of elements with an atomic density greater than 6 g cm^{-3} (Chittum, 1964). But these elements in high contents are toxic for living organisms. Although not completely satisfactory from a chemical point of view, the term "heavy metals" is often used when speaking about the phenomenon of pollution and toxicity.

Emissions of heavy metals into the environment happen through several processes. The emission of heavy metals into the atmosphere is one of the greatest threats to human health. People are directly exposed to the effects of heavy metals through inhalation of airborne micro particles from atmospheric dust (Jarup, 2003). Atmospheric particles affect to the human health when they enter the respiratory system. Depth of penetration and deposition of particles depends on the size of particle, the defense capabilities of the respiratory tract and the respiratory pattern (Godish, 2004). Exposure to heavy metals is possible if there is a coexistence of heavy metals and people. Exposure is usually defined as a function of the pollutant content and time, i.e., that it is "an appearance achieved by direct contact between humans and the environment with the pollutant content in a given interval of time" (Clarke, 1998; Järup, 2003). The health effects of heavy metals distributed with air-dust depend on their mass concentration and where they are deposited in the respiratory tract. The most exposed group in the population to air particles are children, because the particles penetrate deeper in their lungs as compared to adults because children they breathe deeper and faster, they spend more time outdoor and they are more active. The polluted air slows down the development of pulmonary functions in children (Gauderman et al., 2000; Gauderman et al., 2004). Senior citizens, especially those with a weakened cardiovascular and respiratory system are a high risk group too. Another risk group is patients with chronic pulmonary emphysema, asthma or cardiovascular diseases (Vallero, 2008).

Air pollution with heavy metals is a global process that affects every part of the globe. The contribution of the human factor to this phenomenon is very broad and complex, which leads to feedback processes. Rapid increases of heavy metal concentrations in the atmosphere and environment are commonly coupled to the development of exploitative technologies. This kind of sudden change exposes the biosphere to a risk of destabilization, to organisms that developed under conditions whit low concentrations of metal, present have not developed biochemical pathways capable of detoxifying the metal when it is present at high concentrations. Although the human impact on the biosphere dates back to the Neolithic period, the problems of degradation of ecosystems due to the pollution became increasingly acute during the last decades of the 20th century (Barbante et al., 2004). Atmospheric deposition of heavy metals is the main subject of many studies and usually occurs in industrialized areas, in places where exploitation and processing of natural resources (oil, ore, etc.) are performed, in areas with large population centers where traffic and municipal waste are the main sources of metals, etc. (Athar et al., 1995).

Heavy metals in the atmosphere originated mainly from dust dispersion from metal refining, fossil fuel combustion, vehicle exhausts, and other human activities and stay in the atmosphere until they are removed by a variety of cleansing processes.

Particular emphasis is given on ore deposits, mining, and smelter plants as significant anthropogenic sources of dust. Heavy metals emitted in the atmosphere by combustion processes usually have relatively high solubility's and reactivity's; especially under low-pH condition (Hršak et al., 2003; Avila and Rodrigo, 2004; Hou et al., 2005). They can be carried to places far away from the sources by wind, depending upon whether they are in gaseous form or as particulates. Metallic pollutants are ultimately washed out of the air by rain and deposited on the land.

1.1. Atmospheric Dust

Dust is a generic term used to describe fine particles that are suspended in the atmosphere. Atmospheric particles (alternatively referred to as particulate matter - PM) are tiny particles of solids or liquids suspended in air. These particles vary in size and density. They originate as a result of processes on the surface of the earth and in the atmosphere. They may be either emitted directly into the atmosphere or formed there by chemical reactions (Finlayson-Pitts and Pitts, 2000). Particles may be emitted naturally or have an anthropogenic origin. Therefore, some naturally occurring particulates originate from volcanoes, dust storms, forest and grassland fires, living vegetation and sea spray. Human activities, such as the burning of fossil fuels in vehicles, power and smelter plants, mining, ore processing and various industrial processes also generate significant amounts of aerosols.

The main problem of pollution represents anthropogenic sources; however, natural pollution is not negligible. According to the origin and specific characteristics particles differ in size, mass, density, morphology, chemical composition and have various chemical and physical properties (Godish, 2004; Baron and Willeke, 2005).

According to their properties particles settle out from the atmosphere in matter of minutes if they are small and dense or remains suspended for days, even weeks (depend on weather conditions).

Atmospheric dust particles can be primary or secondary according to their origin and formation process. Primary particles in the air are the result of direct emissions in the atmosphere from different natural or anthropogenic sources. However, secondary particles occur in air as a result of different chemical reactions among particles, gases, H_2O and vapor (Godish, 2004). Emission of primary particles can occur from stationary sources (such as factories, power plants etc.) or from moving vehicle (mobile sources). Also non-point or areas sources are known when the sources represent larger areas (construction sites, large open stockpiles, open-pit mining) (Pacyna et al., 2007).

Therefore dust particles suspended in the atmosphere contain hundreds of different chemical compound due to large number of sources of primary particles, formation of secondary particles in air by atmospheric chemistry, growth of atmospheric particles and sorption of gas phase substances (Godish, 2004).

Chemical composition of particles in air differs and is dependent on origin, production processes and atmospheric history of individual particle. However, the chemical composition of particles is very important and highly variable. The chemical composition of tropospheric particles includes inorganic ions, different metals, elemental carbon, organic compounds and crustal (carbonates and compounds of alkali and rare earth elementals) substances (Vallero, 2008). Therefore wind-blown mineral dust tends to be made of mineral oxides and other material blown from the Earth's crust.

1.2. Dust Deposition

Atmospheric dust (large or coarse particles) has traditionally been considered mainly as a cause of nuisance or of health effects due to secondary pathways following its deposition to the ground. Dust deposition is therefore measured using simple passive devices (dry, wet, sticky sampling devices). However, these methods have some deficiencies due to the aerodynamic characteristics of the sampling gauge.

Wind induced flows around the gauge either prevent the entry of particles into the gauge opening or removes them (after deposition) due to a wind generated circulation inside the gauge (Hall et al., 1993). Due to exposure to various disturbances such measurement are less reliable. The strong dependency of gauge performance on wind speed and particle size leads to unrepresentative sampling results (Hall et al., 1993).

Never the less with some improvements of the measuring device results are indicative and therefore useful. The results of such measurement are expressed in g m^{-2} (or mg m^{-2}) in the unit of time; times of collection are usually different (Čačković et al., 2009). Basic principle of such measurements is gravitational force which is proportional to particle mass and gravitational acceleration and force particle to settle. Deposition is the process by which aerosol particles collect or deposit themselves on solid surfaces and thus decreasing the concentration in the air.

The rate of deposition or the deposition velocity is faster for particles of a large size. Very large particles will settle out quickly through sedimentation (settling) or impaction processes, while Brownian diffusion has the greatest influence on small particles.

According to the process of deposition two different types of deposition exist - dry and wet. Dry deposition is caused by gravitational sedimentation, interception, impaction, diffusion or Brownian motion, turbulence and other processes, such as thermophoresis, diffusiophoresis and electrophoresis.

Wet deposition includes some atmospheric phenomena which scavenge aerosol particles - rain, snow, fog, cloud, dew formation and washout the atmosphere (Godish, 2004; Vallero, 2008). The amount of dust released to atmosphere increases with increasing wind speed, with the handling of powdery materials in the open etc. A similar situation occurs with dust carried away from stockpiles.

1.3. Heavy Metals (Dust) Emission

There are many sources and release mechanisms for dust into the environment. It should be emphasized that mining activities, ore processing and processing of waste are potential emitters of heavy metals in air.

As more important are mining activities on surface excavation of ore due to the large number of direct polluting factors that are difficult to control and manage (Pacyna et al., 2001; Pacyna et al., 2007).

Special emphasis is given to the organization of the pit (road movement of dampers carrying ore waste), waste management (ore and flotation tailings) as well as the frequency of explosions for excavation of mining minerals (Sengupta, 1993). The main sources for releasing the pollutants are the extraction, concentration and separation of waste in the open. The level of the environmental pollution will depend on proper control of anthropogenic activities in the mentioned processes (Repley et al., 1996).

Fine dust generated during blasts and excavation of mining minerals is distributed in air by the winds. Rains also contribute to the erosion processes of opencast surfaces and ore tailings. This effect is known as "natural rinsing" (Dudka et al 1997; Cortes-Maramba et al. 2006). Such erosion processes contribute to the increasing content of metals in air and soil. The increased content of heavy metals in the surface layers of soil are carried by the winds in the atmosphere as fine dust (Agarval, 2009). Environmental concerns also include depositional soil contamination, waste/slag dumps and process chemical pollution. This pollution can also contribute to surface and groundwater pollution. Slag's, dust and fumes are common products in almost all extractive metallurgical operations. Dust and fumes may contain toxic metals in fine particle size and as such are a potential environmental and health hazard.

Such anthropogenic activities of using of natural resources and their processing through appropriate technological processes and waste management are a global problem of environmental pollution. Some of the activities that generate dust with high heavy metals content and disperse it in air include (in no particular order of importance):

- movement of waste to and from the facilities;
- storage of waste (under certain conditions) on the facility;
- the handling and processing of the waste materials (e.g. shredding of green waste, turning of windrows and daily cover);
- wind scouring of waste surfaces.

Vehicles driven on the facility can also have a significant impact through the:

- resuspension of deposited particulates on roadways and hard-standing;
- transport of larger particles on vehicle bodies.

External factors influence the degree to which the release and dispersion take place. Meteorological conditions are important, specially the amount of rain and strength of wind. Facilities that process wastes inside buildings are typically affected less by meteorological conditions than wastes processed in the open air.

1.4. Monitoring of Heavy Metals (Dust) Deposition

Atmospheric deposition poses significant ecological concerns. According to the process of deposition particles can be deposited at the earth's surface in two ways, dry deposition or wet deposition, depending on the phase in which a species strikes the earth's surface and is taken up (Finlayson-Pitts and Pitts, 2000). Dry deposition is characterized by direct transfer of gas phase and particulates from air to ground, vegetation, water bodies and other Earth surfaces. This transfer can occur by sedimentation, impaction, and diffusion to surfaces or in case of plants by physiological uptake. In the absence of precipitation dry deposition plays mayor role in removing pollutants from atmosphere (Finlayson-Pitts and Pitts, 2000). Wet deposition includes rain, snow, fog, cloud, dew formation and washout the atmosphere (Godish, 2004; Vallero, 2008). Rain out therefore means process in the clouds, where the occurrence of droplets on solid particles (condensation nuclei) happened. In such way disposal of pollutants by rain, snow etc. occurs. On the other hand, wash out is the process by which pollutants are removed from the atmosphere by being absorbed in or desorbed on water droplets. If two particles in air collide they tend to adhere to each other due to their attractive surface forces.

a) An alternative approach is to use exposed attic dust, because over long periods of time atmospheric particles can accumulate, providing a record of historical local deposition. The determination of historical emissions is based on the data of heavy metals concentration in the attic dust from different measurement sites of the weight of total monthly air deposit. The main idea behind determining past emissions is that heavy metals characterization of deposited dust on a small area is multiplied by the concentration of the elements in that area, the mass of the polluter which has been transported to the place of interest by air can be determinate. Undisturbed attic dusts are potential archive for atmospherically deposited particles and have been shown to be effective across urban areas and in the vicinity of smelters, mines and other potentially emission sources. These have the additional benefit that they are protected from wash and aqueous alteration that may affect other tips of samples (monitors) over long periods of time. Atmospheric emissions attributed to the extraction stage of mining come mainly from the action of wind on disturbed land and stockpiles of ore and waste material. As a result of these processes in the atmosphere permanently introduces dust (Sengupta, 1993, Reply et al., 1996). Continuous monitoring of the content of heavy metals and other toxic components contained in dust is possible thought implementation of monitoring by using samples of dust from attic beams (Ilacqua et al., 2003, Atriola et al., 2004; Tye et al., 2006). Attic dust is derived primary from external sources through aerosol deposition and as a result of soil dusting, and less than household activities (Šajn, 2003). Usefulness of attic dust as a suitable long-term monitor for determination the status and content in air is proven by numerous studies (Citizel, 1998, 1999; Gosar and Šajn, 2001, 2003; Cyrys et al., 2003; Jefrey et al., 2005; Šajn, 2005, 2006; Žibret, 2008).
b) Deposited matter refers to any dust that falls out of suspension in the atmosphere. Solid and liquid particles or dust that falls out of suspension in the atmosphere can get into the environment and lead to its contamination. Atmospheric total deposition (deposited dust) is very useful mechanism for monitoring the fate of anthropogenic

elements introduced into the atmosphere (Tsitouridou et al., 2007; Čačković et al., 2008). Fine powder with a high content of heavy metals is generated as a result of emissions from the processing of ores and metallurgical process and is distributed as a result of wearing the wind. Many investigations have focused on the chemical composition and the content of toxic substances in deposited matter (Morselli et al., 2003; Avila and Rodrigo, 2004; Polkowska et al., 2005; Vike, 2005; Franzaring et al., 2006).

1.5. Study Issue

Monitoring of air pollution has proved as a most useful technique for determining deposition of heavy metals and atmospheric pollution with it's in different geographical areas. The Republic of Macedonia does not deviate from the global framework of air pollution with heavy metals. The results obtained from previous studies on air pollution, suggest that the situation in the Republic of Macedonia is not favorable in terms of air pollution with heavy metals. The main emission sources appear to be power plants, mines, as well as flotation for lead, zinc, copper and metallurgical plants for the production of nickel, steel, lead, zinc and different ferroalloys (Barandovski et al., 2006, 2008; Stafilov et al., 2008, 2010, 2010a, 2010b).

Attic dust and total deposited mater are commonly used as monitors for these purposes. These kind of monitoring programs have been performed as part of a large number of analytical studies for a long time, but their application in recent decades has taken a swing. This is due to the fact that monitoring does not require the use of expensive technical equipment. Analytical results reflect the situation for a long period of time. The monitoring allows analysis of large areas and the types of monitors are chosen according to the analyst and type of analysis.

Within this study a total content of 23 elements was determined (Ag, Al, As, B, Ba, Ca, Cd, Co, Cr, Cu, Ga, Fe, K, Li, Mg, Mn, Mo, Na, Ni, Pb, Sr, V, and Zn) in attic dust, total deposited mater taken from the copper mine and flotation near Radoviš and the deposited dust taken from the ferronickel smelter plant near Kavadarci region. Analyses were performed by the application of atomic emission spectrometry with inductively coupled plasma (ICP-AES) and electrothermal atomic absorption spectrometry (ETAAS) Tables 1 and 2. However for the Kavadarci region samples of attic dust were analysed by the application of mass spectrometry with inductively coupled plasma (ICP-MS). The investigated elements are : Ag, Al, As, Ba, Be, Bi, Ca, Cd, Ce, Co, Cr, Cs, Cu, Dy, Eu, Fe, Ga, Gd, Ge, Hf, Hg, Ho, I, K, La, Li, Lu, Mg, Mn, Mo, Na, Nb, Ni, Pb, Rb, Sb, Sm, Sr, Tb, Th, Ti, U, V, Yb, Zn and Zr. The optimal instrumental parameters for this technique are given in Table 3.

The obtained values for the content of the elements in investigated samples were statistically processed using statistical software for basic descriptive statistics and bivariate analysis, which showed the correlation of the content of chemical elements in samples; multivariate statistic method was used to reveal the associations of the chemical elements.

Distribution maps for each element have been prepared separately based on the results. Statistically processed data and distribution maps for individual elements make it possible to identify narrow areas with their higher content. With that we can determine the scope of the impact of the work of this mine on increasing the content of heavy metals in air, and the direction of their distribution in the area around the town of Radoviš and the Kavadarci region. This is important because of direct human exposure to these pollutants and adverse effects on their health.

2. Study Area

Research was conducted at two locations in the Republic of Macedonia. In the eastern part of the country the appearance of some metals (Au, Mg, Al, Sc, Ti, V, Cu) in the air is related to the presence of a copper mine and flotation plant, "Bučim", near the town of Radoviš (Balabanova et al., 2009; Stafilov et al., 2010b; Balabanova et al., 2010; Balabanova et al., 2011). This was one of the investigations areas. In this area an influence from the former iron mine, Damjan, has also been determined (Serafimovski et al., 2005).

The other study area is located in the south-central part of the country, where the appearance of some metals (Co, Cr, Fe and Ni) in the air is related to a presence of the ferronickel smelter plant near the city of Kavadarci, (Barandovski et al., 2006; 2008; Stafilov et al., 2008, 2010; Bačeva et al., 2009; 2011). As a result of these anthropogenic activities, increase distribution of certain heavy metals in air and their deposition in the environs were expected.

2.1. Radoviš (Copper Mine Case Study)

The study area is located in the eastern part of the Republic of Macedonia with a surface area of 20 km (W–E) × 20 km (S–N), a total of 400 km^2, limited by the coordinates N: 41°32' – 41°44' and E: 22°15' – 22°30' (Figure 1).

The region is characterised by a moderate continental climate (Lazarevski, 1993). The altitude varies between 350 and 1000 m. The average annual temperature is around 10°C. The average annual rainfall amounts to 563 mm with large variations from year to year. The most frequent winds in the region are those from the west with a frequency of 199 ‰ and speed of 2.7 m s^{-1}, and winds from the east with a frequency of 124 ‰ and speed of 2.0 m s^{-1}.

One of the major emission sources of some metals in the eastern part of the R. Macedonia is Bučim copper mine and flotation, near the town of Radoviš. The mine and plant for the preparation of ore are in function from the late seventies of the last century. Ore excavation is from open pit and the ore tailings are stored in the open, in mine vicinity. The produced copper ore from the mine is processed in the flotation plant; after the flotation of copper minerals, the flotation tailings are separated, disposed of and deposited on a dump site in an adjacent valley near the village Topolnica.

The copper mine Bučim is located in the north-west part of the studied area. The Bučim mine territorially and administratively belongs to the municipality of Radoviš, and is located 14 km from the town. In the close vicinity of the mine there are two settlements, the villages Bučim and Topolnica. Mine activities cover 7 km^2 of total mine surfaces, 4 km^2 for the placement of ore tailings and the rest of the land belongs to the open ore pit and to the plant for ore processing. The main ore contents are: 0.3 % Cu, 0.3 g t^{-1} Au, 1 g t^{-1} Ag, 13 g t^{-1} Mo, and 1–4 % pyrite; the igneous rocks have been altered to clays and micas.

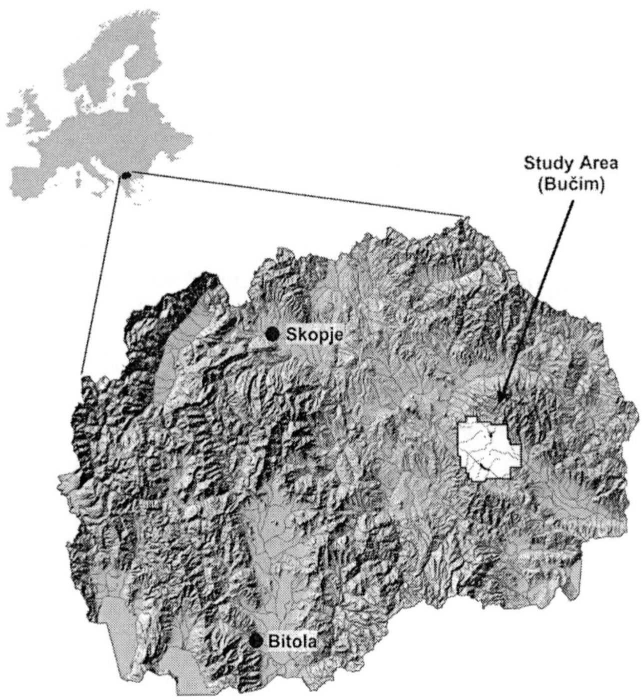

Figure 1. Location of study area of Radoviš region, Republic of Macedonia.

The important metallic minerals are chalcopyrite, pyrite, and bornite, with small amounts of galena, sphalerite, magnetite and hematite (Serafimovski et al., 1995; Stafilov et al., 2010b; Balabanova et al., 2011).

The "Bučim" mine and the ore processing plant have been functioning since 1979 and it is assumed that the mine has about 40 million tons of ore reserves. Ore tailings are dropped out by the dampers from the open ore pit at an open site near the mine. The ore tailings deposit occupies a surface of 0.80 km^2, located southwest of the open ore pit, near the regional road Štip-Strumica. The ore tailings deposit has about 130 million tons of ore tailings. Exposure of this great mass of ore tailings to constant air flow and wind leads to the distribution of fine dust in the air.

The flotation plant produces 4,000,000 tons of copper ore annually. In the process of flotation of copper minerals, the average annual amount of flotation tailings created is approximately 3.95 million tons. These tailings are drained and disposed of on a dump near the mine. The location of this dump is east of the flotation plant, 2.2 km to the dam.

2.1.1. Geological Description

The investigated area represents a part of the Vardar structural zone, separated from the other structural zones during the Caledonian, and subjected to strong tectonic processes, during the Herzynian orogenesis. The structural relations were further complicated by the Alpine orogenesis (Rakićević et al., 1969).

At the study area the following main geotectonic structural units have been identified: (1) the Kriva Lakavica basin (2) the Smrdeš-Gabreš syncline, (3) the Radoviš basin, (4) the Radoviš anticline divided to (4a) the Štip Block and (4b) the Bučim Block (Figure 2). The Radoviš anticline represents the eastern boundary of the Vardar zone towards the Serbian-

Macedonian mass. These two large structural units are separated by a deep NW-SE fault (Hristov et al., 1965).

The Bučim-Damjan-Borov Dol area is divided into two tectonic blocks. The Bučim tectonic block and the southern tectonic block Damjan are a part of the Vardar zone. The blocks are divided by a fault of first order in the SE direction. Despite the disposition in two different tectonic blocks, the metallogenic area is unified based on the similarities of Tertiary magmatism and the analogous ore mineralisations. The Bučim copper-porphyry deposit with additional gold mineralisation is found in the northern block (Stefanova et al., 2004).

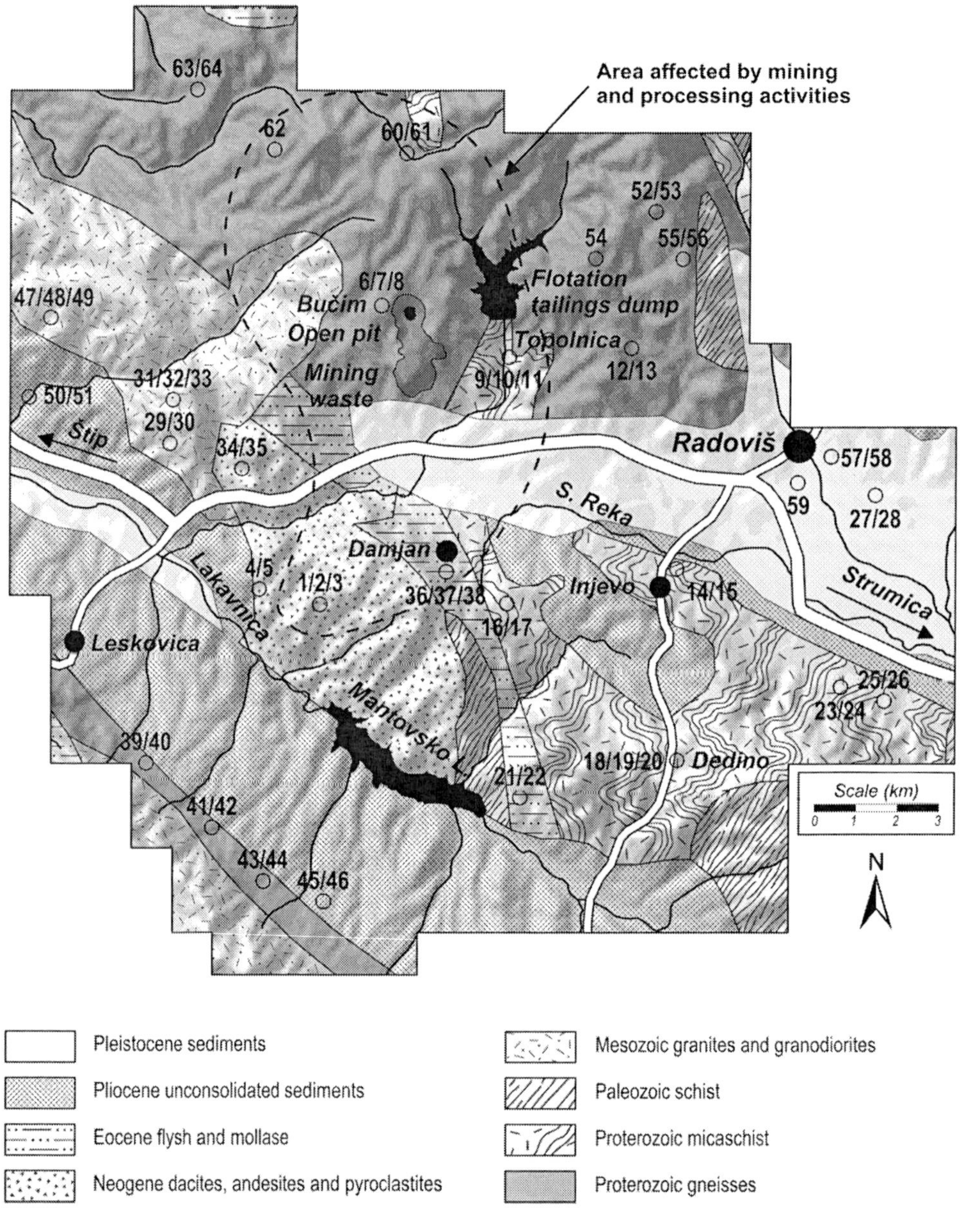

Figure 2. Attic dust sampling network and generalized geological map of Radoviš region.

2.2. Kavadarci (Ferronickel Smelter Plant Case Study)

In the south-central part of the country the appearance of some metals (Co, Cr, Fe and Ni) in the air is related to a presence of the ferronickel smelter plant near the city of Kavadarci, Figure 3 (Barandovski et al., 2006; 2008; Stafilov et al., 2008, 2010; Bačeva et al., 2009; 2011). This smelter plant uses nickel ore from Ržanovo mine, about 30 km south of the plant. In the last 4 years beside ore from Ržanovo mine, ore mainly from Gebe nickel mine, Indonezia reach in nickel (2-2.5 %), is used. It is well known that the nickel ores contain cobalt as well (in average of 0.05 % for Ržanovo ore) (Maksimović, 1982; Boev and Jankovic, 1996). Therefore, the dust from this plant has the similar content like ore used as a raw material including some of the heavy metals like nickel, cobalt and chromium. For that reason, the goal of this work was to determine the total deposition (deposited dust) in the atmosphere in Kavadarci and its surroundings.

2.2.1. Geological Description

The geological description of the investigated area is presented in the Geochemical Atlas of Kavadarci and its Environs (Figure 4) (Stafilov et al., 2008; 2010). The oldest formations have directions NW-SE and belong to the inner parts of the Vardar zone. The Lower Paleozoic (Pz) metamorphic complex is present with two series: amphibole and amphibole chlorite schist with marbles and phylite layers. Serpentinite is present in the form of the narrow belts along the ruptures inside the Vardar zone. The uttermost part in the SW of the study zone is covered with marbles and dolomites probably from Devonian ages. Over the Paleozoic are developed Mesozoic (Mz) formations, mainly form Late Cretaceous ages. Paleozoic and Mesozoic rocks cover approximately 39 km^2 in the SW part of the investigated area.

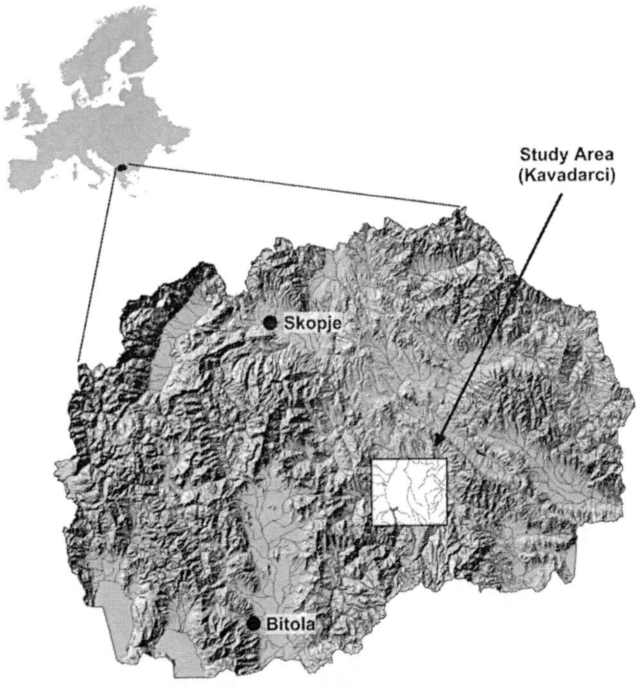

Figure 3. Location of the investigated region of Kavadarci and its environs.

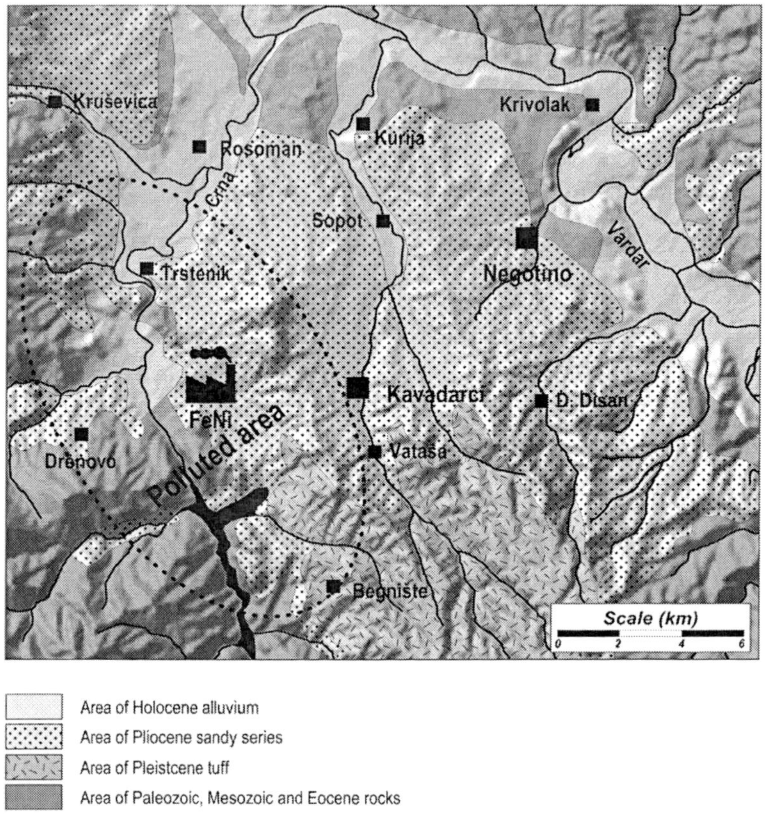

Figure 4. Generalized geological map for the Kavadarci region.

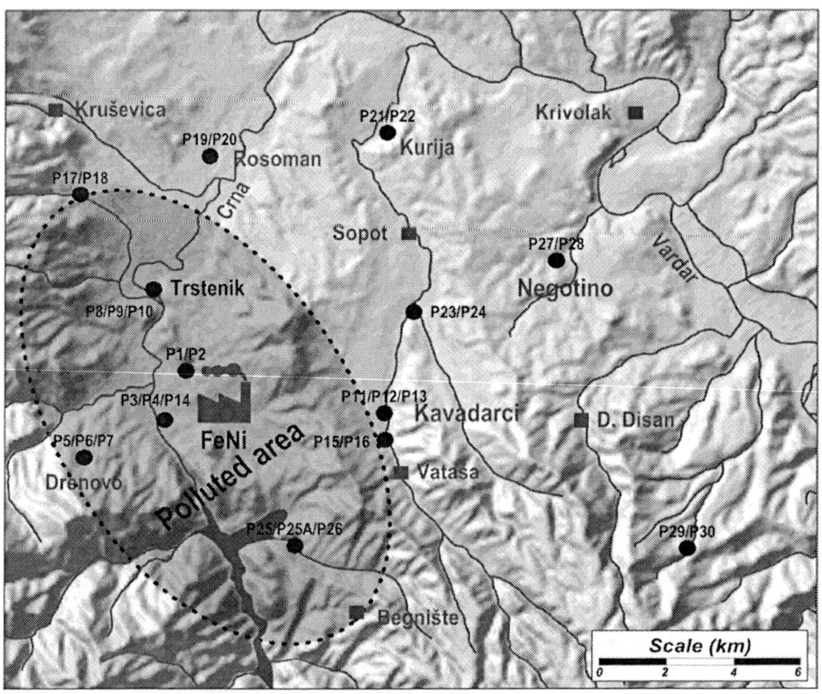

Figure 5. Attic dust sampling network map for the Kavadarci region.

Complexes of Tertiary and Quaternary sediments cover the most of the study area. The Upper Eocene ($^4E_3$) flysch sediments and yellow sandstones are developed along Vardar, Crna Reka and Luda Mara valleys and marginal part of the Tikveš basin. Those sediments with depth up to 3500 m cover approximately 34 km^2 mainly in the N part of investigation area. The Pliocene sediments fill the Tikveš basin, limited with the Vardar on the North, and Paleozoic-Mesozoic formations that have directions NW-SE. This sequence is represented mainly with sandy series. Pliocene (Pl) sediments cover the biggest part (about 182 km^2) in the central part of the investigated area. SE from Kavadarci the Quaternary (Q) pyroclastic vulcanites are found. They're represented by tuffs, breccias and agglomerates and cover approximately 25 km^2.

3. Materials and Methods

3.1. Sampling and Sampling Preparation

Attic dust as sampling media - Samples of attic dust were examined as historical archives of smelter and mine emissions, with the aim of elucidating the pathways of pollution. Dust was collected from the attic of 29 houses in the vicinity of copper mine region (Figure 2) and 31 houses from Fe-Ni smelter plant region, built between 1920 and 1970 (Figure 5). In every settlement, attic dust was collecting from 2 or 3 houses on different sites in the setlements. The collection of attic dust samples was performed according to the adopted protocol (Šajn, 2003). Attic dust was collected with plastic brush in polyethylene bags. The surface from the attic timber was throwing across and the finest dust was collected. The size fraction of attic dust smaller than 0.125 mm was prepared by sieving. In that form samples of attic dust samples were ready for digestion. For digestion of attic dust samples, open wet digestion with mixture of acids was applied. Precisely measured mass of dust samples (0.5000 g) was placed in Teflon vessels and 5 ml concentrated nitric acid, HNO_3 was adding, until the brown vapours came out from the vessels. Nitric acid is very suitable oxidant for digestion of environmental samples. For total digestion of inorganic components was added 5-10 ml hydrofluoric acid. When the digest became clear solution, 2 ml of $HClO_4$ was added. Perchloric acid was used for total digestion of organic matter. After 15 min. cooling the vessels, 2 ml of HCl and 5 ml of H_2O were added for total dissolve of metal ions. Finally the vessels were cooled and digests quantitatively transferred to 50 ml calibrated flasks. For this study reagents with analytical grade or better: nitric acid, trace pure (Merck, Germany), hydrofluoric acid, p.a. (Merck, Germany), perchloric acid, p.a. (Merck, Germany), hydrochloric acid, p.a. (Merck, Germany), and redistilled water were used for preparation of all solutions. Standard solutions of metals were prepared by dilution of 1000 mg l^{-1} solutions (11355-ICP multi Element Standard and Merck VI-ICP multi Element Standard).

Total deposited matter as sampling media - In order to determine daily amount of fine dust contained in the air, samples of total deposited matter (deposited dust) were collected at three locations (hot spots in the area) where were expected large amounts of deposited dust (Figures 6 and 7). The total quantity of the deposited dust was monitored monthly for a period of one year. Monthly samples of deposited dust were collected at three monitoring sites in copper mine environ from one side, and from Fe-Ni smelter plant environ from other side.

The village Vozarci, village Brušani and town Kavadarci (Figure 6) were monitoring spots around the Fe-Ni smelter plant. The town of Radoviš and the villages Bučim and Topolnica (Figure 7) were the three monitoring spots around Bučim copper mine. Samples of total deposited matter (wet and dry deposition) were collected using the dust deposition gauges. This method measures dust deposition rate and involves the passive deposition and capture of dust within a funnel and plastic container.

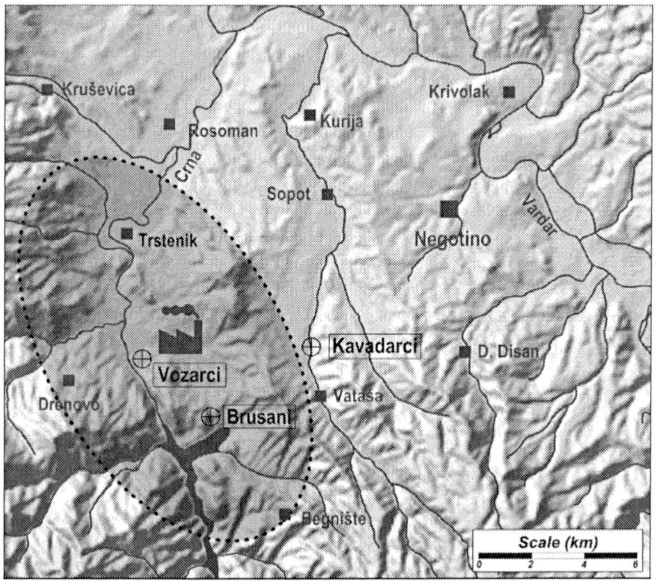

Figure 6. Location of the sampling points for deposited dust for the Kavadarci region.

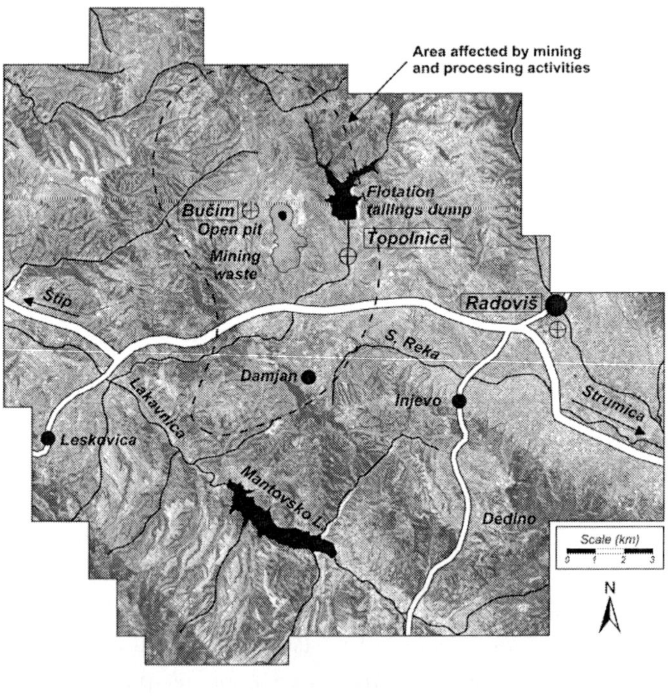

Figure 7. Location of the sampling points for deposited dust for the Radoviš region.

Data is usually collected over monthly periods and results are expressed mg m^{-2} d^{-1} (i.e. the mass of dust deposited per m^2 per day). This method enables determination of the relative 'dustiness' of sampling locations. A deposit gauge, which comprises a 28±1 cm diameter funnel inserted into a plastic container (at least 5-10 liters in size) through a rubber stopper (Holden, 2006). Stand approximately 2 m tall and a canister which holds the plastic container to protect it from sunlight. The plastic container may also collect rainwater and other material such as bugs and leaf litter, etc. This does not contaminate the sample and should not be removed in the field. After 30±2 days, any deposited matter in the funnel was washed into the plastic container using distilled water. The aliquot of each sample was evaporated near dryness and then 3-5 ml of nitric acid, *p.a.* (MERCK, Germany) was added and collected in to the 25 ml volumetric flasks.

3.2. Instrumentation

The investigated elements were analysed by the application of atomic emission spectrometry with inductively coupled plasma (AES-ICP), electrothermal atomic absorption spectrometry (ETAAS) and mass spectrometry with inductively coupled plasma (ICP-MS). The following elements were analyzed: Ag, Al, As, Ba, Be, Bi, Ca, Cd, Ce, Co, Cr, Cs, Cu, Dy, Eu, Fe, Ga, Gd, Ge, Hf, Hg, Ho, I, K, La, Li, Lu, Mg, Mn, Mo, Na, Nb, Ni, Pb, Rb, Sb, Sm, Sr, Tb, Th, Ti, U, V, Yb, Zn and Zr.

The heavy metals content was determined using ICP-AES, Varian 715-ES, for Al, Ca, Cr, Cu, Fe, K, Li, Mg, Mn, Na, Ni, Pb, Sr, V and Zn, and ETAAS determination for As, Cd and Co (Varian SpectrAA-604Z). The optimal instrumental parameters for these techniques are given in Tables 1 and 2. For AES-ICP instrument calibration and quantitative determination of each element in deposited matter, a commercial standard mix solution (11355-ICP Multi Element Standard IV, Merck) was used. The correlation coefficient of calibration curve for each element was 0.999. In order to check for possible background contamination, blank samples were used and processed simultaneously with field samples. The method detection limit was calculated based on average measuring of the blank sample ($\gamma \pm 3\delta$). For all laboratory samples and standard solutions, treated ultra pure water was used. The QC of the applied techniques was performed by standard addition method, and it was found that the recovery for the investigated elements ranges for ICP-AES between 98.5–101.2 % and for ETAAS, between 96.9 % – 103.2 %

For ICP-MS measurements a SCIEX Perkin Elmer Elan DRC II (Canada) inductively coupled plasma mass spectrometer (with quadruple and single detector setup) was used (Table 3). The instrument's running parameters were checked and adjusted before every batch of measurements, using a solution with 1 μg ml^{-1} In, 1 μg ml^{-1} Ce, 10 μg ml^{-1} Ba and 1 Th and Mg μg ml^{-1}. Oxide levels and double ionized levels were kept under 3%, background for both low and high mass was under 1 cps and all the other parameters were chosen considering the best signal/noise ratio. The dynamic reaction chamber (DRC) was used in RF-only mode (no gas) and its parameters optimization have been optimised elsewhere (Tănăselia et al., 2008). For sample introduction system, a classic set-up was used, consisting in a peristaltic pump, a Meinhard nebuliser and a cyclonic spray chamber, were the fine aerosols are formed that goes directly into plasma. All other reagents were supplied by Merck. 18 MΩ cm^{-1} DI water was prepared in the laboratory, using a Millipore-Milli-Q® ultrapure water purification system.

Table 1. Instrumentation and operating conditions for ICP-AES (Varian 715-ES) system

RF Generator	
Operating frequency	40.68 MHz free-running, air-cooled RF generator.
Power output of RF generator	700–1700 W in 50 W increments
Power output stability	Better than 0.1%
Introduction Area	
Sample Nebulizer	V- groove
Spray Chamber	Double-pass cyclone
Peristaltic pump	0-50 rpm
Plasma configuration	Radially viewed
Spectrometer	
Optical Arrangement	Echelle optical design
Polychromator	400 mm focal length
Echelle grating	94.74 lines/mm
Polychromator purge	0.5 L min^{-1}
Megapixel CCD detector	1.12 million pixels
Wavelength coverage	177 nm to 785 nm

Conditions for program					
RFG Power		1.0 kW	Pump speed	25 rpm	
Plasma Ar flow rate		15 l min^{-1}	Stabilization time	30 s	
Auxiliary Ar flow rate		1.5 l min^{-1}	Rinse time	30 s	
Nebulizer Ar flow rate		0.75 l min^{-1}	Sample delay	30 s	
Background correction		Fitted	Number of replicates	3	
Element	Wavelength, nm	Element	Wavelength, nm	Element	Wavelength, nm
Al	396.152	Cu	324.754	Na	589.592
As	188.980	Fe	238.204	Ni	231.604
Ca	370.602	K	766.491	Pb	220.353
Cd	226.502	Li	670.783	Sr	407.771
Co	238.892	Mg	279.553	V	292.401
Cr	267.716	Mn	257.610	Zn	213.857

All measurements were done using the semiquantitative method (TotalQuant) supplied by Elan 3.4 software that uses a response factor calibration curve which was obtain by a calibration in multiple points, low, medium and high mass, for optimum set-up, using a multi-element Merck VI standard solution, diluted to mimic real sample composition. The drawback is that the accuracy tends to be worse than a proper quantitative method for some elements, however the main advantage is the large mass interval that can be studied (up to 65 elements per each sample during a single run), a good choice for screening type measurements that requires a large throughput of samples with many elements of interest.

For this study, a NIST 2709 and NIST 1643e certified reference materials was used to check method accuracy and for all considered elements, the difference between measured and certified values was within 15%.

Table 2. Optimal instrumental parameters for ETAAS determination by Varian SpectrAA-604Z

Parameter	As	Cd	Co
Wavelength, nm	193.7	288.8	242.5
Spectral width slit, nm	0.2	0.5	0.2
Calibration mode	Peak height		
Lamp current, mA	10.0	4.0	7.0
DRY			
Temperature, °C	120	120	120
Ramp time, s	55	55	55
Hold time, s	-	-	-
PYROLYSIS			
Temperature, °C	1400	250	400
Ramp time, s	10	5	5
Hold time, s	35	15	22
ATOMIZING			
Temperature, °C	2600	1800	2300
Ramp time, s	1	1	1
Hold time, s	2	2	2
CLEANING			
Temperature, °C	2600	1800	2650
Time, s	2	2	5
SHEATH GAS	Argon		

Table 3. Spectrometer's running parameters for ICP-MS

Parameter	Value
Plasma	
Power	1350 W
Plasma gas flow	12.00 l min^{-1}
Auxiliary gas flow	1.20 l min^{-1}
Nebuliser gas flow	1.05 l min^{-1}
Sample/Skimmer cone	Platinum
Quadrupole	
Quadruple rod offset (QRO)	0.00 V
Cell rod offset (CRO)	- 8.00 V
Cell path voltage (CPV)	- 20.00 V
Measurement mode	Peak hopping
Dwell time/ms	Varying
Integration time/ms	Varying
Reading per point	300
Reading per replicate	1
Replicate measurements	4
DRC	
Reaction Gas	None
Lens voltage	11.00 V

The theoretical limit for ICP-MS methods are in ppt (ng l^{-1}) range for the majority of the elements. Matrix effects above 1 ppb (μg l^{-1}) threshold while using TotalQuant were not observed during our study. For some elements, values between these two levels were further investigated using more complex quantitative methods.

Table 4. Lower detection limits of determinations by ETAAS and ICP-AES

Elements	Detection Limit
ETAAS	
Cd, Co	0.1 mg kg^{-1}
As	0.2 mg kg^{-1}
ICP-AES	
Ba, Mo	0.1 mg kg^{-1}
Cu, Mn	0.5 mg kg^{-1}
B, Cr, Ga, Fe, Mg, Ni, Sr, V, Zn	1 mg kg^{-1}
Al, K, Li, Pb	2 mg kg^{-1}
Ag, Na	5 mg kg^{-1}
Ca	10 mg kg^{-1}

Table 5. Lower detection limits of determinations by ICP-MS

Elements	Detection Limit
Hg	0.01 mg kg^{-1}
Ag, Bi, Cd, Ce, Co, Cs, Cu, Dy, Eu, Gd, Ge, Hf, Ho, I, Li, Mo, Nb, Ni, Pb, Rb, Sb, Sm, Tb, Th, Yb, U, Zr	0.1 mg kg^{-1}
As	0.5 mg kg^{-1}
Ba, Be, Cr, Ga, La, Mn, Sr, Zn	1 mg kg^{-1}
V, Lu	2 mg kg^{-1}
Na, Ti	0.001 %
Al, Ca, Fe, K, Mg	0.01 %

3.3. Data Processing

For the statistical analysis of data parametric and nonparametric statistical methods were used (Hollander and Wolfe, 1999). The obtained values for the contents of the investigated elements were statistically processed using basic descriptive statistics. Data distribution was examined with the application of normality tests.

The application of bivariate statistics showed the correlation between the content of the investigated elements in deposited dust and attic dust samples. In order to analyse this, the linear coefficient of correlation was used, and the absolute values $r > 0.50$ indicated a good association between variables.

Multivariate statistical methods (cluster and R-mode factor analyses) were used to reveal the associations of the chemical elements. The factor analysis was performed on variables standardised to zero mean and unit standard deviation (Reimann et al., 2002; Filzmoser et al., 2005; Šajn, 2006; Žibret and Šajn, 2010). As a measure of similarity between variables, the product-moment correlation coefficient (*r*) was applied. For orthogonal rotation the varimax method was used. In the final multivariate factor solution some elements were retained. The elements with low communalities were excluded because of lack of significant associations.

The universal method kriging with linear variogram interpolation was applied for the construction of spatial distribution maps of each factor score. Seven classes of the following percentile values were selected: 0–10, 10–25, 25–40, 40–60, 60–75, 75–90 and 90–100. Line and bar/colon plots were used for better visibility of elements content and trends of dust deposition through the year.

4. Results and Discussion

4.1. Attic Dust as Sampling Media

4.1.1. Copper Ore Processing Case Study

For each observation there were few variables: sample identification number, locality, geographic coordinates and concentration level for total of 19 elements. Basic statistical analyses were done.

Minima and maxima mean calculation methods of median, geometrical mean, standard deviation, geometric standard deviation, skewness, kurtosis and arithmetical mean were calculated and histograms were drawn to show the distribution features. Based on the normality tests and distribution histograms of the independent variables, the normality for some elements was determined in the distribution of data.

These skewed distributions are confirmed by relatively high values of standard deviation, which is a measure of the degree of data deviation (dispersions) around the mean. On the other hand, significant differences between mean and median value were not obtained for the variables (Table 6).

On the basis of the matrix of correlation coefficients, factor analysis was done. Principal component factor analysis was used to identify and characterise element associations. From 19 analyzed variables, seven elements (Ag, Al, Ba, Fe, K, Na, and Zn), had low factor loading, or tendency to form independent factor and therefore do not belong to any factor group. But these elements generally present as naturally born in environment, so they can be considered as geogenic elements. The matrix of dominant rotated factor loadings is presented in Table 8. Tree factors were identified, two geogenic and one anthropogenic, interpreted as Factor 1, Factor 2 and Factor 3, which including 67 % of variability of treated elements.

Factor 1 (Ca, Li, Mg, Mn and Sr). These elements are naturally found in soil as macro elements (Figure 8). Contents of these elements are variable and are not related to any anthropogenic activities. Their sources are mainly natural phenomena such as rock weathering and chemical processes in soil. Occurrence of this Factor is typical for the oldest formations in the Republic of Macedonia (Proterozoic micaschist and Proterozoic gneisses).

Factor 2 (Co, Cr, Ni) presents geogenic factor (Figure 9). These determined elements are considered as 'natural' because their origin is primarily crustal, soil particles suspended and transported by wind. High factor loadings are related to some old formation as Neogene dacites, andesites and pyroclastites and Pliocene unconsolidate sediments.

Table 6. Descriptive statistics for elements content values in attic dust samples (n=64, values are given in mg kg^{-1})

Element	n	Dis.	X_a	X_g	Md	Min	Max	P_{10}	P_{90}	s	CV	A	E
Ag	64	log	0.24	0.10	0.05	0.009	1.12	1.09	13.4	0.83	17.8	0.90	-0.62
Al	64	N	11040	10292	10886	1787	21957	6407	16085	3797	34.4	0.26	0.39
As	64	log	11.3	8.34	9.58	1.16	51.6	2.13	18.8	9.29	82.2	-0.29	-1.42
Ba	64	log	4635	361	146	3.95	38676	24.2	21121	10533	227	0.43	-0.83
Ca	64	N	6049	4681	5950	342	18833	1380	10609	3624	59.9	0.79	1.30
Cd	64	log	2.02	1.97	1.98	1.09	3.11	1.39	2.64	0.46	22.8	-0.24	-0.07
Co	64	log	6.26	5.95	5.63	2.42	11.3	4.03	9.95	2.10	33.6	0.19	0.05
Cr	64	log	38.7	34.8	31.1	16.7	110	23.8	78.1	21.3	55.2	1.19	0.80
Cu	64	log	52.1	33.4	26.8	10.9	415	15.5	116	75.8	142	1.51	2.18
Li	64	log	4.82	3.98	4.10	1.20	15.9	1.61	8.72	3.09	64.1	-1.31	1.48
Fe	64	N	11966	10525	13228	1131	18350	5986	16080	4120	34.4	0.10	-0.34
K	64	N	8846	8741	8613	5560	11612	7161	10876	1362	15.4	0.02	-0.79
Mg	64	log	1321	642	715	24.6	5859	94.3	3414	1399	106	-0.43	-0.66
Mn	64	log	211	201	195	113	454	143	302	70.4	33.3	0.40	-0.17
Na	64	log	6846	5024	4784	326	28682	2348	15810	6336	92.5	-0.24	1.95
Ni	64	log	23.1	20.2	17.2	8.94	58.7	11.5	41.4	12.5	54.3	0.43	-1.11
Pb	64	log	28.8	21.4	20.4	3.12	121	7.49	53.3	23.9	83.1	-0.20	-0.07
Sr	64	log	36.1	28.4	32.1	3.64	141	10.3	60.9	25.2	69.8	-0.74	0.92
Zn	64	log	50.9	48.7	49.5	21.4	92.9	34.3	72.9	15.1	29.8	-0.20	0.11

Table 7. Matrix of correlation coefficients

Ag	1.00																		
Al	0.12	1.00																	
As	0.34	-0.11	1.00																
Ba	0.29	-0.21	**0.75**	1.00															
Ca	0.28	-0.19	-0.03	0.28	1.00														
Cd	**0.66**	0.18	-0.17	0.11	0.18	1.00													
Co	**0.67**	0.03	0.33	**0.59**	0.32	**0.76**	1.00												
Cr	**0.52**	-0.09	-0.17	0.02	0.33	**0.67**	**0.64**	1.00											
Cu	-0.01	0.28	-0.06	0.17	0.06	0.26	0.39	-0.11	1.00										
Li	**0.70**	0.09	0.57	**0.57**	0.44	0.29	**0.51**	0.31	-0.23	1.00									
Fe	**0.67**	0.29	-0.10	0.13	0.14	**0.92**	**0.77**	**0.63**	0.36	0.22	1.00								
K	0.24	0.21	-0.12	-0.44	-0.27	-0.02	-0.14	0.24	-0.41	-0.06	0.07	1.00							
Mg	0.42	-0.15	-0.20	0.05	**0.74**	0.45	0.46	**0.72**	-0.18	**0.53**	0.31	0.06	1.00						
Mn	**0.66**	-0.27	0.35	**0.60**	0.30	0.59	**0.78**	**0.57**	-0.08	**0.69**	0.49	-0.14	**0.57**	1.00					
Na	0.12	-0.21	**0.81**	**0.89**	0.09	-0.08	0.43	-0.11	0.13	**0.54**	-0.09	-0.44	-0.04	0.48	1.00				
Ni	0.36	-0.03	-0.28	-0.10	0.05	**0.74**	0.48	0.82	-0.20	0.13	**0.69**	0.12	0.46	0.48	-0.20	1.00			
Pb	0.03	**0.74**	-0.44	-0.19	0.11	0.30	0.08	0.19	0.18	0.08	0.36	0.09	0.18	-0.12	-0.27	0.23	1.00		
Sr	**0.69**	0.13	0.24	0.46	0.20	0.72	**0.70**	0.43	-0.06	**0.72**	**0.66**	-0.08	0.45	**0.83**	0.33	**0.53**	0.19	1.00	
Zn	-0.06	-0.21	0.38	0.59	0.49	-0.16	0.12	-0.18	0.20	0.34	-0.16	-0.77	0.05	0.20	**0.56**	-0.23	-0.21	0.12	1.00
	Ag	Al	As	Ba	Ca	Cd	Co	Cr	Cu	Li	Fe	K	Mg	Mn	Na	Ni	Pb	Sr	Zn

Factor 3 (As, Cu, Pb) associates chemical elements that indicate anthropogenic influence in the studying area (Figure 10). This association of elements was expected because of the studying area geology and the open pit mining and flotation activities. The acid drainage rapidly dissolved the elements, providing increasing content in soil. The open ore pit and flotation tailings dam allows direct exposure of the finest ore particles to the atmosphere.

Corpuscle dust from the surface layer of ore body and soil is spread in the atmosphere by the winds, through which is performed atmospheric distribution of these elements in the vicinity of the mine. High loading value for this factor was attempt for Cd (0.47). That is because higher contents from this element introduces to the environment as a result of anthropogenic activity.

Table 8. Matrix of dominant rotated factor loadings (F>0.60)

	F1	F2	F3	Com
As	-0.08	-0.06	**0.78**	62
Ca	**0.80**	-0.03	-0.09	70
Cd	0.27	0.51	0.47	67
Co	0.04	**0.89**	0.09	89
Cr	-0.07	**0.90**	-0.23	89
Cu	-0.18	-0.07	**0.70**	57
Li	**0.81**	-0.18	0.04	80
Mg	**0.78**	0.21	-0.22	79
Mn	**0.76**	0.34	0.11	80
Ni	0.19	**0.85**	0.04	65
Pb	0.27	0.09	**0.70**	81
Sr	**0.72**	0.11	0.19	66
Var	31.26	19.38	16.33	66.98

F1, F2, F3-Factor loading; Var-Variance (%); Com-Communality (%).

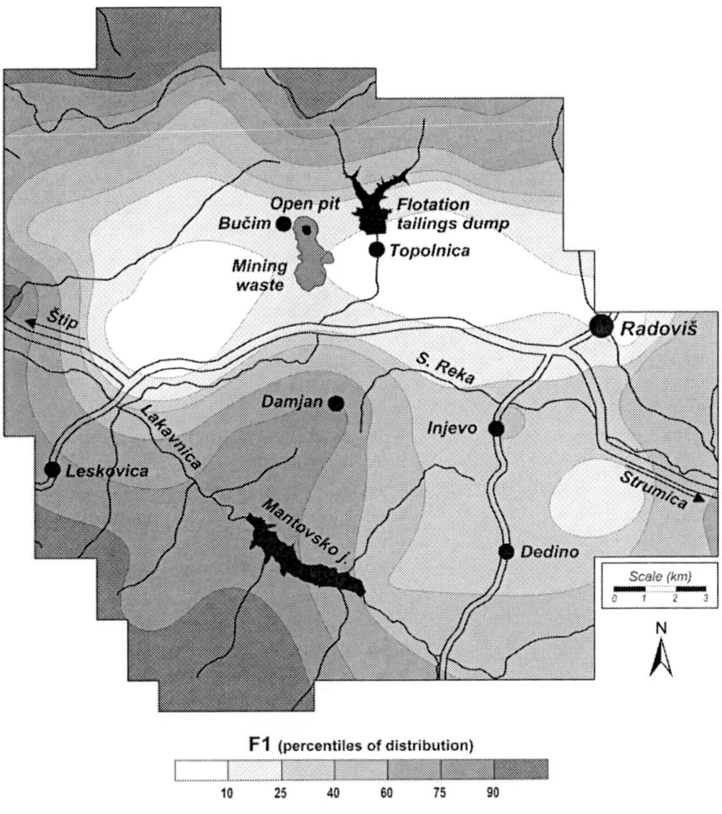

Figure 8. Spatial distribution of Factor 1 scores (Ca, Li, Mg, Mn, Sr) in attic dust samples.

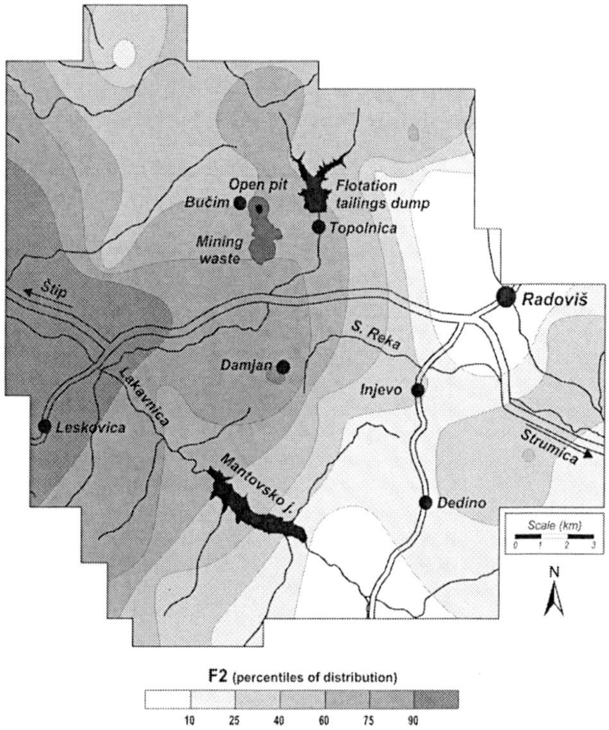

Figure 9. Spatial distribution of Factor 2 scores (Co, Cr, Ni) in attic dust samples.

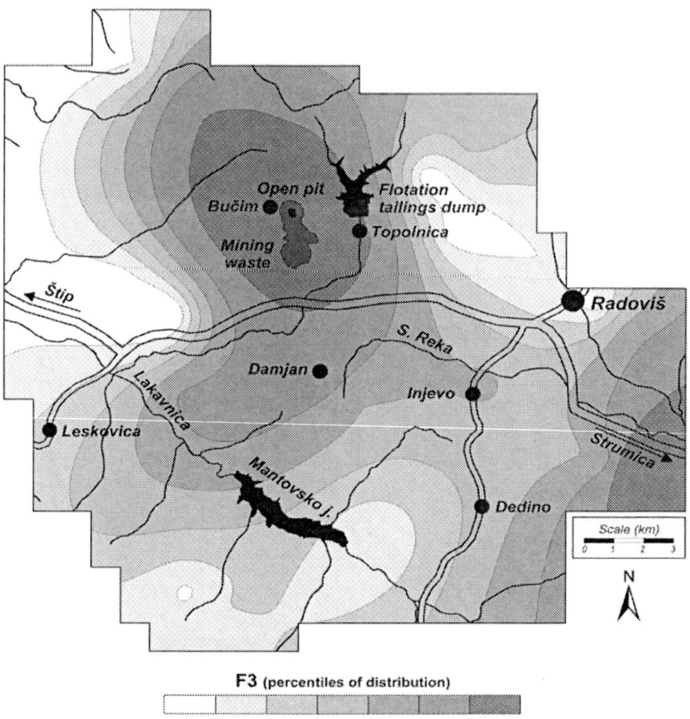

Figure 10. Spatial distribution of Factor 3 scores (As, Cd, Cu, Pb) in attic dust samples.

Urban elements and those related with mine activities, in dust are associated with fine particles, which are preferentially mobilized by wind and trapped in attics, and because old attics have had the longest time to collect this material (Miguel, et al. 1999).

4.1.2. Ferronickel Smelter Plant Case

The descriptive statistics of analysed elements are shown in Table 9 along with the results of 46 chemical elements (Ag, As, Al, Ba, Be, Bi, Ca, Cd, Ce, Co, Cr, Cs, Cu, Fe, K, Mg, Na, Ti, Dy, Eu, Ga, Gd, Ge, Hf, Hg, Ho, I, La, Li, Lu, Mn, Mo, Nb, Ni, Pb, Rb, Sb, Sm, Sr, Tb, Th, U, V, Yb, Zn and Zr) in 31 samples of attic dust. Values of Al, Ca, Fe, K, Mg, Na and Ti are given in % and remaining elements in mg kg^{-1}.

The distributions of most of the elements in attic dust samples are closely dependent on the lithogenesis and are slightly affected by anthropogenic activities. Thus, the highest contents for Ba, La, Th and U were found in areas of the Pleistocene tuff, Holocene alluvium (W of the town of Kavadarci) and Holocene alluvium of the river Luda Mara, and their lowest values in areas of the Eocene upper flysch zone and Paleozoic and Mesozoic rocks (Stafilov et al., 2008).

Anthropogenic distribution represents the pollution when trace elements are introduced into the environment through human activities. In the case of anthropogenically produced trace elements, they concentrations can increase several times compared to the background that consists of minor elemental fluctuations. The ferronickel smelter plant uses ore that contains between 1 % and 2.5 % Ni, about 0.05 % Co, 1-3 % Cr etc. Therefore, we expect these elements to have significantly higher contents in samples of attic dust compared to the other elements.

As it can be seen from the data presented in Table 9, the median value of nickel in samples of attic dust taken from the Kavadarci area is 220 mg kg^{-1}. However, the range of values shows a high content of nickel in the samples taken from the vicinity of the ferronickel smelter plant ranges from 89 to 1200 mg kg^{-1} (Figure 11). Cobalt and chromium results show significantly higher contents in the samples of attic dust taken from the Kavadarci area, as well (Figs. 12 and 13). The median value for Co in attic dust samples taken from the studied area is 17 mg kg^{-1} (ranges from 10 to 52 mg kg^{-1}), and the median value for Cr is 140 mg kg^{-1} (ranges from 72 to 510 mg kg^{-1}). For the remaining elements there are no such significant distortions in the distribution of values.

These findings were confirmed also by the application of continuous probability distribution and enrichment ratio for the attic dust samples studied (Table 10). According to the sampling locations, attic dust samples from the polluted areas around the ferronickel smelter plant (14 samples) are compared with corresponding data with the rest of the samples of the investigated region (unpolluted area, 17 samples).

Based on the values of the data in Tables 9-10 it can be seen that the averages for samples that are around ferronickel smelter plant (polluted area) for Co, Cr, Mo and Ni are separated by significantly higher contents of these elements, in the values for averages in the rest (unpolluted area) of the samples of the investigated region.

Samples around the ferronickel smelter plant are characterised with the highest contents of those elements. Thus, the average value for the content of nickel in attic dust samples from the polluted area is 354 mg kg^{-1} and from the unpolluted area is 156 mg kg^{-1}, having an enrichment ratio of 2.27. Distribution maps for Ni, Co and Cr are presented in Figures 11–13.

Table 9. Descriptive statistic of chemical analyses of attic dust samples from Kavadarci area ($n = 31$, 46 elements) Average values of Al, Ca, Fe, K, Mg, Na and Ti are in %, remaining elements in mg kg^{-1}

Element	N	Dis	X, Xg	Md	P_{10}	P_{90}	Min	Max	A	E
Al	31	Log	2.5	2.6	1.4	3.5	0.58	7.5	-1.25	3.23
Ca	31	N	2.9	3.0	1.8	3.5	1.2	3.8	-0.99	0.64
Fe	31	N	3.7	3.5	2.4	4.9	0.53	6.5	0.43	1.79
K	31	Log	1.3	1.3	1.1	1.5	0.83	2.0	-0.39	1.88
Mg	31	N	0.44	0.45	0.17	0.74	0.10	0.83	0.13	-1.25
Na	31	N	0.84	0.84	0.41	1.3	0.20	1.9	0.62	1.61
Ti	31	N	0.38	0.38	0.32	0.46	0.019	0.58	-1.59	6.63
Ag	31	Log	2.7	2.3	1.4	6.7	1.1	9.0	0.62	-0.85
As	31	Log	18	16	11	31	4.7	190	1.68	4.56
Ba	31	Log	320	350	210	460	36	1800	-1.16	7.81
Be	31	Log	2.4	2.5	1.7	3.3	1.3	5.6	0.31	1.47
Bi	31	N	3.7	3.4	1.8	5.7	0.77	7.6	0.39	-0.28
Cd	31	Log	8.6	8.2	6.1	14	3.8	16	-0.14	0.10
Ce	31	N	38	39	17	54	3.8	99	0.96	4.15
Co	31	Log	18	17	13	29	10	52	1.05	1.18
Cr	31	Log	140	140	81	290	72	510	0.85	0.37
Cs	31	Log	1.4	1.4	0.71	2.9	0.35	5.5	-0.01	0.06
Cu	31	N	50	52	40	64	32	66	-0.09	-0.68
Dy	31	N	2.3	2.5	0.32	3.4	0.15	5.5	0.02	1.06
Eu	31	N	0.90	0.90	0.30	1.2	0.15	1.7	0.03	1.28
Ga	31	N	15	15	12	18	0.69	32	0.81	7.46
Gd	31	N	3.4	3.6	0.55	4.6	0.23	8.4	0.24	2.25
Ge	31	N	0.96	1.0	0.28	1.7	0.010	2.2	0.30	-0.95
Hf	31	N	1.0	0.89	0.47	1.8	0.030	2.4	0.74	-0.07
Hg	31	Log	0.093	0.18	0.005	0.60	0.005	4.6	-0.38	-1.17
Ho	31	N	0.44	0.45	0.090	0.66	0.030	1.0	-0.03	0.91
I	31	N	0.47	0.40	0.13	1.1	0.020	1.2	0.98	0.44
La	31	N	17	17	3.6	25	0.90	49	1.07	4.59
Li	31	N	16	17	12	21	12	22	0.32	-0.91
Lu	31	N	0.15	0.15	0.040	0.23	0.010	0.34	0.25	0.35
Mn	31	N	510	500	450	650	410	680	0.93	0.19
Mo	31	Log	3.0	4.0	0.99	6.5	0.23	21	-0.79	0.96
Nb	31	N	14	14	9.3	18	0.17	27	0.11	4.01
Ni	31	Log	230	220	110	560	89	1200	0.63	-0.22
Pb	31	Log	180	180	110	310	66	390	-0.33	0.11
Rb	31	N	56	52	34	81	11	130	1.11	3.29
Sb	31	Log	2.4	2.3	1.6	4.1	1.0	5.2	-0.04	-0.37
Sm	31	N	3.1	3.2	0.61	4.2	0.13	7.0	-0.09	1.41
Sr	31	N	170	160	110	210	51	230	-0.68	0.64
Tb	31	N	0.45	0.46	0.080	0.65	0.020	1.1	0.16	1.66
Th	31	Log	5.3	7.0	0.79	17	0.26	25	-1.05	0.81
U	31	Log	3.4	3.4	2.2	8.6	0.54	12	-0.44	3.16
V	31	N	110	100	63	170	9.8	200	0.16	-0.14
Yb	31	N	1.0	1.1	0.20	1.4	0.040	2.5	0.12	1.55
Zn	31	Log	350	350	260	510	230	600	0.24	-0.57
Zr	31	N	46	40	15	70	0.47	160	1.78	4.14

Data round at two digits; Dis. – distribution (Log - lognormal); X – mean; Xg – geometric mean; Md – median; P_{10} and P_{90} – percentile; Min - minimum; Max – maximum; A – skewness; E – kurtosis.

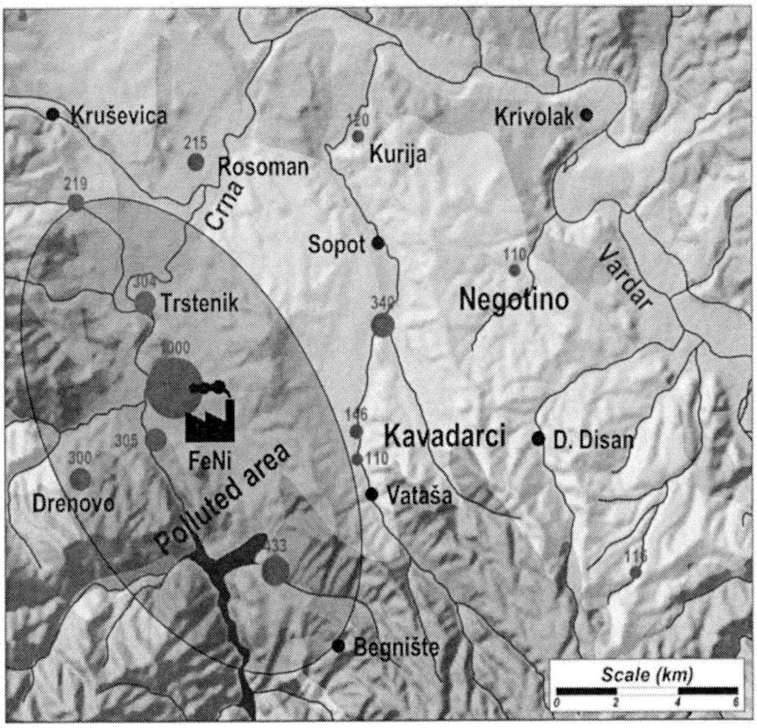

Figure 11. Distribution of Ni in attic dust samples (in mg kg^{-1}).

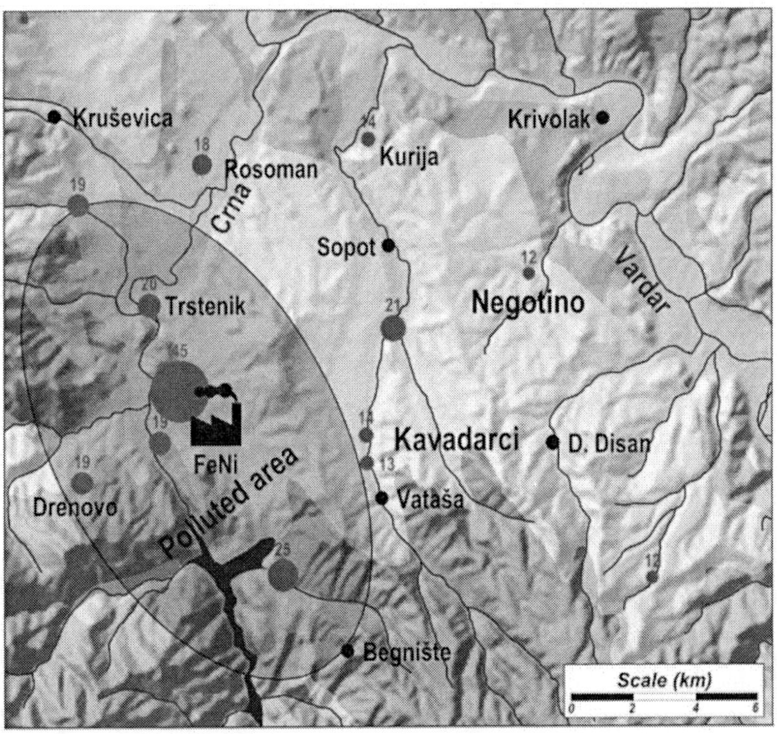

Figure 12. Distribution of Co in attic dust samples (in mg kg^{-1}).

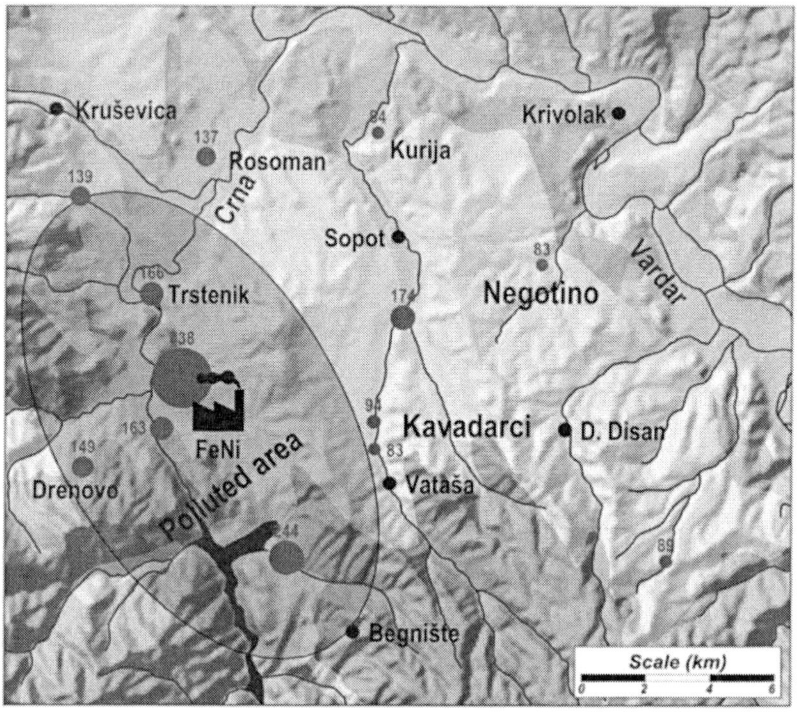

Figure 13. Distribution of Cr in attic dust samples (in mg kg^{-1}).

Table 10 also presents the values of the continuous probability distribution – F, which is acquired by a running variation analysis on the attic dust samples between two variables (samples in the vicinity of the smelter plant and samples from the rest of the investigated area). This data confirms that in this area an anthropogenic group consisting of the elements Co, Cr, Mo and Ni distinguishes itself.

From the results shown in Table 10 it is clear that the values for the continuous probability distribution – F, for Co, Cr, Mo and Ni in the attic dust samples originating from the vicinity of the smelter plant and the ones from the remainder of the investigated area are significantly higher compared to the F values of the rest of the elements. Namely, if these values are compared with the tabular critical value for F, which for the corresponding number of samples and freedom degrees is 2.32 (Miller and Miller 1993) it is clearly noted that the values for some of these elements are remarkably high (17.90 for nickel, 18.67 for chrome, 12.86 for cobalt and 5.15 for molybdenum).

The presented data makes it clear that the source of some of the heavy metals in the air in the investigated region is the dust that comes from this smelter plant (Figs. 11-13). Moreover, the content of these elements is considerably higher in the attic dust collected from this region than their natural content in the surrounding rocks and soil (Stafilov et al., 2010).

Data presented in Tables 9 and 10 and in Figs. 11–13 undoubtedly show that we can separate two populations that are influenced by activities of the ferronickel smelter plant, polluted area, which spreads about 120 km^2, and the unpolluted area. Distribution of these elements follows the wind direction (Lazarevski 1993), with the dominant winds from N and NW and SE (Figures 11–13). This fact assuredly confirms the influence of the air pollution with dust from the ferronickel plant in the closest region with the population of about 3500.

Table 10. Averages and enrichment ratios of 46 chemical elements in attic dust samples according to the sampling locations. Values of Al, Ca, Fe, K, Mg, Na and Ti are in %, remaining elements in mg kg^{-1}

No	Element	Average for Kavadarci region	Samples around the FeNi smelter	Samples from the rest of the region	ER (Feni/Rest)	F (around Feni smelter)
1	Al	2.5	2.0	2.9	0.69	4.71
2	Ca	2.9	3.0	2.8	1.09	1.05
3	Fe	3.7	3.8	3.5	1.10	0.59
4	K	1.3	1.3	1.3	1.03	0.23
5	Mg	0.44	0.44	0.44	1.00	0.00
6	Na	0.84	0.83	0.85	0.98	0.02
7	Ti	0.38	0.38	0.38	1.00	0.00
8	Ag	2.7	2.4	2.9	0.85	0.52
9	As	18	18	17	1.08	0.08
10	Ba	320	297	342	0.87	0.45
11	Be	2.4	2.4	2.5	0.96	0.11
12	Bi	3.7	4.5	3.2	1.41	5.18
13	Cd	8.6	7.9	9.2	0.86	1.77
14	Ce	38	32	43	0.74	3.48
15	**Co**	18	**22**	**15**	**1.50**	**12.86**
16	**Cr**	140	**193**	**105**	**1.84**	**18.67**
17	Cs	1.4	1.2	1.5	0.81	0.90
18	Cu	50	48	52	0.92	1.52
19	Dy	2.3	2.0	2.7	0.73	3.10
20	Dy	0.90	2.0	2.7	0.73	3.10
21	Eu	15	0.80	0.98	0.81	2.16
22	Ga	3.4	15	15	1.00	0.00
23	Gd	0.96	3.1	3.6	0.86	0.71
24	Hf	1.0	0.91	1.1	0.80	1.19
25	Hg	0.093	0.076	0.11	0.69	0.24
26	Ho	0.44	0.41	0.47	0.86	0.69
27	I	0.47	0.43	0.50	0.86	0.35
28	La	17	16	18	0.90	0.28
29	Li	16	17	15	1.12	2.88
30	Lu	0.15	0.13	0.16	0.82	1.10
31	Mn	510	522	508	1.03	0.25
32	**Mo**	3.0	**4.4**	**2.1**	**2.07**	**5.15**
33	Nb	14	14	13	1.02	0.04
34	**Ni**	230	**354**	**156**	**2.27**	**17.90**
35	Pb	180	167	182	0.92	0.30
36	Rb	56	59	54	1.08	0.32
37	Sb	2.4	2.5	2.4	1.07	0.23
38	Sm	3.1	2.8	3.2	0.87	0.64
39	Sr	170	155	173	0.90	1.48
40	Tb	0.45	0.41	0.49	0.83	1.06
41	Th	5.3	4.4	6.1	0.72	0.63
42	U	3.4	3.7	3.3	1.12	0.29
43	V	110	112	110	1.02	0.02
44	Yb	1.0	0.93	1.1	0.85	0.73
45	Zn	350	337	368	0.92	0.87
46	Zr	46	40	51	0.78	0.75

n – number of samples; FeNi – Area around ferronickel smelter (polluted area); Rest – unpolluted area; F – continuous probability distribution (analysis of variance) for attic dust between polluted area and unpolluted area, significant values ($p=0.05$) are underlined; ER – enrichment ratios of attic dust - Kavadarci area (polluted area vs. unpolluted area). Average values of Al, Ca, Fe, K, Mg, Na and Ti are in %, remaining elements in mg kg^{-1}. Data round at two digits.

4.2. Total Deposited Dust as Sampling Media

4.2.1. Copper Mining and Ore Processing Case

The obtained values for 17 elements content were proceed using descriptive statistics (Table 11). As it can be seen, median values for the copper in deposited dust samples taken from the studied area is 158 mg kg^{-1} and the min/max range of values shows much higher content of this element in the samples from this area (ranges from 52 to 1182 mg kg^{-1}). Similar results were obtained for the distribution of Fe, Pb and Zn thereby the median value and the min/max range for these elements indicate increase content of these elements.

Table 11. Descriptive statistics of measurements (N=36, 18 elements) Values of Al, Ca, Fe, K, Mg and Na are in %, remaining elements in mg kg^{-1}

Element	N	X	Xg	med	min	max	P_{10}	P_{90}	s	A	E
Al (%)	36	0.2	0.16	0,17	0.02	0.57	0.07	0.3	0.13	1.51	2.16
Ca (%)	36	6.20	4.01	4.17	0.52	21.9	1.01	14.6	5.65	1.30	1.19
Fe (%)	36	0.32	0.26	0.27	0.05	1.00	0.14	0.66	0.20	1.59	2.77
K (%)	36	1.21	0.84	0.77	0.12	6.21	0.28	2.20	1.23	2.62	7.14
Mg (%)	36	0.48	0.44	0.44	0.15	1.36	0.24	0.81	0.24	1.56	3.71
Na (%)	36	0.51	0.35	0.35	0.06	2.42	0.08	1.06	0.48	2.06	5.71
B	36	443	195	188	8.91	3085	38.3	1427	654	2.52	7.02
Cr	36	8.61	6.60	6.69	0.96	41.2	3.17	17.8	7.36	2.75	10.4
Cu	36	310	210	158	52.5	1182	84.1	861	313	1.70	1.95
Li	36	3.70	1.99	2.06	0.27	60.3	0.95	3.83	9.77	5.89	35.1
Mn	36	183	162	151	50.8	459	95.7	362	98.9	1.37	1.53
Mo	36	4.25	2.73	3.59	0.50	15.7	0.50	9.57	3.70	1.37	1.91
Ni	36	21.7	13.5	12.9	1.78	148	5.39	35.3	28.8	3.24	11.5
Pb	36	49.3	35.1	28.6	7.19	189	14.6	122	47.3	1.83	2.61
Sr	36	212	128	132	16.9	887	34.0	572	223	1.59	1.79
V	36	9.35	8.01	7.24	2.31	27.1	4.34	16.8	5.47	1.27	1.76
Zn	36	259	176	149	38.9	1045	72.7	665	275	2.04	3.29

The amount of total deposited dust that is spread in the air is presented in Figure 14. From the results obtained in this investigation it is evident that large amounts of deposited dust were recorded in the close vicinity of the mine (villages Bučim and Topolnica). In some period of the year the values are much above the maximum permitted amount of dust powder (300 mg m^{-2} d^{-1}).

The first months of the year (January to February), mining activities were placed on hold due to the decline in the price of copper. Such suspension of mining activities for a short period indicates the impact of mining on the amount of fine dust spread in the air. Maximum value for the total deposited dust (815 mg mg m^{-2} d^{-1}) was obtained in August in the Bučim village. This is the highest value for the amount of total deposited dust compared to the other two places.

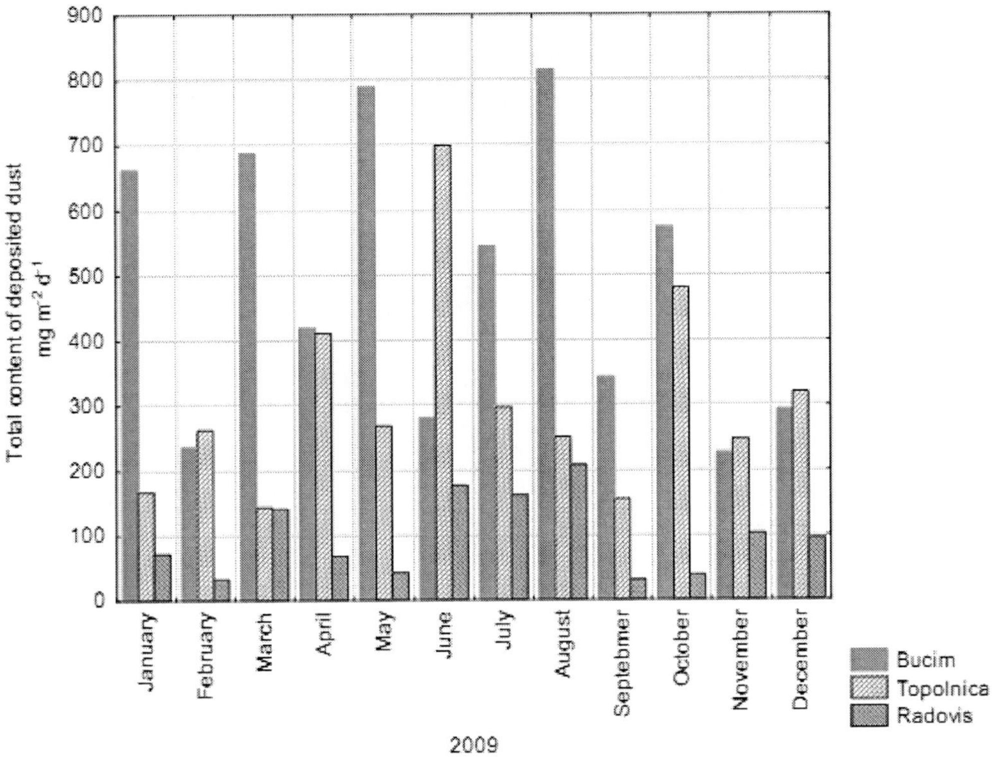

Figure 14. The total content of deposited dust.

The annual average for the total deposited dust in the vicinity of the Bučim village is 489 mg mg m^{-2} d^{-1}, for Topolnica the 309 mg mg m^{-2} d^{-1} and accounted for Radoviš is 97 mg mg m^{-2} d^{-1}. These values indicate that in the vicinity of the Bučim village the population is exposed to high amounts of dust in the air.

A lower value was obtained for the amount of the total deposited dust in village Topolnica environ, while for Radoviš was obtained a much lower value (Figure 14). The town of Radoviš was not affected by the dust deposition from the mine activities. Bučim village is exposed to large amounts of dust almost through the whole year.

Because of high ore content of Cu (0.3 %) and large amount of copper in flotation tailing (Serafimovski et al., 2005) was expected this element to have significantly higher content in samples of deposited dust compared to the other elements.

As it can be seen from the data presented in Table 12. the median values for Cu in samples of deposited dust taken from the Radoviš area is 396 mg kg^{-1} and the ranges (from 94.8 to 1171 mg kg^{-1}) for the Topolnica village the median values in samples of deposited dust is 150 mg kg^{-1} with ranges (from 52.5 to 1183 mg kg^{-1}) and for the Bučim village the median values in deposited dust samples is 145 mg kg^{-1} and the ranges (from 85.3 to 317 mg kg^{-1}).

From these results can be seen that the maximum value for the content of Cu was obtained from Topolnica village (settlement near by the flotation tailings landfill). In the town of Radoviš was not found deposited dust above the maximum permitted amount of dust powder, but high content of Cu content was worrying.

Table 12. Statistical parameters for annual values for the content of chemical elements in samples of deposited dust measured in Radoviš and is environs Values of Al, Ca, Fe, K, Mg and Na are in %, remaining elements in mg kg^{-1}

Element	N	Sampling site					
		Vill. Bučim		Vill. Topolnica		Radoviš	
		Md	**Range**	**Md**	**Range**	**Md**	**Range**
Al %	12	0.13	0.02-0.19	0.15	0.07-0.58	0.25	0.16-0.56
Ca %	12	9.12	1.74-2.19	1.32	0.53-4.74	7.25	1.56-20.9
Fe %	12	0.23	0.05-0.34	0.27	0.10-1.02	0.36	0.14-0.77
K %	12	0.76	0.12-4.71	0.86	0.28-6.21	0.69	0.23-2.12
Mg %	12	0.43	0.15-0.84	0.34	0.21-0.84	0.48	0.19-1.36
Na %	12	0.16	0.06-0.92	0.31	0.08-0.87	0.77	0.24-2.41
B	12	203	8.91-1427	93.4	34.7-305	454	101-3085
Cr	12	4.22	0.96-7.78	6.32	1.74-20.1	12.6	6.42-41.2
Cu	**12**	**145**	**85.3-317**	**150**	**52.5-1183**	**396**	**94.8-1171**
Li	12	1.17	0.27-2.11	2.13	1.20-5.16	2.79	0.95-60.3
Mn	12	128	50.8-459	148	69.8-411	197	98.4-419
Mo	12	3.53	0.50-6.91	2.51	0.50-12.3	4.18	0.50-15.7
Ni	12	6.94	2.81-14.1	13.5	6.49-27.8	26.9	1.79-148
Pb	12	25.7	12.1-56.6	26.8	7.20-184	69.5	20.1-189
Sr	12	303	70.7-888	46.4	16.9-269	170	34.1-572
V	12	5.49	2.31-8.23	8.63	4.34-27.1	13.1	5.23-18.5
Zn	12	217	92.5-1045	89.0	38.9-240	176	86.6-665

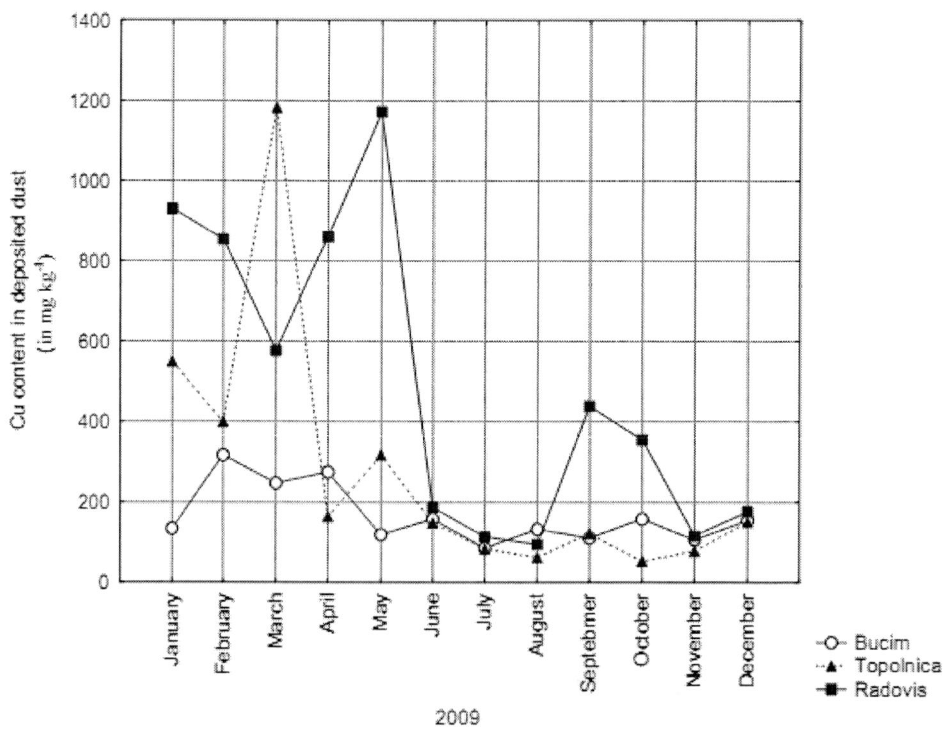

Figure 15. Trends of copper content in deposited dust through the whole year.

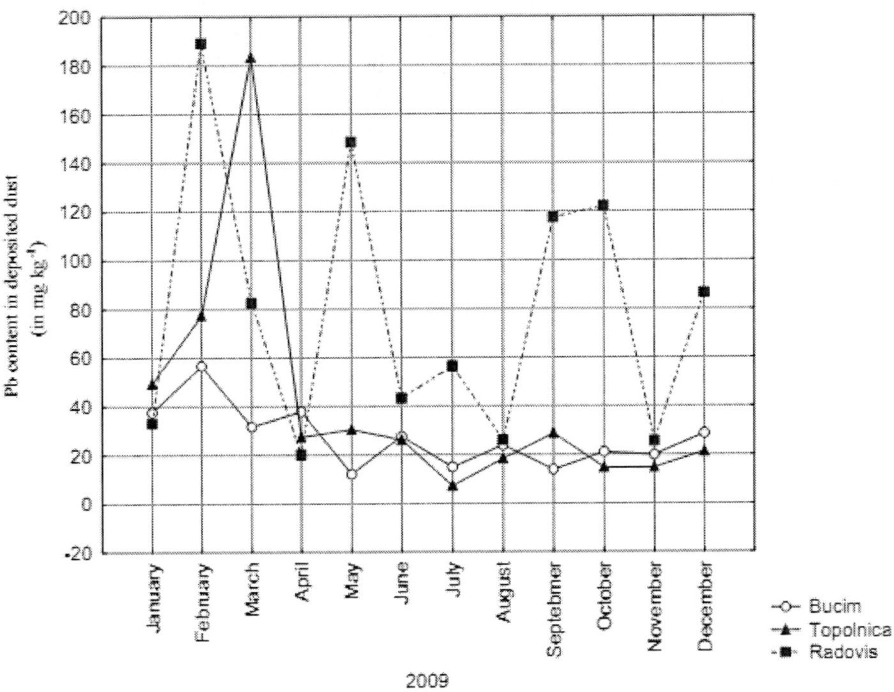

Figure 16. Trends of lead content in deposited dust through the whole year.

Similar results were obtained for the content of Pb, where clearly is visible that the Bucim village was not affected from high content of lead in the deposited dust from air. Maximum value was obtained from the town of Radoviš (189 mg kg^{-1} in February) and varies during the whole year. Despite the large amounts of total deposited dust from Bučim village, there were not found high values for the lead contend. Maximum value (183 mg kg^{-1}) for the lead content in deposited dust from the Topolnica village was obtained in March, when starts' decreasing of lead content and the trend is retained until the end of the year.

The ultimate effect is that despite large amounts of deposited dust, population in Bučim village is not affected with high content of anthropogenic introduced metals (Cu and Pb) due to copper mining works. Unexpectedly the town of Radoviš, despite little amounts of deposited dust, was affected by large content of Cu and Pb. This suggests that there is Cu and Pb distribution on large distance from mine, which was not concluded with the historical monitoring using attic dust samples.

4.2.2. Ferro-Nickel Smelter Plant Case

For each observation there were few variables: sample identification number, locality, geographic coordinates, sample type and concentration level for 18 elements. For statistical analysis of data was used parametric and nonparametric statistic. Minima and maxima mean calculation methods of median, geometrical mean, standard deviation, geometric standard deviation, skewness, kurtosis and arithmetical mean were calculated and histograms were drawn to show the distribution features. Table 13 shows the basic statistics for samples of deposited dust for the period from (January-December-2009). As it can be seen from the obtained data presented in Table 13, the median values for the nickel in deposited dust samples taken from the studied area is 101 mg kg^{-1} and the min/max range of values shows much higher content of this element in the samples from this area (ranges from 5.24 to 1289 mg kg^{-1}). Similar results are obtained for the distribution of Co and Cr, thereby the median value and the min/max range for these elements indicate increase content of these elements. The median value for Cr in deposited dust samples taken from the studied area is 11.96 mg kg^{-1} (ranges from 0.73 to 139 mg kg^{-1}), and for Co the median value in deposited dust samples taken from the studied area is 3.91 mg kg^{-1} (ranges from 0.23 to 75.58 mg kg^{-1}).

The amount of total deposited dust that is spread in the air is presented in Figure 17. From the results obtained in this investigation it is evident that a large amount of deposited dust were recorded in summer in 2009 when the values are close to or above the maximum permitted amount of dust powder (300 mg m^{-2} d^{-1}).

Table 13. Descriptive statistics of measurements (N=36, 18 elements) Values of Al, Ca, Fe, K, Mg and Na are in %, remaining elements in mg kg^{-1}

Elements	N	X	Xg	Md.	Min.	Max.	P_{10}	P_{90}	s	A	E
Al %	36	0.13	0.10	0.12	0.01	0.41	0.03	0.27	0.09	1.23	1.86
Ca %	36	6.24	5.23	6.06	1.17	14.09	1.85	11.23	3.33	0.37	-0.53
Fe %	36	0.26	0.13	0.16	0.01	1.42	0.02	0.59	0.33	2.24	5.19
K %	36	2.09	1.28	1.23	0.32	15.69	0.49	2.88	3.23	3.52	12.34
Mg %	36	0.74	0.62	0.60	0.19	1.96	0.31	1.55	0.48	1.06	0.05
Na %	36	0.74	0.54	0.59	0.07	2.04	0.18	1.66	0.56	0.92	-0.23
As mg kg^{-1}	36	10.55	6.47	7.98	0.24	39.19	0.45	21.45	8.33	1.28	2.55
Cd mg kg^{-1}	36	0.93	0.63	0.79	0.10	4.94	0.20	1.55	0.97	2.69	8.46
Co mg kg^{-1}	36	9.80	3.92	3.91	0.23	75.58	0.31	25.18	15.57	3.05	10.14
Cr mg kg^{-1}	36	24.02	11.93	11.96	0.73	139	2.13	47.11	31.97	2.64	7.48
Cu mg kg^{-1}	36	227	115	82.1	17.2	1093	23.0	703	270	1.63	2.11
Li mg kg^{-1}	36	6.47	3.16	2.54	0.37	42.21	1.09	20.17	9.92	2.71	7.18
Mn mg kg^{-1}	36	180	110	114	11.48	795	26.81	513	188	1.76	2.77
Ni mg kg^{-1}	36	180	89.38	101	5.24	1289	13.62	428	243	3.15	12.31
Pb mg kg^{-1}	36	354	127	121	6.66	2306	14.20	1079	525	2.40	6.03
Sr mg kg^{-1}	36	160	135	144	31.21	325	47.59	287	84.9	0.30	-0.88
V mg kg^{-1}	36	16.01	6.60	7.14	0.81	136	0.81	41.75	26.52	3.30	12.43
Zn mg kg^{-1}	36	754	282	227	59.88	7346	61.77	1392	1550	3.31	10.86

N-number of samples; X – mean; Xg – geometric mean; Md – median; Min - minimum; Max – maximum; P_{10} - 10 percentile; P_{90} - 90 percentile; s – standard deviation; A – skewness; E – kurtosis. Data round at two digits.

The ferronickel smelter plant uses ore that contains between 1 % and 2.5 % Ni, about 0.05 % Co, 1-3 % Cr etc. Therefore, we expect these elements to have significantly higher contents in samples of deposited dust compared to the other elements. As it can be seen from the data presented in Table 14, the median values for nickel in samples of deposited dust taken from the Kavadarci area is 28 mg kg^{-1} and the ranges (from 5.23 to 140.9 mg kg^{-1}) for the Brušani village the median values in samples of deposited dust is 196.3 mg kg^{-1} with ranges (from 24.7 to 290 mg kg^{-1}) and for the Vozarci village the median values in deposited dust samples is 144.9 mg kg^{-1} and the ranges (from 20.7 to 393 mg kg^{-1}). From this results can be seen that the values for the content of Ni in Vozarci village which is near the ferro-nickel smelter plant (Figure 18), has the highest content of Ni compared with the other two locations where are collected monthly samples of deposited dust from the Kavadarci area in 2009.

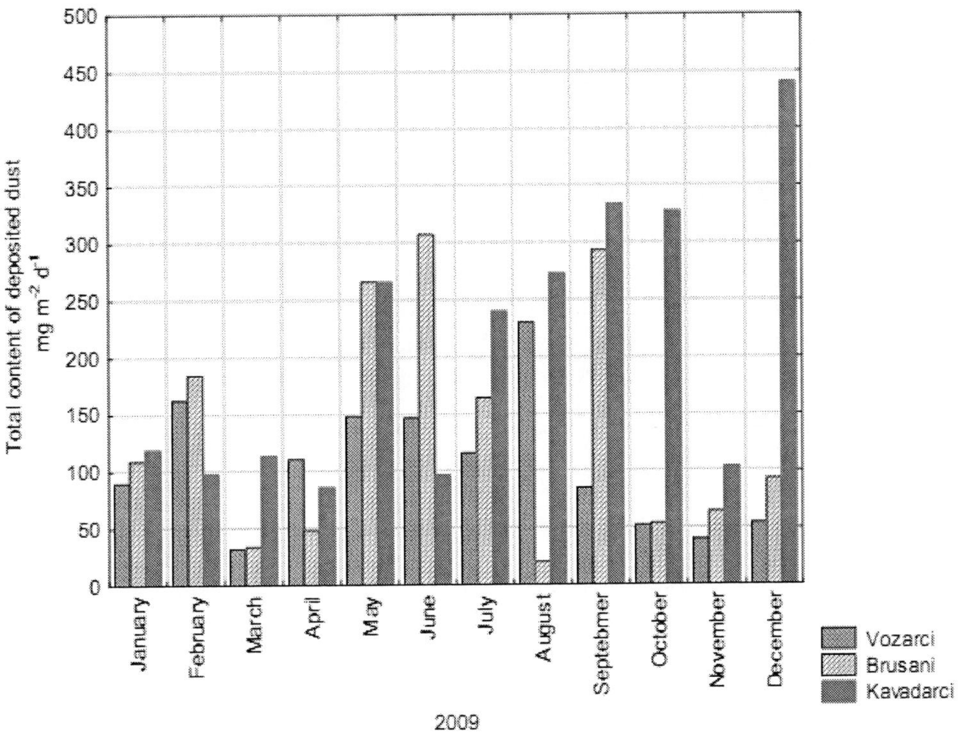

Figure 17. The total content of deposited dust.

Figures 18–20 has shown the values for the content of Ni, Cr and Co in deposited dust collected at all three locations. The obtained values for the content of nickel in deposited dust collected in the Vozarci village (near ferro-nickel smelter) in January-March 2009 is very high (477 to 1290 mg kg^{-1}) while the remainder content is relatively low (25 to 222 mg kg^{-1}) with the exception of the samples collected in July (301 mg kg^{-1}) and September (428 mg kg^{1}), Figure 18.

This trend of increased content of nickel in the deposited dust is followed by increased content of chromium (Figure 19) and cobalt (Figure 20). The content of chromium in deposited dust collected in the Vozarci village is high for the first four months of January-April 2009 (65 to 140 mg kg^{-1}) and for the cobalt (Figure 20). The obtained values in deposited dust collected in the Vozarci village in January-March 2009 (28 to 76 mg kg^{-1}). This suggests that in these months had increased emission of dust from the smelter plant. These elements (Co, Cr and Ni) can enter the atmosphere from refining plants. Once it is airborne, they can attaches to dust in the air.

Over time, the dust falls to the ground or mixes with rain or snow, where it enters the soil. The exposure to nickel for the general population takes place via inhalation of dust, ingestion of food and water. Inhaled nickel is absorbed by the body depending on the solubility of the nickel compound in the air.

The median value for Co in deposited dust samples taken from the Kavadarci is 1.43 mg kg^{-1} (ranges from 0.23 to 7.5 mg kg^{-1}), for the Brušani village the median values is 10.8 mg kg^{-1} and the ranges (from 1.7 to 75.5 mg kg^{-1}) and for the Vozarci village the median values is 3.9 mg kg^{-1} and the ranges (from 0.99 to 25 mg kg^{-1}). The median value for Cr for Kavadarci is 4.3 mg kg^{-1} with ranges (from 0.7 to 33.5 mg kg^{-1}), the median for Brušani village is 31.9 mg kg^{-1} with ranges (from 1.83 to 139 mg kg^{-1}) and for Vozarci village the median is 14 mg kg^{-1} and ranges (from 4.3 to 47 mg kg^{-1}).

In Brušani village the content of cobalt and chromium in deposited dust samples is higher compared to content of these elements in village Vozarci and Kavadarci. From here it can be conclude that the dust distribution follows the wind rose (Lazarevski 1993) with the dominant winds from N and NW and SE. These results indicate that content for these elements (Co, Cr and Ni) are significantly higher, compared with the other elements that are investigated.

Table 14. Statistical parameters for annual values for the content of chemical elements in samples of deposited dust measured in Kavadarci and is environs Average values of Al, Ca, Fe, K, Mg and Na are in %, remaining elements in mg kg^{-1}

Element/	Sampling sites								
	Kavadarci			Vill. Brušani			Vill. Vozarci		
Statistical parameters	**N**	**Md**	**Range**	**N**	**Md**	**Range**	**N**	**Md**	**Range**
Al %	12	0.08	0.02-0.306	12	0.13	0.05-0.41	12	0.13	0.01-0.20
Ca %	12	4.71	1.17-14.09	12	6.35	1.69-11.3	12	7.14	1.22-11.8
Fe %	12	0.06	0.008-0.86	12	0.28	0.01-1.42	12	0.18	0.03-0.59
K %	12	1.27	0.43-15.68	12	0.86	0.51-2.52	12	1.33	0.32-2.88
Mg %	12	0.40	0.18-1.34	12	0.77	0.31-1.96	12	0.67	0.28-1.69
Na %	12	0.42	0.07-2.03	12	0.55	0.18-1.88	12	0.82	0.28-1.76
As mg kg^{-1}	12	2.85	0.24-25.17	12	8.97	3.55-39.2	12	14.73	7.43-21.7
Cd mg kg^{-1}	12	0.28	0.10-1.220	12	0.90	0.30-4.94	12	0.83	0.20-2.99
Co mg kg^{-1}	**12**	**1.43**	**0.23-7.51**	**12**	**10.86**	**1.74-75.5**	**12**	**3.91**	**0.99-25.18**
Cr mg kg^{-1}	**12**	**4.3**	**0.73-33.53**	**12**	**31.92**	**1.83-139**	**12**	**14.11**	**4.38-47.1**
Cu mg kg^{-1}	12	46.2	17.2-319	12	83.78	37.6-839	12	438.83	40-1093
Li mg kg^{-1}	12	1.81	0.36-10.2	12	2.45	0.86-39.8	12	5.98	1.32-42.2
Mn mg kg^{-1}	12	64.1	21.5-289	12	153.1	11.4-608	12	145.8	26.8-795
Ni mg kg^{-1}	**12**	**28.13**	**5.23-140.9**	**12**	**196.3**	**24.7-290**	**12**	**144.9**	**20.7-393**
Pb mg kg^{-1}	12	53.3	6.66-485	12	229.2	14.9-306	12	236.9	30.6-1163
Sr mg kg^{-1}	12	122.5	31.2-286	12	132.2	40.6-323	12	215.4	46.9-325
V mg kg^{-1}	12	3.6	0.81-24	12	7.14	0.81-41.7	12	9.91	0.81-135
Zn mg kg^{-1}	12	189	59.8-7346	12	235	60.3-5140	12	272.66	61.7-1113

N-number of samples; Md-median.

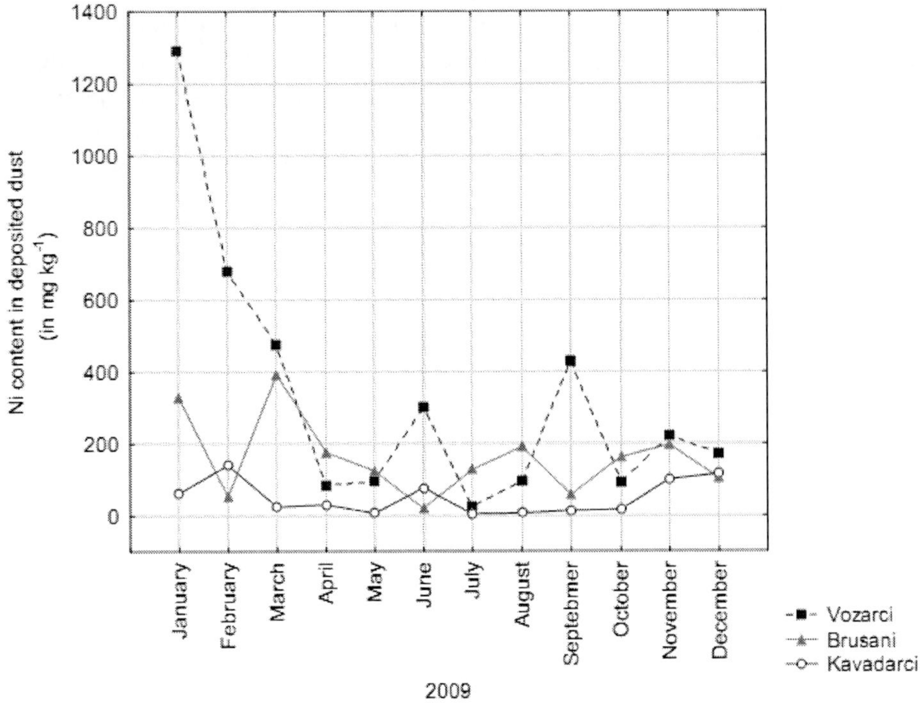

Figure 18. Trends of nickel content in deposited dust trough the whole year (in mg kg^{-1}).

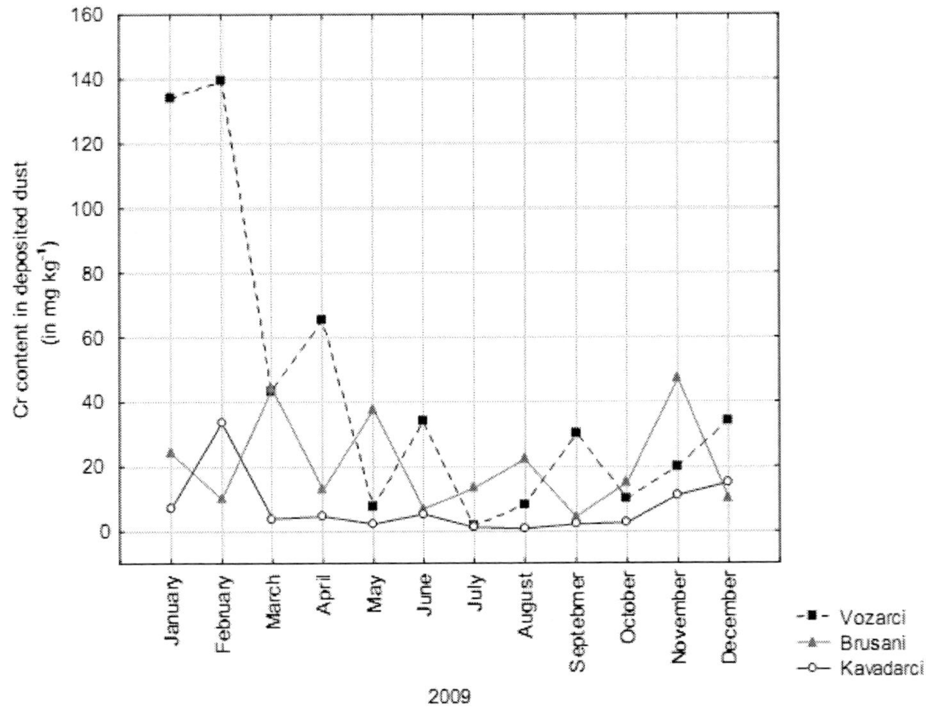

Figure 19. Trends of chrome content in deposited dust trough the whole year (in mg kg^{-1}).

Atmospheric deposition of heavy metals can pollute soils and sediments and can have severe ecological impacts such as bioaccumulation or loss of biodiversity (Mulder and Breure, 2006; Norr and Riepert, 2007; Morvan et al., 2008; Kwaterczak and Helios-Rybicka, 2009; Schröder et al., 2010). Therefore, deposition measurements are key to soil monitoring (Schröder et al., 2004) and are needed for spatial estimations of soil contamination through atmospheric deposition (Zheng et al., 2008).

The same region has been investigated for the soil pollution (Stafilov et al., 2008, 2010), and did not identify a statistical significance between the soil samples. For the nickel content in the soil in Kavadarci and its environs, it is found that the median and average values are very similar (72 mg kg^{-1} and 74 mg kg^{-1}, respectively) and that the content of nickel in the whole investigated area ranges between 42 and 150 mg kg^{-1} (Stafilov et al., 2010). Moreover, it was found that nickel content in subsoil and topsoil is almost the same, which shows that the presence of nickel in soil is from natural origin.

The average value for the content of nickel in moss samples from the investigated region is 80 mg kg^{-1}, ranges from (14 to 340 mg kg^{-1}) (Bačeva et al., 2009). This value is much higher than the average of 6.6 mg kg^{-1} for the content of nickel in moss samples collected from the territory of the Republic of Macedonia, (Barandovski et al., 2006, 2008), which show direct connection with the pollution from the ferronickel smelter plant situated near the city of Kavadarci. Therefore, main goal was to follow the influence on the elements' distribution in the air caused by the ferronickel smelter plant and its connection to the some other natural conditions' (climate conditions, lithological, dust, etc).

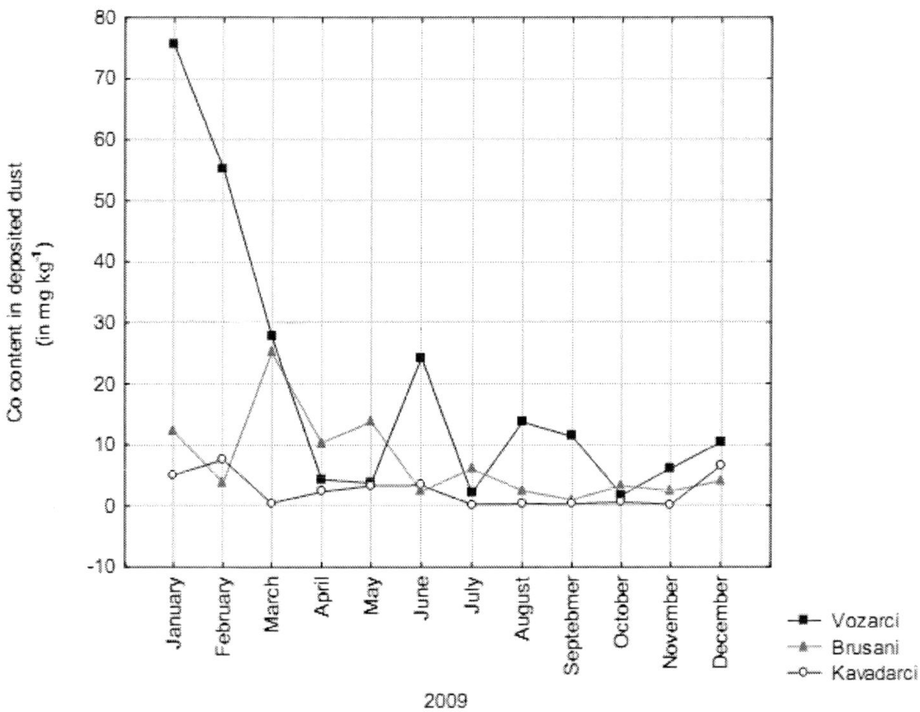

Figure 20. Trends of cobalt content in deposited dust trough the whole year (in mg kg^{-1}).

Our previous investigation on the distribution of Ni, Co and Cr, in attic dust shows an increased content in the samples from the smelter plant surroundings compared to the content in the rest of the samples of the same region (Bačeva et al., 2011). This also confirms that the dust that is discharged from this plant is connected with the ferronickel ore that is being processed in this smelter plant.

CONCLUSION

Attic and total deposited dust has proved to be very effective environmental samples for monitoring historical and present distribution of heavy metals in a potentially contaminated area. In this Chapter data about the air pollution with heavy metals in the vicinity of copper mine and ferro-nickel smelter plant are given.

The historical (long period) distribution was followed by the analysis of the content of heavy metals in attic dust collected from the houses in the investigated areas. The presence (since the operation starts) of an open copper ore pit, ore processing and flotation plant (Bučim mine, near the town of Radoviš, Republic of Macedonia) leads to increased content of copper in attic dust in its surroundings (max. content of Cu 415 mg kg^{-1}). Apart from the copper, increased content of As and Pb in attic dust (max. contents 51 mg kg^{-1} and 121 mg kg^{-1}, respectively) were also found, singled out as anthropogenic association with copper. The distribution of the anthropogenic association of elements (As, Cu and Pb) at distant regions was not determined. The distribution of the rest of the elements relies on the geology of the region. The distributions of most of the elements in ferronickel smelter plant environ in attic dust samples are closely dependent on the lithogenesis and are slightly affected by anthropogenic activities. Historical record (attic dust) of nickel distribution showed high contents deposited in Kavadarci environ (max. content 1200 mg kg^{-1}). Cobalt and chromium results show significantly higher contents in the samples of attic dust taken from the Kavadarci area, as well (max. contents 52 mg kg^{-1} and 510 mg kg^{-1}, respectively).

The distribution of dust with heavy metals was also followed by the present distribution (total deposited dust), were it assumed that anthropogenic elements although deposit in higher content at distant regions from their hot spots. In close vicinity of the copper mine environ, there was deposition of large amounts of dust, values much above the maximum permitted amount of dust powder (300 mg m^{-2} d^{-1}); annual average for the total deposited dust in the vicinity of the Bučim village - 489 mg mg m^{-2} d^{-1}, for Topolnica - 309 mg mg m^{-2} d^{-1} and for Radoviš - 97 mg mg m^{-2} d^{-1}. Maximum value for the Cu content was obtained from Topolnica village (settlement near by the flotation tailings landfill). In the town of Radoviš was not found deposited dust above the maximum permitted amount of dust powder, but high content of Cu content was worrying (1171 mg kg^{-1}). The population from mentioned villages is continuously exposed to a high content of copper and lead, which has a negative effect on human health and toxic effects on the environment.

It was found that ferronickel smelter plant (the second case study) affects on the distribution of large amounts of dust in it environ trough the whole year. From the results obtained it is clear that a large amount of deposited dust were recorded in summer when the values are close to or above the maximum amount of dust powder (300 mg m^{-2} d^{-1}). The values for the content of nickel in deposited dust collected in the village Vozarci (near smelter plant) in the winter period is very high while in the remaining period of its content is relatively low (max. content 1290 mg kg^{-1}). This trend of increased content of nickel in the dust monitor is followed by the content of chromium and cobalt in deposited dust. The dust distribution followed the wind rose with the dominant winds from N and NW and SE. These results indicate that content for these elements (Co, Cr and Ni) are significantly higher, compared with the other elements that are investigated. The population of the settlements near smelter plant was affected with the high contents of these elements due to large amounts of dust distribution from ferronickel smelter plant.

REFERENCES

Agarval, S. K. (2009). *Heavy metal pollution*. New Delhi, SB: A P H Publishing Corporation: Nangia.

Aleksander-Kwaterczak, U. and Helios-Rybicka, E. (2009). Contaminated sediments as a potential source of Zn, Pb and Cd for river system at the historical metalliferous ore mining and smelting industry area in South Poland. *Journal of Soils and Sediments*, *9*, 13–22.

Athar, M. and Vohora, S. (1995). *Heavy metals and environment*, New Delhi, New Age International Publishers.

Artiola, J. F. Pepper, I. and Brussean, L. (2004). *Environmental Monitoring and Characterization*. San Diego, Elsevier Academic Press.

Avila, A. and Rodrigo, A. (2004). Trace metal fluxes in bulk deposition, through fall and stemflow at two evergreen oak stands in NE Spain subject to different exposure to the industrial environment, *Atmospheric Environment*, *38*, 171–180.

Bačeva, K. Stafilov, T. and Šajn, R. (2009). Biomonitoring of nickel air pollution near the city of Kavadarci, Republic of Macedonia. *Ecology and Protection of the Environment*, *12*, 57–69.

Bačeva, K. Stafilov, T, Šajn, R, Tănăselia, C. and Ilić Popov, S. (2011). Distribution of chemical elements in attic dust in the vicinity of ferronickel smelter plant, *Fresenius Environmental Bulletin*, *20*(9), 2306–2314.

Balabanova, B.Stafilov, T. Bačeva, K. and Šajn, R. (2009). Atmospheric pollution with copper around the copper mine and flotation, Bučim, Republic of Macedonia, using biomonitoring moss and lichen technique. *Geologica Macedonica*, *23*, 35–41.

Balabanova, B. Stafilov, T. Bačeva, K. and Šajn, R. (2010). Biomonitoring of atmospheric pollution with heavy metals in the copper mine vicinity located near Radoviš, Republic of Macedonia. *Journal of Environmental Science and Health, Part A*, *Toxic/Hazardous Substance and Environmental Engineering*, *45*, 1504–1518.

Balabanova, B. Stafilov, T. Šajn, R. and Bačeva, K. (2011). Distribution of chemical elements in attic dust as reflection of lithology and anthropogenic influence in the vicinity of copper mine and flotation. *Archives of Environmental Contamination Toxicology*, *61*, 173-184.

Barandovski, L. Cekova, M. Frontasyeva, M. V. Pavlov, S. S. Stafilov, T. Steinnes, E. and Urumov, V. (2006). *Air pollution studies in Macedonia using the moss biomonitoring technique, NAA, AAS and GIS Technology*. Preprint E18-2006-160. Dubna, Joint Institute for Nuclear Research.

Barandovski, L. Cekova, M. Frontasyeva, M. V. Pavlov, S. S. Stafilov, T. Steinnes, E. and Urumov, V. (2008). Atmospheric deposition of trace element pollutants in Macedonia studied by the moss biomonitoring technique. *Environmental Monitoring and Assessment, 138*, 107-118.

Barbante, C., Schwikowski, M., Döring, T., Gäggeler, H. W., Schotterer, U., Tobler, L., Van de Velde, K., Ferrari, C., Cozzi, G., Turetta, A., Rosman, K., Bolshov, M., Capodaglio, G., Cescon, P. and Boutron, C. *(2004)*. Historical Record of European Emissions of Heavy Metals to the Atmosphere Since the 1650s from Alpine Snow/Ice Cores Drilled near Monte Rosa. *Environmental Science and. Technology, 38,* 4085–4090.

Baron, P. A. and Willeke, K. (2005). Aerosol Fundamentals. In: Baron PA; Willeke K. editors. *Aerosol Measurement: Principles, Techniques, and Applications*. 2nd edition. New Jersey, Wiley Interscience, 1111.

Boev, B. and Jankovic, S. (1996). *Nickel and nickeliferrous iron deposits of the Vardar zone (SE Europe) with particular references to the Rzanovo-Studena Voda ore-bearing series.* Special issue No. 3. Štip: Faculty of Mining and Geology.

Čačković, M. Šega, K. and Vaidjic, Bešli. I. (2008). Characterisation of major acidic anions in TSP and PM10 in Zagreb Air. *Bulletin of Environmental Contamination and Toxicology, 80*, 112-4.

Čačkovič, M. Kalinič, N. and Vadjič, V. G. (2009). Pehnec heavy metals and acidic components in total deposited matter in Šibenik and National park Kornati, Croatia. *Archives of Environmental Contamination and Toxicology, 56*, 12–20.

Chittum, J. W. (1964). Van Nostrand's International Encyclopaedia of Chemical Science, *Journal of Chemical Education, 41*, 517.

Citizel, J. V. (1998). *Plutonium anomalies in attic dust and soils at locations surrounding the Nevada Test Site*. PhD thesis.University of Nevada, Reno.

Citizel, J. Hodge,V. and Faller, S. (1999). Resolving Nevada Test Site and global fallout plutonium in attic dust and soils using (super 137)Cs/(super 239+240)Pu activity ratios. *Health Physics, 77*, 67-75.

Clarke, A. G. (1998). *Industrial air pollution monitoring*. London, Chapman and Hall Press.

Cortes-Maramba, N., Reyes, J. P., Francisco-Rivera, A. T., Akagi, H., Sunio, R. and Panganiban, L. C. (2006). Health and environmental assessment of mercury exposure in a gold mining community in western Mindanao, Philippines. *Journal of Environmental Management, 81*, 126–134.

Cyrys, J. Stölzel, M. Heinrich, J. Kreyling, W. G. Menzel, N. Wittmaack, K. Tuch, T. and Wichmann, H. E. (2003). Elemental composition and sources of fine and ultrafine ambient particles in Erfurt, Germany. *Science of the Total Environment, 305*, 143-156.

Dudka, S. and Adriano, C. D. (1997). Environmental impacts of metal ore mining and processing: A review. *Journal of Environmental Quality, 26*, 590–602.

Filzmoser, P. Garrett, R. G. and Reimann, C. (2005). Multivariate outlier detection in exploration geochemistry. *Computers and Geosciences, 31*, 579-587.

Finlayson-Pitts, B. and Pitts, J. N. (2000). *Chemistry of the Upper and Lower Atmosphere*, San Diego, CA: Academic Press.

Gauderman, W. J., McConnell, R., Gilliland, F., London, S., Thomas, D., Avol, E., Vora, H., Berhane, K., Rappaport, E. B., Lurmann, F., Margolis, H. G., and Peters, J. (2000). Association between air pollution and lung function growth in southern California children. *American Journal of Respiratory and Critical Care Medicine, 162*, 1383-1390.

Gauderman, W. J., Avol, E., Gilliland, F., Vora, H., Thomas, D., Berhane, K., McConnell, R., Kuenzli, N., Lurmann, F., Rappaport, E., Margolis, H., Bates, D., and Peters, J. (2004). The effect of air pollution on lungd from 10 to 18 Years of age. *The New England Journal of Medicine, 351*, 1057-1067.

Godish, T. (2004). *Air quality*. 4th edition. Boca Raton, Lewis publishers.

Gosar, M. and Šajn, R. (2001). Mercury in soil and attic dust as a reflection of Idrija mining and mineralization (Slovenija). *Geologija*, *44*, 137-159.

Gosar, M. and Šajn, R. (2003). Geochemical soil and attic dust survey in Idrija, Slovenija, *Journal du Physique*, *107*, 561-564.

Gosar, M. Šajn, R. and Harald, B. (2006). Binding of mercury in soils and attic dust in the Idrija mercury mine area (Slovenia). *Science of the Total Environment*, *369*, 150-162.

Hall, D. J. Upton, S. L. and Marsland, G. W. (1993). Improvements in dust gauge design. In: Couling S, editor. *Measurement of airborne pollutants*. Oxford: Butterworth-Heinemann Ltd.

Holden, T. (2006). *Air Quality – Dust Monitoring*. NSW: Environmental Defender's Office Ltd.

Hollander, M.and Wolfe, D. A. (1999). *Nonparametric Statistical Methods, 2nd Edition*. New York, John Wiley and Sons.

Hou, H. Takamatsu, T. Koskiwa, M. K. and Hosomi, M. (2005). Trace metal sin bulk precipitation and through fall in a suburban area of Japan. *Atmospheric Environment, 39*, 3583–3595.

Hristov, S. Karajovanovič, M. and Stračkov, M. (1965). *Basic geological map of SFRJ, sheet Štip,*. Beograd, Federal Geological Survey.

Hršak, J. Škrbec, A. Balagovič, I. and Šega, K. (2003). Thallium content in Zagreb air. *Bulletin of Environmental Contamination and Toxicology, 71*, 131–134.

Ilacqua, V. Freeman, N. C. Fagliano, J. P. and Lioy, J. (2003). The historical record of air pollution as defined by attic dust. *Atmospheric Environment*, *37*, 2379–2389.

Jarup, L. (2003). Hazards of heavy metal contamination. *British Medical Bulletin, 68*, 167-182.

Jeffrey, J. Davisa, B. and Gulson, B. L. (2005). Ceiling (attic) dust: A ''museum'' of contamination and potential hazard. *Environmental Research*, *99*, 177-194.

Lazarevski, A. (1993). *Climate in Macedonia*. Skopje, Macedonia : Kultura.

Maksimović, Z. (1982). *Mineralogical Study of the Ore from Ržanovo Mine*. First Edition. Belgrade, Faculty of Geology and Mining.

Miguel, E. Lamas, J. F. Chacon, E. and Mazadiego. L. F. (1999). Sources and pathways of trace elements in urban environments: a multi-elemental qualitative approach. *Atmospheric Environment*, *31*, 2733- 2740.

Miller, J. C. and Miller, J. N. (1993). *Statistics for Analytical Chemistry*. 3rd Edition. London, Ellis Horwood PTR Prentice Hall.

Morselli, L. Olivieri, P. Brusori, B. and Passarini, F. (2003). Soluble and insoluble fractions of heavy metals in wet and dry atmospheric depositions in Bologna, Italy. *Environmental Pollution*, *124*, 457–469.

Morvan, X. Saby, N. P. Arrouays, D. Le Bas, C. Jones, R. J. Verheijen, F. G. Bellamy, P. H. Stephens, M. and Kibblewhite, M. G. (2008). Soil monitoring in Europe: a review of existing systems and requirements for harmonisation. *Science of the Total Environment, 391*, 1–12.

Mulder, C. and Breure, A. M. (2006). Impact of heavy metal pollution on plants and leaf-miners. *Environmental Chemistry Letters, 4*, 83–86.

Norr, C. and Riepert, F. (2007). Bioaccumulation studies with *Eisenia fetida* using an established degradation test system. *Journal of Soils and Sediments, 7*, 393–397.

Pacyna, J. M. and Pacyna, E. G. (2001). An assessment of global and regional emissions of trace metals to the atmosphere from anthropogenic sources worldwide. *Environmental Reviews, 9*, 269-298.

Pacyna, E. G. Pacyna, J. M. Fudala, J. Strzelecka-Jastrzab, E. Hlawiczka, S. Panasiuk, Nitter, D. S. Pregger, T. Pfeiffer, H. and Friedrich, R. (2007). Current and future emissions of selected heavy metals to the atmosphere from anthropogenic sources in Europe. *Atmospheric Environment, 41*, 8557-8566.

Polkowska, Z., Astel, A., Walna, B., Małek, S., Mędrzycka, K., Górecki, T., Siepak, J. and Namieśnik, J. (2005). Chemometric analysis of rain water and throughfall at several sites in Poland. *Atmospheric Environment, 39*, 837–855.

Rakićević, T., Dumurdzanov, N. and Petkovski, M. (1969). *Basic geological map of SFRJ, sheet Štip, M 1:100,000 (map and interpreter)*. Beograd, Federal Geological Survey.

Reimann C., Filzmoser, P. and Garrett R. G. (2002). Factor analysis applied to regional geochemical data: problems and possibilities. *Applied Geochemistry, 17*, 185-206.

Repley, E. A., Redmann, R. E. and Crowder, A. A. (1996). *Environmental Effects of Mining*. Delray Beach, FL: St. Lucie Press.

Šajn, R. (2003). Distribution of chemical elements in attic dust and soil as reflection of lithology and anthropogenic influence in Slovenia. *Journal de Physique, 107*, 1173–1176.

Šajn, R. (2005). Using attic dust and soil for the separation of anthropogenic and geogenic elemental distributions in an old metallurgic area (Celje, Slovenia), *Geochemistry, 5*, 59–67.

Šajn, R. (2006). Factor analysis of soil and attic-dust to separate mining and metallurgy influence, Meza valley, Slovenia. *Mathematical Geology, 38*, 735–746.

Schröder, W. Holy, M. Pesch, R. Harmens, H. Ilyin, I. Steinnes, E. Alber, R. Aleksiayenak, Y. Blum, O. Coşkun, M. Dam, M. Temmerman, L. Frolova, M. Frontasyeva, M. V. Miqueo, L. Grodzińska, K. Jeran, Z. Korzekwa, S. Krmar, M. Kubin, E. Kvietkus, K. Leblond, S. Liiv, S. Magnússon, S. Maňkovská, B. Piispanen, J. Rühling, Å. Santamaria, J. Spiric, Z. Suchara, I. Thöni, L. Urumov, V. Yurukova, L. and Zechmeister, H. G. (2010). Are cadmium, lead and mercury concentrations in mosses across Europe primarily determined by atmospheric deposition of these metals. *Journal of Soils and Sediments, 10*, 1572 – 1584.

Sengupta, M. (1993). *Environmental Impacts of Mining: Monitoring, Restoration and Control*. Boca Raton, Lewis Publishers.

Serafimovski, T. Jankovic, S. and Cifliganec, V. (1995). Alpine metallogeny and plate tectonics in the SW flank of the Carpatho-Balkanides. *Geologica Macedonica, 9*, 3-14.

Serafimovski, T. Alderton, D. H. M. Dolenec, T. Tasev, G. and Dolenec, M. (2005). Metal pollution around the Bučim Mine. In: Boev, B. and Serafimovski, T. *Anthropogenic Effects on the human Environment in Tertiary Basins in the Mediterranean*, Štip, Faculty of Mining and Geology, 36-56.

Stafilov, T. Šajn, R. Boev, B. Cvetković, J. Mukaetov, D. and Andreevski, M. (2008). *Geochemical atlas of Kavadarci and the environment*. Skopje, Sts. Cyril and Methodius University, Faculty of Natural Sciences and Mathematics.

Stafilov, T. Šajn, R. Pančevski, Z. Boev, B. Frontasyeva, M. V. Strelkova, L. P. (2010). Heavy metal contamination of surface soils around a lead and zinc smelter in the Republic of Macedonia. *Jornal of Hazardous Materials*, *175*, 896-914.

Stafilov, T. Šajn, R. Boev, B. Cvetković, J. Mukaetov, D. Andreevski, M. and Lepitkova, S. (2010a). Distribution of some elements in surface soil over the Kavadarci region, Republic of Macedonia. *Environmental Earth Science*, *61*, 1515-1530.

Stafilov, T. Balabanova, B. Šajn, R. Bačeva, K. and Boev, B. (2010b). *Geochemical atlas of Radoviš and the environs and the distribution of heavy metals in the air*. Skopje, Faculty of Natural Sciences and Mathematics.

Stefanova, V. Neviealkov, R. and Moritz, R. (2004). Magmatism of the Borov Dol copper occurrence, In Kovachev, V. editor. *Proceedings on the Annual Scientific Conference of the Bulgarian Geological Society*. Sofia, House of Science and Technology.

Tănăselia, C. Miclean, M. Roman, C. Cordoş, E. and David, L. (2008). Determination of lead isotopic ratio in organic and soil materials using a Quadrupole mass spectrometry method with fast inductively coupled plasma. *Optoelectronics and Advanced Materials: Rapid Communications*, *2*, 299-302.

Tsitouridou, R. and Anatolaki, C. (2007). On the wet and dry deposition of ionic species in the vicinity of coal-fired power plants, northwestern Greece. *Atmospheric Research*, *83*, 93-105.

Tye, A. M. Hodgkinson, E. S. and Rawlins, B. G. (2006). Microscopic and chemical studies of metal particulates in tree bark and attic dust: evidence for historical atmospheric smelter emissions, Humberside, UK. *Journal of Environmental Monitoring*, *8*, 904–912.

Vallero, D. (2008). *Fundamentals of air pollution*. 4th edition. Amsterdam, Boston, Heidelberg, London, New York, Oxford, Paris, San Diego, San Francisco, Singapore, Sydney, Tokyo, Academic Press Elsevier.

Vike, E. (2005). Uptake, deposition and wash off fluoride and aluminum plant foliage in the vicinity of an aluminum smelter in Norway. *Water Air and Soil Pollution*, *160*, 145–159.

Zheng, Y. M. Chen, T. B. and He, J. Z. (2008). Multivariate geostatistical analysis of heavy metal in topsoils from Beijing, China. *Journal of Soils and Sediments*, *8*, 51–58.

Žibret, G. (2008). Determination of historical emission of heavy metals into the atmosphere: Celje case study. *Environmental Geology*, *56*, 189-196.

Žibret, G. and Šajn, R. (2008). Hunting for Geochemical Associations of Elements: Factor Analysis and Self-Organising Maps. *Mathematical Geosciences*, *42*, 681-703.

Reviewed by Prof. Katarina Čundeva, Professor at the Institute of Chemistry, Faculty of Science, Sts. Cyril and Methodius University, POB 162, 1000 Skopje, Macedonia (e-mail: kcundeva@pmf.ukim.mk).

In: Dust: Sources, Environmental Concerns and Control
Editors: Laurent B. Wouters and Michel Pauwels
ISBN: 978-1-61942-547-7

Chapter 3

METHODS FOR THE CHARACTERISATION AND CONTROL OF DUST IN MATERIALS HANDLING APPLICATIONS WITH A SPECIFIC FOCUS ON PASSIVE DUST CONTROL IN TRANSFER CHUTES

T. J. Donohue, C. A. Wheeler, A. W. Roberts, X. L. Chen and A. Katterfeld
TUNRA Bulk Solids, University of Newcastle, NSW, Australia

ABSTRACT

Dust control is a widespread problem affecting the mining industry in Australia and most parts of the world where the mining of natural resources is taking place. Due to the ever increasing focus on the health and safety of personnel, as well as environmental concerns, the problem of dust control is certainly an important one that needs total attention. Within the mining industry, the majority of dust problems arise from the transportation of the material being mined. However there are also problems associated with product that is exposed to the environment during storage, such as in large stockpiles. In this chapter current technology is presented and discussed that addresses both the characterisation of dust and different avenues for classifying a material in terms of "dustiness". In addition, analysis techniques currently available for use in the design of materials handling equipment are also presented. The particular emphasis is on passive dust control measures with the focus on transfer chutes, as it is well known the transfer of material from one conveyor to another is a major cause of dust problems. The work presented herein includes a series of case studies that seek to highlight the challenges involved in the design of plant equipment to handle dusty materials.

INTRODUCTION

The control of dust emissions in industrial operations is a matter of high importance. In the materials handling chain, belt conveyors have been used for the transportation of bulk material over short distances for many years. Recent times have seen improvements in technology leading to longer belt conveyors as well as increased belt speeds. The need to improve the efficiency of the handling chain is being driven by significant increases in global demand for natural resources. With this in mind, it is not only important that belt conveying systems be designed, operated and maintained for optimum efficiency and economy, but also ensuring there is minimal impact on the environment, and the health and safety of personnel and the surrounding community. An important aspect of the transportation phase, and one of the key contributors to dust generation, is the bulk material loading, unloading and transfer operations. The magnitude of this problem is identified by the U.S. Department of Labor [1] which cites these areas of belt conveyor operations as the most frequent sources of dust emissions. This is also the case in Australia where the control of dust emissions is a major problem. For instance, a Coal mine in Australia uses more than 85×10^6 litres of water per annum to control dust in only seven conveyor transfer stations. On a national scale, and indeed worldwide, the annual use of water for dust control in mining operations would be very significant indeed. In addition, the application of water for dust control, almost invariably, increases the cohesive strength of the bulk material leading to flow blockages and, therefore, increased running costs.

The use of water to control dust is an example of an active dust control method where external energy is used. Apart from the use of water for dust suppression, a great many industries rely on the extensive use of energy driven, dust extraction systems. The energy costs associated with these active dust suppression techniques is substantial. In view of the high cost and complexity of active dust control systems, there is a great need for passive dust control which does not depend on external energy or equipment, or the use of water or chemical surfactants. Current technology in belt conveyor transfer chutes is such that passive dust control is mainly achieved by using an enclosure of sufficient volume so that the dust laden air within the enclosure eventually settles, and re-enters the exiting bulk material stream. Such systems are mainly used for low speed, in-plant installations; a typical example being is illustrated in Figure 1 (a). These types of installations are typically limited to slow speed conveyors with their design being largely empirically based.

A more sophisticated approach to passive dust control in conveyor transfers is to employ a “hood” and “spoon” chute system as illustrated in Figure 1 (b) and, by appropriate analysis, integrate the dust control within the streamlined flow of bulk material. The aim is to ensure that the dust remains encapsulated in the material stream so it is unable to separate out and disperse. This is one of the main underlying themes of the material presented in this chapter.

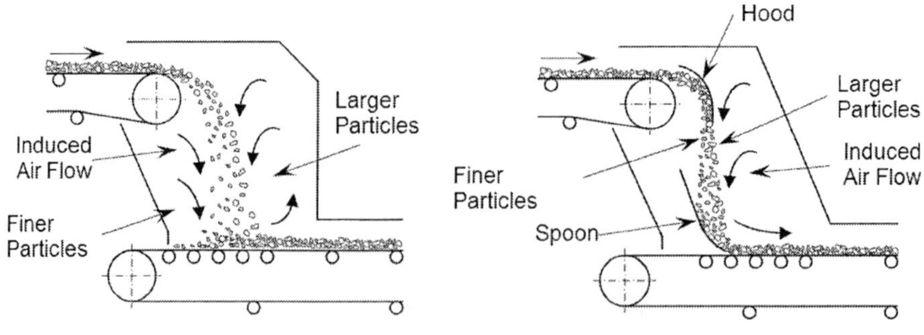

Figure 1. (a) Transfer chute showing entrained dust; (b) Hood and spoon transfer chute system.

Based on the analytical and experimental research of Roberts [3,4], hood and spoon chutes can now be designed with a view to achieving optimum performance with minimum chute wear as well as minimum belt wear at the loading point.

Characterisation Techniques

In the design of materials handling equipment to satisfactorily handle material whilst ensuring a minimum amount of dust is emitted, it is essential that characterisation procedures are performed on the material being handled.

These procedures are aimed at classifying the material in terms of its likelihood to generate dust while being transported as well as identifying the options that are available to reduce dust generation by active or passive means. A discussion of common characterisation techniques, and classification standards, are now presented.

Particle Size Distribution

A particle size distribution (PSD) is a simple characterisation test that can be carried out to reveal valuable information with regards to dust generation. Generally speaking, a material with a large proportion of fines will be more susceptible to producing dust than a material that consists of mostly coarse particles.

For many bulk solids that are handled, a PSD is performed by mechanical sieving of the sample with sieve opening sizes ranging from 100mm to 45μm. A standard PSD performed by sieving can be seen in Figure 2, which is shown for a Coal sample. For bulk solid samples that are finer and powder-like in nature, laser diffraction is a common method for measuring the PSD. This is performed by automated machines, such as the well known Malvern Mastersizer, with these devices being guided by the international standard ISO 13320-1.

In general, particle size ranges that can be analysed in one of the devices are in the range of 0.1-1000μm, with a typical graph shown in Figure 3. The analysis provided by a PSD provides valuable information on the size of particles in a sample and also the range of particle sizes. From the range of particle sizes, calculations can be made of the corresponding range of particle terminal velocities based on a single particle analysis.

This analysis can then be used to make some initial observations regarding the likely characteristics of the material with respect to dust generation. The case of a single particle undergoing freefall motion in air is now considered.

From the free body diagram shown in Figure 4, the sum of forces acting on the particle can be expressed as shown in Equation (1).

$$\sum F = F_W - F_D \tag{1}$$

where F_D is the drag force and F_W is the weight force. The weight force is simply the mass multiplied by the gravity, whereas the drag force is calculated through Equation (2).

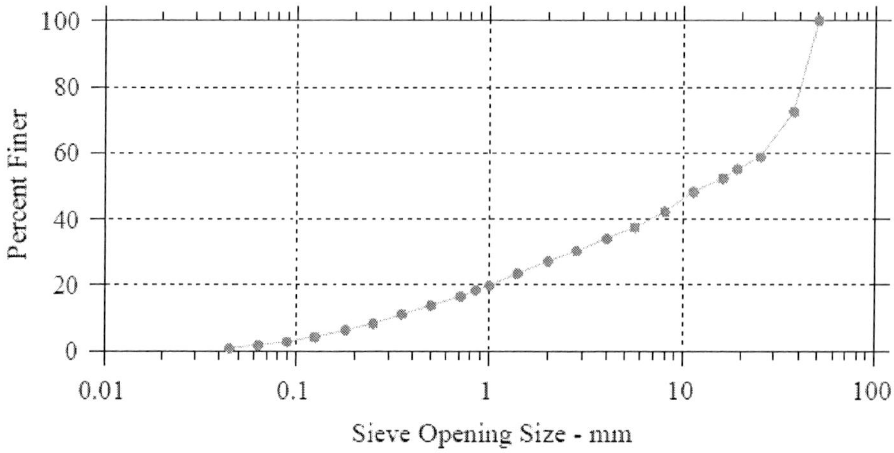

Figure 2. Typical PSD performed by sieving (Coal Sample).

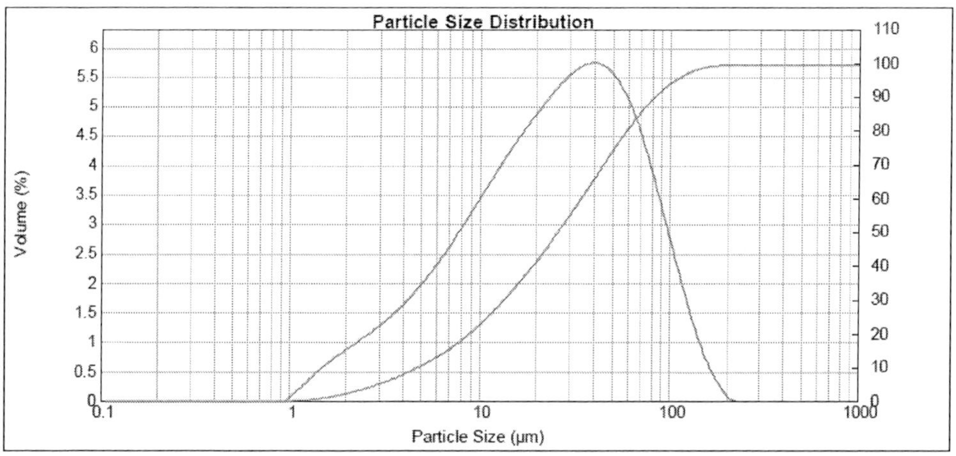

Figure 3. Typical PSD performed by laser diffraction (Fly ash sample).

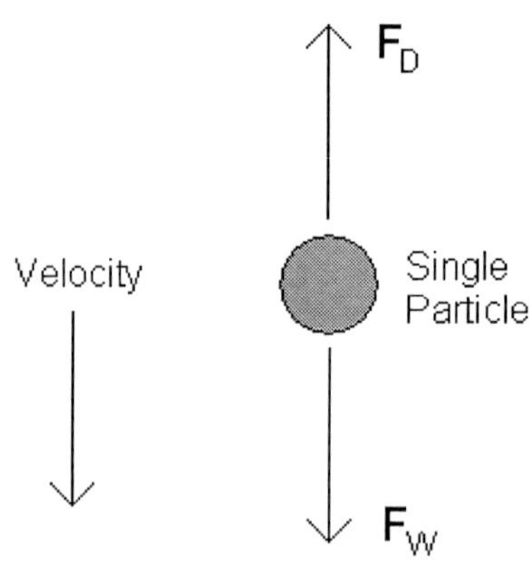

Figure 4. Forces acting on a freefalling individual particle.

The calculation for the drag coefficient (C_D) appears in Equation (3).

$$F_D = C_D \, v^n \tag{2}$$

$$C_D = C_d \, \rho_a \frac{A}{2} \tag{3}$$

In Equations (2) and (3), ρ_a is the air density, v is the particle velocity, n is the drag index (normally taken as 2), and A is the frontal area of the particle. The terminal velocity of a single particle can be calculated through Equation (1), as terminal velocity is characterised by zero acceleration. If we make the assumption of the particle being approximately spherical in shape with a particle diameter d, the equation for the terminal velocity can shown as in Equation (4), with ρ_s being the solids density of the particle.

$$v_t = \left[\frac{4 \, g \, \rho_s \, d}{3 \, \rho_a \, C_d}\right]^{1/n} \tag{4}$$

Equation (4) for the terminal velocity of a particle can be used to calculate an approximate terminal velocity for particles of varying diameter. Figure 5 shows the terminal velocity of particle diameters of varying particle size calculated assuming a drag coefficient of 0.45 (based on an approximate spherical shape) and a solids density of 1500 kg/m^3 (typical solids density for Coal). This simple calculation for the terminal velocity is representative of the air velocity required to suspend the particles. For example, for particles of nominal size 0.1mm, an air velocity of approximately 2m/s is required to keep these particles suspended in air. This information can be translated to transfer chute design by considering the velocity of the bulk solid as it travels through the transfer chute (analysis method presented in a later section) and the likely corresponding air velocity. Performing this kind of empirical analysis can provide an indicative estimate of dust generation, however more detailed information can be found through experimental means. These are discussed in the following section.

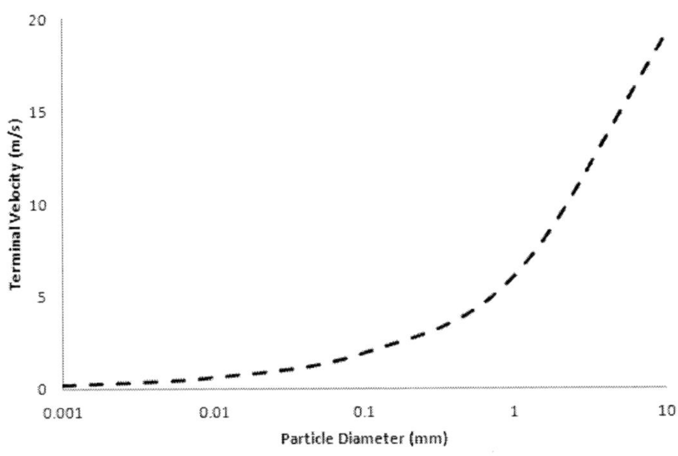

Figure 5. Plot of terminal velocity vs particle diameter.

Minimum Pick-Up Velocity

A test that can be performed to yield experimental information on a sample of material is to test for the minimum air velocity that is needed to "pick-up" the particles and suspend them in air. This is a similar principle to the analysis appearing in the previous section, however no assumptions need to be made for the particle shape.

A test to determine the minimum pick-up velocity can be carried out in a number of ways, with the simplest method involving a sample of material placed inside a wind tunnel and increasing the air velocity until the sample begins to lift-off. With this type of test the results are often obtained visually, without rigorous analysis. However, as so often is the case, materials have a range of particle sizes each with their own pick-up velocity, and so a more detailed experimental process is necessary. One method of carrying out such tests is to employ the apparatus shown in Figure 7. The apparatus consists of a series of chambers through which air is drawn using a vacuum system. The lower most chamber forms a venturi with the highest air velocity within the apparatus. Following this venturi is a long (4 diameters) section of constant diameter with the lowest air velocity. The final chamber converges to a smaller diameter which leads to a vacuum system.

The method involves starting the vacuum system at a prescribed air flow rate (found from the removable Pitot tube) which draws a sample of the material into the test chamber through the venturi. The lower two chambers aim to separate the material into individual particles and classify the sample in terms of terminal velocity. The high air velocities in the lower section of the apparatus serve two purposes, the separation of material into individual particles, and the retention of particles that are not drawn through the system. The constant diameter section forms a barrier to those particles that have a higher terminal velocity than the fixed air velocity due to the vacuum system. Particles with lower terminal velocities rise up through the chambers and exit the apparatus. The sample is repeatedly drawn into the constant velocity section and the smaller particles (lower terminal velocity) are removed.

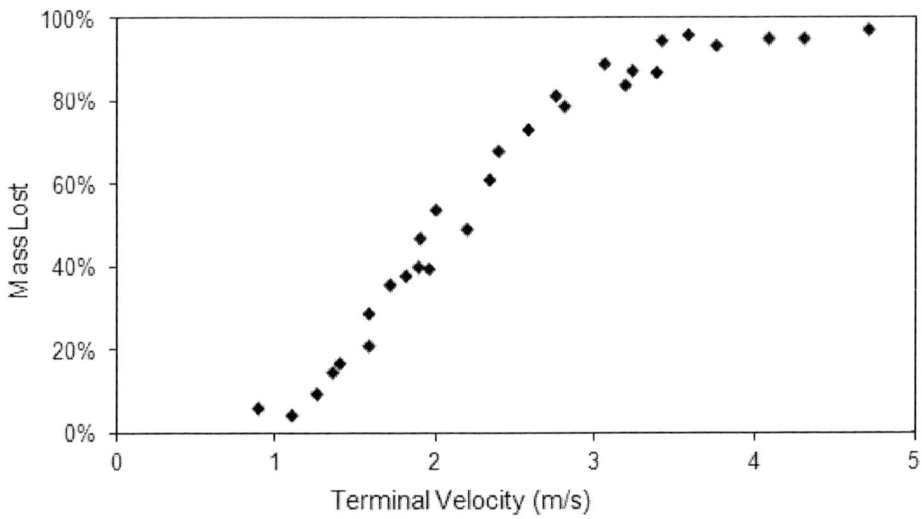

Figure 6. A sample result for the minimum pick-up velocity test.

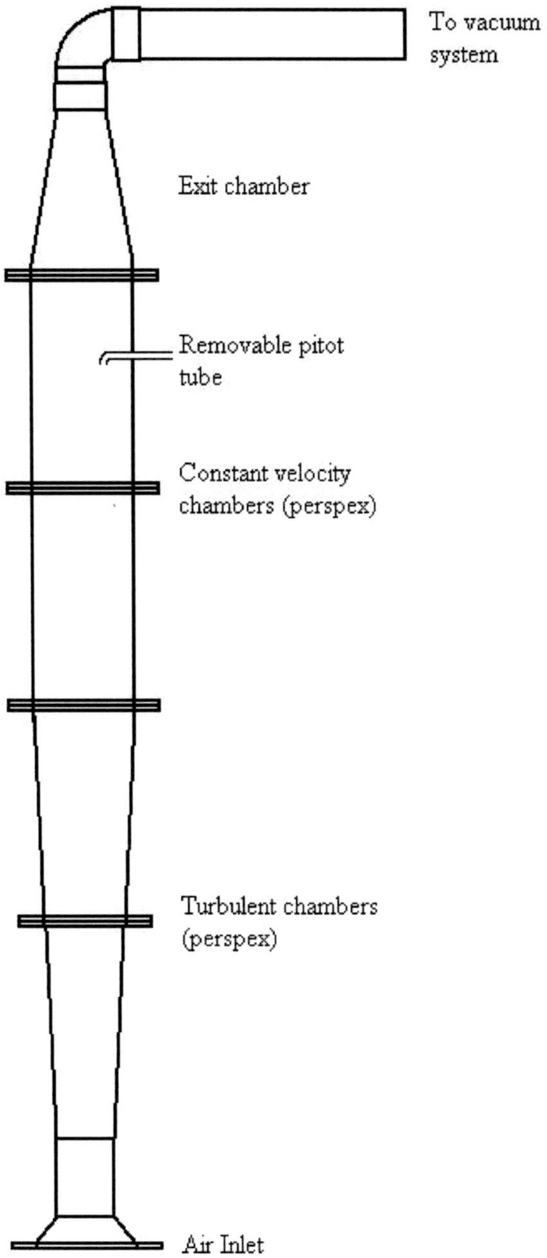

Figure 7. Experimental apparatus used to determine the minimum pick-up velocity.

After a period of time steady state is achieved whereby all of the particles that have a lower terminal velocity than the air velocity in the constant velocity chamber have been drawn out of the system and the vacuum system is switched off. The remaining material is weighed to determine the percentage of the mass lost to the vacuum system. By conducting this process over a range of air flow rates a distribution of terminal velocities can be found. A sample result for this test is shown in Figure 6. The advantage of this test is that it can be used for a wide range of materials including even those materials with a wide range of particle sizes and particle shape. The resulting distribution from the experiment, as shown in Figure 6, can be used estimate the likely amount of fine particles, or dust, which will become suspended in the air for a given velocity. As discussed previously, this can be cross referenced with velocity profile calculations that will be used in the design of a transfer chute.

Dustiness Tester

An established "dustiness" test involves the use of an apparatus incorporating a rotating drum in which a sample is placed at a premeasured moisture level. A photograph of the apparatus can be seen in Figure 8. The drum is rotated around its horizontal axis at a speed of 30 RPM for a period of 5 minutes while an air flow rate of 170 l/m is drawn through a hole in the drum lid, through a hollow drive shaft and a paper filter bag which collects the dust generated in the drum. The weight of the filter bag is measured before and after the test to determine the quantity of dust collected. A dust number is then calculated using the formula given in Equation 5.

$$Dust\ Number = \frac{M_b - M_a}{M_s}\ x\ 100{,}000 \qquad (5)$$

where M_b is the mass of the filter bag and dust, M_a is the mass of the filter bag, and M_s is the mass of the sample placed in the drum (all masses in grams).

Using this dustiness tester, a dust extinction moisture content can be found that is defined as the total moisture at which a dust number of 10 is attained. The above test is carried out at a number of moisture contents which will then allow a dust/moisture curve to be drawn. An example of this type of curve can be seen in Figure 9.

Australian Standard AS4156.6-2000 outlines this process in detail, and although the standard is specifically for Coal, the test can also be applied to other minerals such as Iron Ore.

Figure 8. Dustiness Tester.

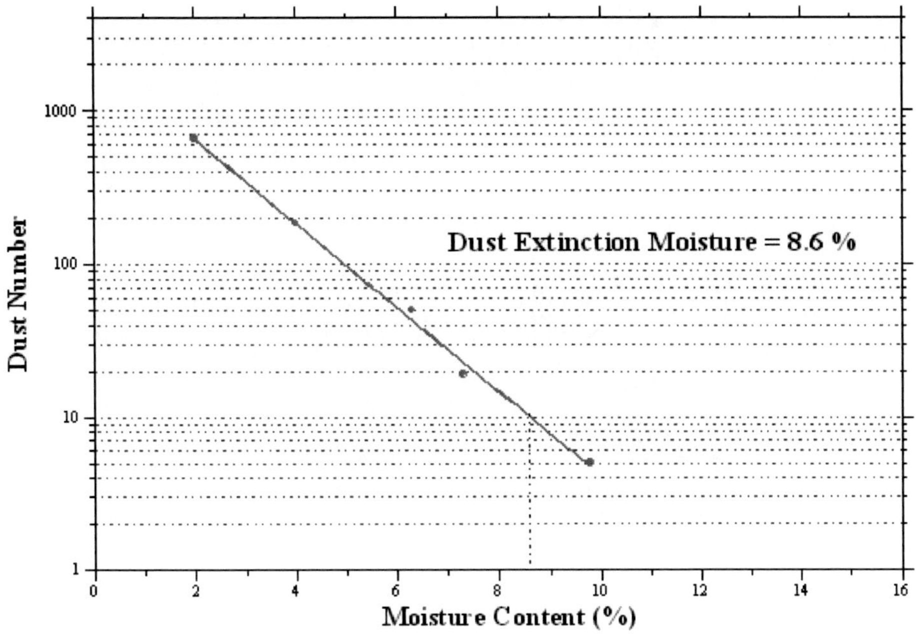

Figure 9. Dust extinction moisture level shows relationship between dustiness and moisture content for typical coal sample.

Determining the dust extinction moisture content for each material that is to be handled allows operators to nominate the required moisture content for each material type to reduce the dustiness. Furthermore, by knowing this information for each material type, several measures can be implemented in order to minimise fugitive dust emission during loading, conveying and discharging.

This standardised test to find a moisture content to reduce dust emissions can be used to reduce the dustiness of a material at almost any location in the materials handling chain. However, it is commonly applied to materials to be stored in stockpiles in an effort to reduce dust generation and associated pollution resulting from prevailing winds. The following section discusses this in more detail along with other methods to prevent dust lift off in stockpiles such as surface veneer treatments.

Wind Tunnel Testing

A wind tunnel apparatus can be used to simulate the flow of wind across a material surface and in particular the flow of air across a stockpile. A wind tunnel designed and built for this purpose at the University of Newcastle [5] can be seen in Figure 10.

This particular wind tunnel makes use of a suction unit, where the air is drawn through an opening in the square section at the end of the tunnel and passes over the material sample surface. The wind tunnel unit was designed to achieve a consistent cross sectional flow over the surface of the test sample. A suitable length working chamber with a small wall roughness coefficient was chosen in the design stage to create a uniform laminar boundary layer along the wind tunnel base and in particular across the test sample surface. Perspex was, therefore, selected for the wind tunnel unit, which also allowed an opportunity to visualise the test work.

The two main sections of the wind tunnel are the working chamber and the contraction chamber, with the working chamber having a cross sectional area of 550mm x 650mm and a length of 1200mm. The main purpose of this working chamber is to create a smooth air flow before it reaches the test sample. Inside the working chamber a tray holder is positioned to hold up to two test sample trays at a nominated angle of repose.

Figure 10. Wind tunnel used to replicate air flow over a stockpile.

Figure 11. Wind tunnel with sample trays inclined at 37.

The air drawn through the working chamber will, prior to reaching the test samples, achieve laminar boundary layer conditions. The flow continues along the chamber base towards the sample and across the sample surface with increasing pick-up velocity. The shearing between the air stream and the sample surface will create surface drag forces picking up loose participles, creating dust emission which with time will significantly increase. The main purpose of the contraction chamber is to join the working chamber to the suction fan, with its design being based on avoiding any vortex and back pressure being created which would affect the flow in the working chamber.

In general, the wind tunnel test is performed on a -6.3mm size range sample of the bulk material, since this size range is deemed to target the most likely particles vulnerable to lift-off. In the sample preparation, the screening is performed in accordance with Australian Standard AS-4156.6-2000. A typical test in the wind tunnel begins with ensuring the moisture content of the sample corresponds to that occurring on site, and corrected for the smaller particle size distribution. The sample is then placed inside the working chamber on the sample trays. The inclination of the sample trays is then set to the angle of repose of the material. Figure 11 shows the wind tunnel with two sample trays inclined at 37°, with this sample of material being Coal. Using a Pitot tube to monitor air velocity, the air speed in the wind tunnel is set to match typical conditions expected on site, the test then being run for a suitable duration. Depending on the sample the test duration may be 4-8 hours. After completion of the test the sample tray is carefully extracted from the wind tunnel and weighed to calculate the mass lost due to dust lift-off. In this process, the sample moisture content is again determined so as to allow for any reduction in sample weight due to the sample drying out during the test. This test gives a useful estimate of the expected dust lift-off from a stockpile over time. Furthermore, multiple moisture contents can be studied if necessary to find an optimum level in regards to dust lift-off.

However, a more common method of reducing dust lift-off from stockpiles is the use of a surface veneer treatment. This method involves spraying either water or a chemical treatment, or a mixture of both, to the surface of stockpiles which causes the fine particles to agglomerate thus creating a protective layer or crust. The wind tunnel test can also be used to evaluate the performance of a range of surface veneer treatments. As an example, the data in Figure 12 is presented.

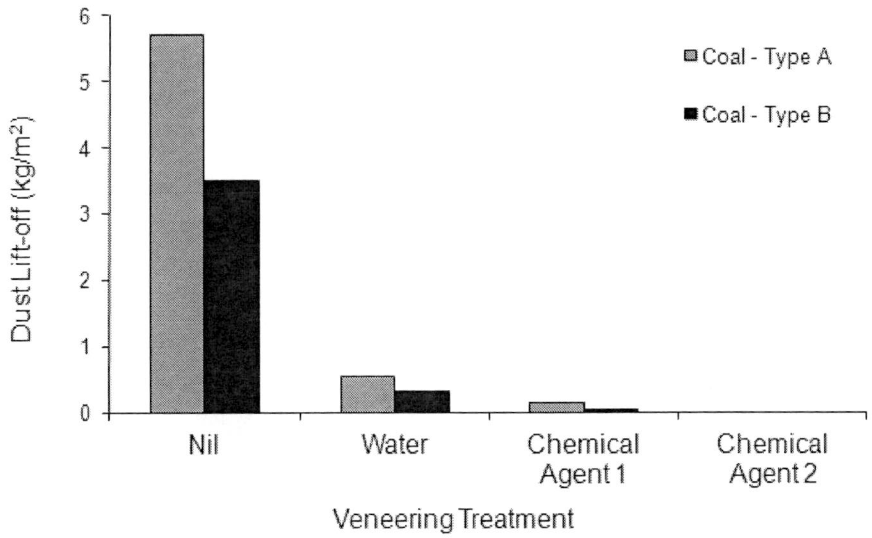

Figure 12. Performance of different surface veneer treatments.

This graph shows the results of two samples of Coal with three different types of surface veneer treatments. As can be seen, the addition of water to the surface has a large effect on reducing the dust lift-off from the stockpile, while the addition of the two types of chemical veneer treatments, denoted as Chemical Agent 1 and Chemical Agent 2, also reduces the dust lift-off. In the case of the Chemical Agent 2 veneer treatment, the test showed that no measurable dust was lost in the time period of the test. The application of these surface veneer treatments must not significantly affect the quality of the bulk material or the handling characteristics. It has also been shown that the surface veneer treatment needs to be applied at regular intervals due to evaporation and surface drying.

Occupational Health and Safety Standards

There are two main Australian Standards for measuring dust in the workplace, namely AS2985 and AS3640. AS3640 is concerned with inhalable, also known as inspirable dust, which is defined as the dust fraction consisting of those particles that are taken in through the nose or mouth during breathing.

AS2985 is concerned with respirable dust, which is defined as the proportion of airborne particulate matter that penetrates to the non-ciliated airways when inhaled. In terms of health and safety, it is the respirable dust that poses the greatest threat to the safety of workers. AS 2985 defines respirable dust as a cumulative log-normal distribution with a median equivalent aerodynamic diameter of 4.25μm and a geometric standard deviation of 1.5μm. A table showing respirable dust as defined by AS2985 is shown in Table 1.

The previously mentioned Australian Standards are guidelines defining the definitions of respirable and inhalable dust, and in other countries similar standards exist. In terms of allowable dust levels in the environment, this depends to an extent on the local regulatory body.

Table 1. Respirable dust as defined by AS2985

Equivalent Aerodynamic Diameter (μm)	Respirability (%)
0	100
1	100
2	97
3	80
4	56
5	34
6	20
7	11
8	6
10	2
12	0.5
14	0.2
16	0.1
18	0

Table 2. Dust Concentration Limits for the State of NSW, Australia

Pollutant	Averaging Period	Concentration ($\mu g/m^3$)	Set By
TSP	Annual	90	NSW DEC
PM_{10}	1 Day (24 hr) 1 Day (24 hr) Annual	50 50 30	NSW DEC NEPM (allows exceedance 5 times/year) NSW DEC
$PM_{2.5}$	1 Day (24 hr) Annual	25 8	NEPM (advisory standard) NEPM (advisory standard)
Dust Deposition	Annual (total) Annual (increase)	$4g/m^2$/month $2g/m^2$/month	NSW DEC NSW DEC

For example, in the State of New South Wales in Australia, outdoor air quality is regulated by The National Environment Protection Measure (Air NEPM) and The NSW Department of Environment and Conservation (NSW DEC). Air NEPM provides air quality standards that are applied in cities and large towns across Australia. These standards apply to average concentrations across a region. The NSW DEC has regulatory criteria that are consistent with Air NEPM, but are more comprehensive and can apply to localised areas such as those areas in the vicinity of mine sites. The relevant dust concentration limitations are summarised in Table 2.

In Table 2, TSP refers to the Total Suspended Particulate Matter which is defined as the total amount of all particles suspended in the air. PM_{10} is a subset of TSP, and refers to all particles suspended in the air smaller in size than 10μm, but larger in size than 2.5μm. $PM_{2.5}$ is also a subset of TSP, but refers to all particles suspended in the air smaller in size than 2.5μm. The dust deposition is a measurement that collects the dust particles deposited in a collection device over a yearly period. These types of regulatory standards exist to maintain safe levels of dust in the atmosphere. While the example given here applies to the State of NSW Australia, similar standards exist elsewhere in the world. Strict controls are typically in place in those areas directly surrounding mine sites, with regular testing undertaken by the regulatory bodies to ensure safe levels of air borne dust, with harsh penalties being in place to deal with any offenders. The existence of these types of standards illustrates once again the ever increasing need for effective dust control.

Analysis Techniques

There are a number of analysis procedures available to study the dust problem. In this chapter a discussion is presented for both classical continuum methods as well as computational techniques. For each of the procedures presented case study examples are included. A brief summary of each of the procedures is as below;

1) Continuum Method – this method relies on a lumped mass approach of the bulk solid through the transfer chute and the prediction of dust is qualitative
2) Computational Fluid Dynamics – this category can be further classified into sub-categories;

- Single phase flow – this modelling is of the air only with assumptions being made for the granular flow
- Multiphase flow – this modelling is of the air as well as a granular phase

3) Distinct Element Modelling – discrete modelling of particles with assumptions being made regarding the flow of air
4) Computational Fluid Dynamics coupled with Distinct Element Modelling – this category can be further classified into sub-categories;
   - Indirect coupling – this procedure utilises DEM to estimate the granular flow and porosity of the material in different regions and then uses a porous single phase CFD simulation
   - Direct coupling – This method involves coupling the DEM simulation directly to the CFD simulation.

Continuum Method

As illustrated previously, the terminal velocities of dust particles are quite low when compared to typical air velocities encountered in the handling of bulk solids. As an example of this, particles in the range of 0.1 – 0.7 mm correspond to minimum pick-up velocities in the range of 2-5 m/s.

As harmful dust particles are below 10μm in size, as outlined in the Australian Standards, then the minimum pick-up velocity to eliminate any ejection of dust particles is very small and in the order of 0.6 m/s. As this is the case, in the design of transfer chutes, it is impractical to reduce velocities through the chute such that no dust particles are discharged in the surrounding air. With this in mind, a successful design of a transfer chute must be one that contains any dust that is generated and if possible reintegrate that dust back into the granular stream. The continuum method, developed by Roberts [3,4], is a design tool developed for the analysis of streamlined flow of the bulk solid through a transfer chute. It has been shown that streamlined flow is a very effective flow mode for passive dust control. The continuum method primarily deals with hood sections (Figure 13a) and spoon chutes (Figure 13b) and is based on a lumped parameter model. The basis for the continuum method is shown in Figure 13. As an example, the following analysis applies to a spoon chute, however a similar set of equations can be determined for the hood as well. The governing differential equation for the velocity of the bulk material as a function of the angular position θ for the spoon is shown in Equation (6).

$$\frac{dv}{d\theta} + \mu_e v - \frac{gR}{v}(\cos\theta - \mu_e \sin\theta) = 0 \tag{6}$$

where μ_e = equivalent drag friction factor which is given by Equation (7), and the parameter C_1 is given by Equation (8).

$$\mu_e = \mu\left(1 + \frac{C_1}{v}\right) \tag{7}$$

$$C_1 = \frac{K_v B_o H_o v_o}{(B_o - 2s\tan\lambda)^2}\left(1 + \frac{\tan\lambda}{\mu}\right) \tag{8}$$

In the above set of equations, B_o is the initial chute width, v_o is the initial velocity of the bulk solid in the hood or spoon, λ is the convergence angle (shown in Figure 13d), and μ is the boundary friction coefficient.

For the case of a parallel chute, λ=0 and Equation (8) reduces to the form of Equation (9).

$$C_1 = K_V \frac{v_o H_o}{b} \tag{9}$$

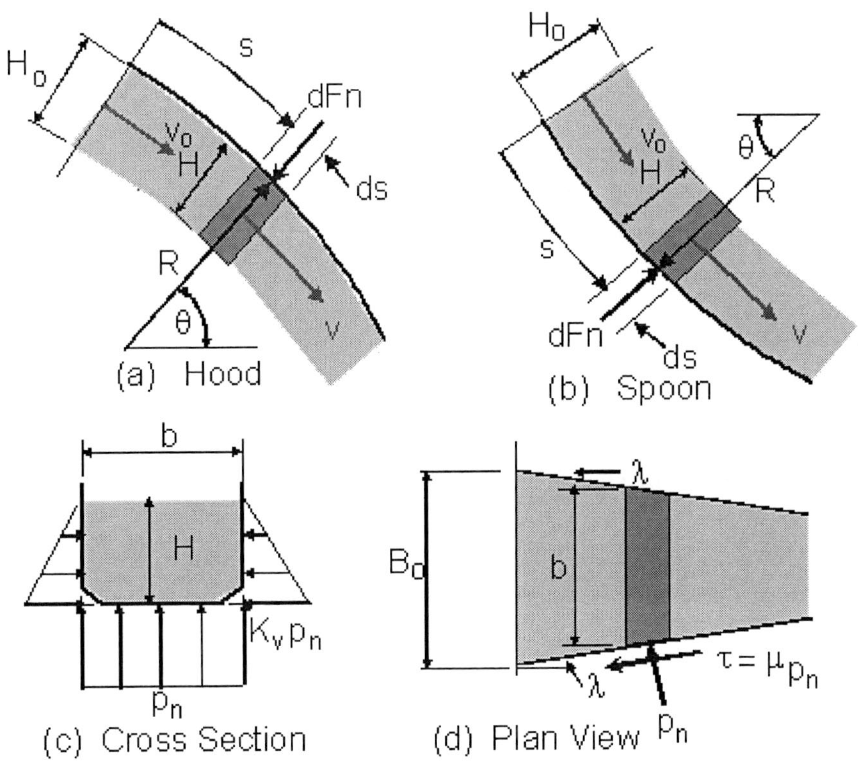

Figure 13. Chute Flow Models (Roberts [4]).

The pressure ratio K_V is based on the stress field of the bulk solid in the chute and the measured value of the effective angle of internal friction. The equations described above allow the velocity as a function of position, as defined by "θ" or "s", to be determined for the spoon. From the continuity of flow, the stream thickness variations around the chute profiles denoted by "H" can then be determined. Furthermore, the combination of the velocity and normal pressure distributions, combined with the measured boundary friction characteristics allow the chute wear profiles to be determined. For effective dust control, severe impact zones and unconstrained or dispersed flow of the bulk solid in its passage through the chute is to be avoided at all costs. Fundamentally, the objective is to achieve a streamlined flow of bulk material throughout a transfer chute. The analytical procedures described previously allow this performance objective to be met. By maintaining streamlined flow with minimal impact losses ensures that the dust particles are substantially entrained with the stream of larger particles. In effect, the dust is encapsulated by the main stream.

Past research has indicated that this continuum method approach can be very successful in reducing dust emissions [6]. A recent project where this approach was taken was the loading of grains into a ship hold. This particular project evolved primarily due to unacceptable dust levels arising from a poorly designed load-out chute. Due to environmental restrictions and operational requirements of the existing ship loader, several constraints were imposed on the solution, including:

- Modifications to the ship loader should be able to be undertaken in a minimum period of time and if possible not involve any significant structural work to the ship loader.
- Ability to be retrofitted to the existing booms of the ship loaders.
- Existing filling levels of ships are to be maintained and the ship loader should be able to load grain beneath the top of the ship's cargo hold.
- Due to strict export quality requirements no additives, including water, were to be added to the grain as a dust suppression agent.

This project involved a systematic review of the existing grain handling terminal, with the result of this review being the identification of several key dust generation points throughout the terminal. However, the main dust generation point was the final stage of loading where the grain discharged directly into the ship hold. As this was the focus of the dust reduction project, a pilot scale test facility was constructed to experimentally quantify the relative change in dust generation likely through the use of different loading methods. The test facility is shown in Figure 14.

The test facility consisted of an enclosure to simulate, to scale, the comparatively large size of the ship's cargo hold relative to the size and diameter of the chute. The test enclosure consisted of a rectangular section with a lower wedge shaped hopper for ease of emptying. The test enclosure was supported on four legs, which rested on four load cells to monitor the flow rate of grain into the enclosure. The front face of the enclosure was made from Perspex to enable visual observation of the spoon/chute performance and dust concentration. The test enclosure was well sealed to prevent dust from escaping during conveying trials. A vacuum system was attached to the bin in order to evacuate the dusty air from the enclosure after each test. The testing procedure involved fitting a number of different chutes and discharge spoon designs to the inclined screw conveyor. The system facilitates the recirculation of the grain as it discharges from the inclined screw conveyor. The product flow rate is controlled by a slide gate on the storage bin (not shown for clarity) which feeds onto the belt conveyor.

A relative measure of the effectiveness of each solution to reduce dust was undertaken by measuring the dust emission within the enclosure during each test. Dust concentration levels inside the enclosure were measured using a DustTrak aerosol monitor at two second intervals for the duration of the tests and for approximately 5 minutes after.

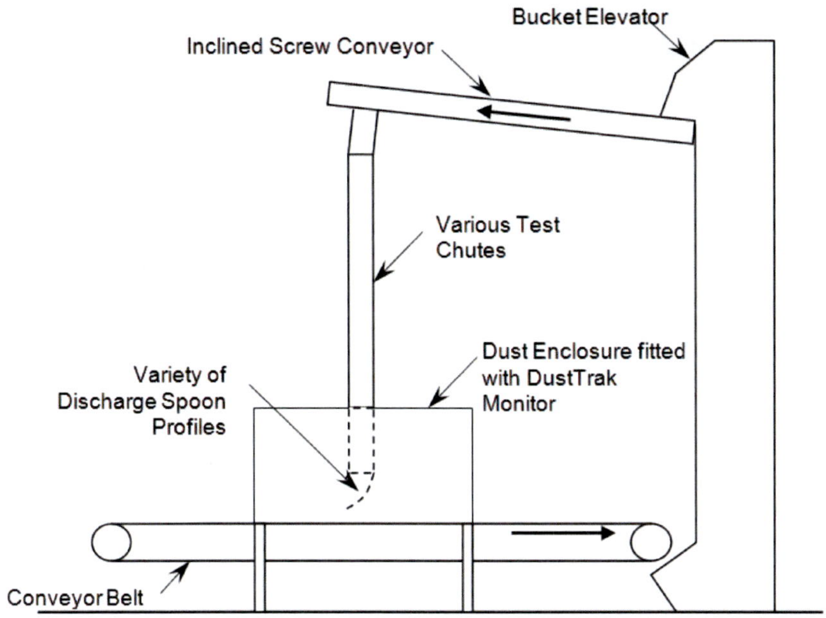

Figure 14. Pilot scale test facility.

During the emptying stage, the vacuum system is switched on to evacuate the dusty air from the enclosure. After each test the dust collected by the vacuum system is returned to the grain sample.

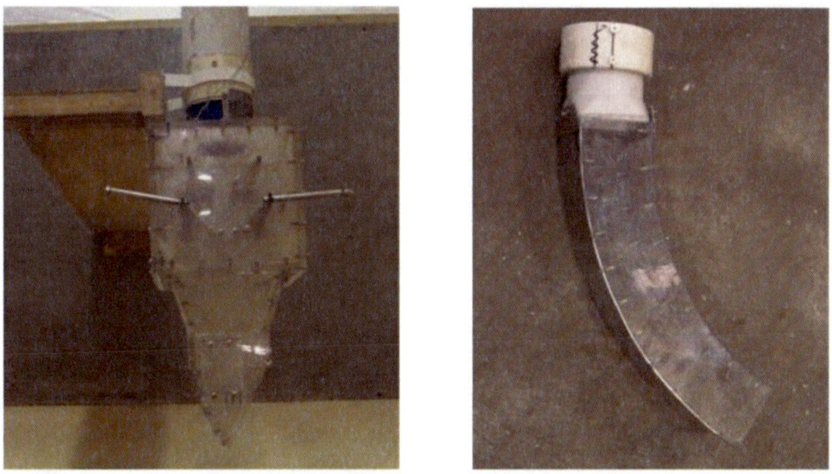

Figure 15. (a) Original design with spring loaded gates (b) Constant radius spoon revised design.

Using the previously described continuum method, a constant radius spoon was designed to replace the original chute used. Photos of the chutes used in the scale modelling can be seen in Figure 15. For benchmarking, the original chute was modelled and tested and then various chute profiles focussing on constant radius geometry were trialled in the test facility. The dust emissions were measured using the method described previously.

Initial observations showed a significant reduction in airborne dust within the enclosure. As opposed to the original spoon, the product stream exiting the constant radius spoon was very well defined with a sharp boundary layer between the air and grain. Knowing the product flow rate and the height of the grain stream the exit velocity of the grain was estimated to be approximately 4.5 to 5 m/s.

The increased component of the exit velocity in the horizontal direction compared to the original spoon improved the loading reach of the spoon and therefore reduces the required inclination angle of the vertical chute. Using the DustTrak aerosol monitor, the average dust concentration level for the series of tests showed more than a 50% reduction in dust compared to the original loading spoon. The reduction in dust levels were attributed to the concentrated product stream entraining the dust.

During this course of experimental testing, further spoon optimisations were undertaken in which the product stream was further streamlined through the use of a tapered chute. This use of a tapered chute further reduced dust emissions by a further 20%, with a detailed study of this appearing in a later section of this chapter on modelling.

Computational Fluid Dynamics

Computational Fluid Dynamics (CFD) is a numerical modelling technique that can be used for a wide variety of applications. A great deal of literature exists covering the fundamentals of CFD and the associated set of equations that are solved. The focus of this chapter is on the use of CFD and how it can assist in analysing transfer chute flows, so the reader is referred elsewhere to the great wealth of information that exists for a detailed explanation on the fundamentals of CFD.

Single Phase Flow

Single phase flow in CFD can be approached from a quite simple perspective in that it can model the flow of air through a certain geometry which can lead to valuable insights into the flow of dust and any corresponding dust generation or concentration. One particular project in which this approach was applied was for an electrostatic dust collection system which was used for dust suppression. In this particular instance, the environment prevented the application of any moisture additive solutions. Instead of the more common dust extraction fans being employed, the electro-static collection system was used instead, as illustrated in Figure 16.

The operation of this particular dust suppression arrangement relies on the airflow through the system to be turbulent in nature so that the airflow will be directed towards the electrodes, which are located at the top of the dust suppression system just below the fans shown in Figure 16.

The electrodes attract the dust due to an electrostatic force, and then at certain time periods the system is turned off and a vibration of the electrodes causes the dust to fall back down onto the belt. The fans shown in the figure operate whenever the dust suppression system is on, and their role is to ensure the power connection to the electrodes remains clean and free of dust build up. The general idea of these systems is for them to be installed in a location that is directly after a transfer chute where the flow of air surrounding the bulk solid on the conveyor belt will still be transient.

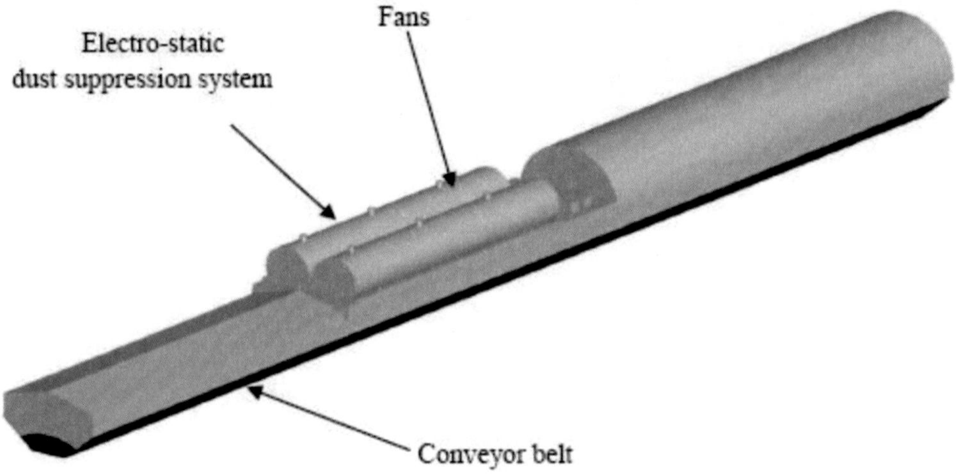

Figure 16. Electro-static dust suppression geometry.

However, in this case this was not possible due to space constraints of the existing structure, the system having been retrofitted as a solution to the dust problem being experienced on site. For this particular installation of the dust suppression system, it was determined that the system was not reducing the dust present in the air directly surrounding the transfer, as determined by air sampling that was carried out before and after the installation.

CFD simulations were then undertaken in an effort to firstly understand the problem, and then investigate any possible solutions. The initial investigation involved the mine site using a Pitot tube to measure the velocity of the air at the entrance of the dust suppression system in the region directly above the bulk solid travelling on the conveyor belt. Measurements taken from site indicated an air velocity ranging from a value of 1.6m/s measured directly above the belt to a value of 0.1 m/s towards the top of the inlet. The conveyor belt and bulk solid being transported on the belt was travelling at 7.7m/s in comparison. The flow rate of the fans into the dust suppression system was also known.

An initial single phase simulation was first run in which only the air passing through the dust suppression system was simulated. For this simulation, the inlet boundary conditions were set to be equal to the measurements taken from site, and only the volume above the bulk solid on the belt was used. The interface that represented the profile of the bulk solid was set to be a moving wall at the same speed as the conveyor belt (to represent the moving bulk solid). The results of the CFD simulation can be seen in Figure 17. As the figure shows, the greater air velocities through the dust suppression plant are in the region close to bulk solid. In other words, the velocity profile at the entrance to the dust suppression plant remains approximately the same as the air passes through the plant.

As Figure 17 shows, very little air travels up into the region where the electrodes are located. The figure also shows that even with the presence of the fans injecting air into the dust suppression system at velocities approaching 12 m/s, only minimal turbulence is caused to force the dust laden air towards the electrodes. In summary, this simulation shows that very little of the air passing through the dust suppression plant is being directed towards the electrodes and hence the system is not performing as well as it could.

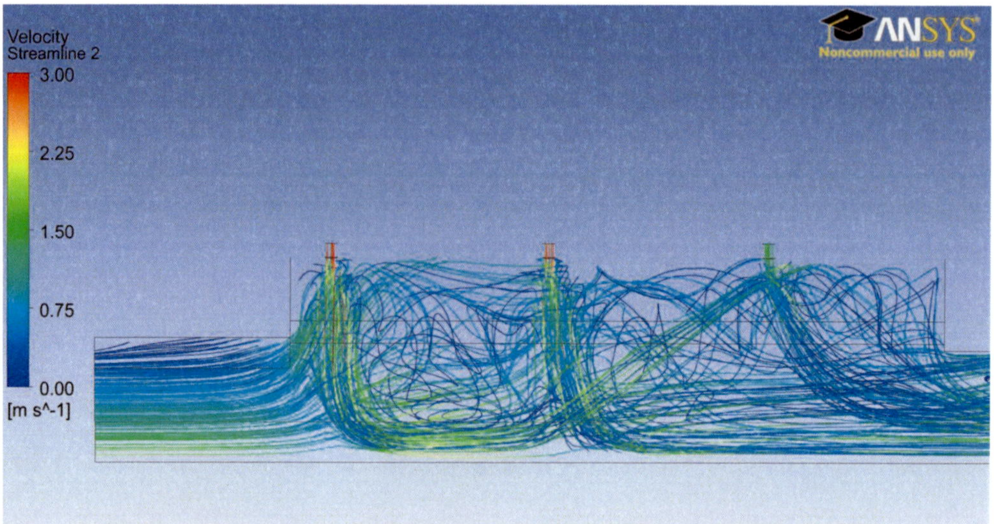

Figure 17. Single phase simulation of air passing through the dust suppression system.

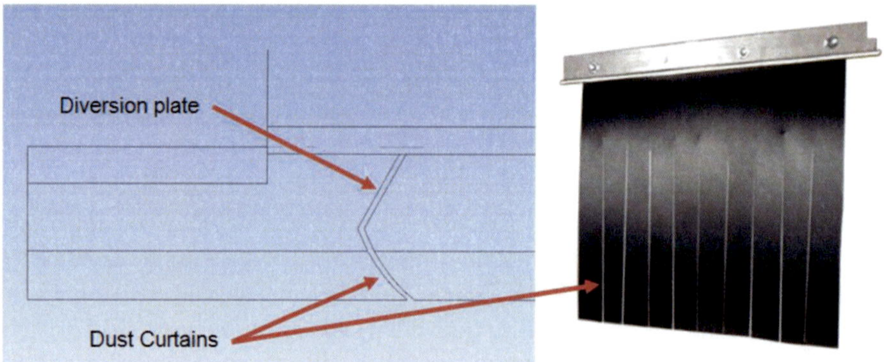

Figure 18. Modification of geometry to include a diversion plate and dust curtains.

A simulation was then undertaken in which a deflector plate was used to redirect the incoming air in conjunction with dust curtains. The modification to the geometry can be seen in Figure 18.

The role of the deflector plate was to direct the air flow into the upper region of the dust suppression system so the electrodes would have an opportunity to extract the dust particles from the air. The dust curtains were necessary as a flexible interface with the bed of bulk material on the conveyor was necessary to account for differential heights in the burden depth as it passes through the dust suppression plant on the conveyor belt.

The results of the simulation are shown in Figure 19. The results show that the incoming air is successfully diverted upwards into the region of the dust suppression plant where the electrodes are located at the entry to the dust suppression plant. However, as the figure indicates, the presence of the fans blowing on the electrodes shows the majority of the air stream is forced back down into the region close to the bulk solid and away from the electrodes. This indicates that, in this particular case, the presence of multiple sets of deflector plates and dust curtains may lead to this dust suppression system becoming more efficient.

Figure 19. Simulation results for the modified geometry.

The dimensions associated with the geometry used in the modification of the dust suppression plant are only representative in nature, and can possibly be optimised further. The assumptions made in the modelling of this geometry are also idealised in that the model assumed that the dust curtains made constant contact with the bed of bulk material and also had no gaps in them across the width of the belt. However, from the results of these single phase simulations it is possible to quickly analyse a materials handling plant and also easily investigate the effectiveness of a range of options to improve the characteristics of the plant.

Multiphase Flow

Multiphase flow in CFD software is a challenging area and is associated with much longer times to obtain solutions than for a typical single phase flow problem. However, the advantage of multiphase flows is that it is possible to model both the granular phase as well as the air phase. There are two main approaches that can be used when modelling a multiphase simulation, with the first approach being known as the Eulerian-Lagrangian approach where the fluid phase is treated as a continuum and the dispersed phase is solved by tracking a large number of particles. This method is limited to small volume fractions, as computationally the process for large volume fractions is too intensive. However, this method does model the real behaviour of the particles where a coefficient of restitution is defined between particle to particle and also particle to wall. The other method is the Eulerian-Eulerian approach where all of the phases are treated as interpenetrating continua. This approach requires the solid dispersed phase to have a low value of dynamic viscosity so that it can be modelled as a fluid. This method can be used to solve flow fields for densely packed solids. However the disadvantage of this method is that there is no direct method to account for the friction between the dispersed solid phase and a wall. The two CFD solvers presented in this chapter, Ansys CFX and Ansys Fluent, have slightly different ways of handling this boundary condition and will now be discussed.

A multiphase simulation can be employed to include a number of phases. This is discussed in the following sections, where results are presented from Ansys Fluent for a two phase simulation, as well as results from Ansys CFX in which a three phase simulation was modelled. The two phase simulation, which included modelling the granular and air phases, was carried out for an experimental laboratory scale model transfer chute. The study formed part of a research project aimed at improved dust control in an industrial conveyor transfer. The 3 phase simulation is a continuation of the case study presented earlier on the grain loading project and focuses only on the spoon section of the transfer. This simulation includes the modelling of a granular phase, an air phase, as well as a dust phase. The dust phase was defined as a solid dispersed phase, similar to the granular phase, but was assigned an average particle diameter much smaller than for the granular phase.

Two Phase Simulation

A two phase simulation can be carried out to include a granular phase and an air phase, with the behaviour of the air phase used to predict the behaviour of the dust since the dust is not directly modelled.

In modelling an entire transfer chute with multiphase CFD, as is the case in the example presented in this section, the simulation times become relatively long. In addition, the solver becomes more sensitive to the simulation parameters making it more difficult to achieve a converged solution. For the example considered in this section, only a two phase simulation was considered as the addition of a third phase for the dust would have translated to very long simulation times and a reduced solver capacity. In the following section the focus is on a specific aspect of a transfer chute in which a three phase simulation is carried out to achieve a direct representation of the dust phase.

The work presented in this section uses scale model testing to verify results obtained from a series of two phase simulations. In particular, Ansys Fluent was used to analyse air and particle flow in the transfer chute enclosure. This work originated from a particularly dusty transfer chute where it was decided to investigate a number of possible options to retrofit to the existing chute to reduce dust emissions. This was done by evaluating the relative performance of a number of transfer chute designs aimed at minimising dust emission.

This comparison was performed by analysing air velocities and the impact zones for each chute configuration. The validity and accuracy of CFD to predict dust emissions was verified by comparing the simulation results with scale model tests.

For this particular transfer chute, a total six configurations were analysed. Each of these configurations are shown in Figure 20 (a) to (c) and Figure 21 (a) to (c). The six configurations are labelled A to F in each of these figures which also highlight the relative differences between each of the chutes.

Scale model testing was carried out in the laboratory to evaluate the performance of the six transfer chute designs. The transfer chute was placed in a larger control enclosure to contain the dust expelled from the transfer chute. The enclosure measured 2.5 m wide, 2.5 m long and 3 m high, and is shown in Figure 22. The scale model transfer chute was constructed predominantly out of 10 mm clear Perspex in order to allow for visualisation of the bulk material flow.

The bulk material used in the testing was -11.2 mm Iron Ore, with the same sample used for testing in each of the six configurations to ensure the moisture content and dust fraction remained the same between each of the chutes. As a means to evaluate the dust emission from each of the six configurations, a plastic drop sheet was used inside the enclosure to collect the fugitive dust that escaped from the granular stream of material. After each test, and after the dust had settled inside the enclosure, the plastic drop sheet was carefully extracted from the enclosure and the dust collected was weighed. A total of five tests were completed for each chute configuration.

In comparison to the grain loading example previously presented, the DustTrak monitor could not be used to measure the dust in the air as it was found that there was a significant proportion of the Iron Ore particles in the air greater in size than the maximum detectable size of the DustTrak monitor.

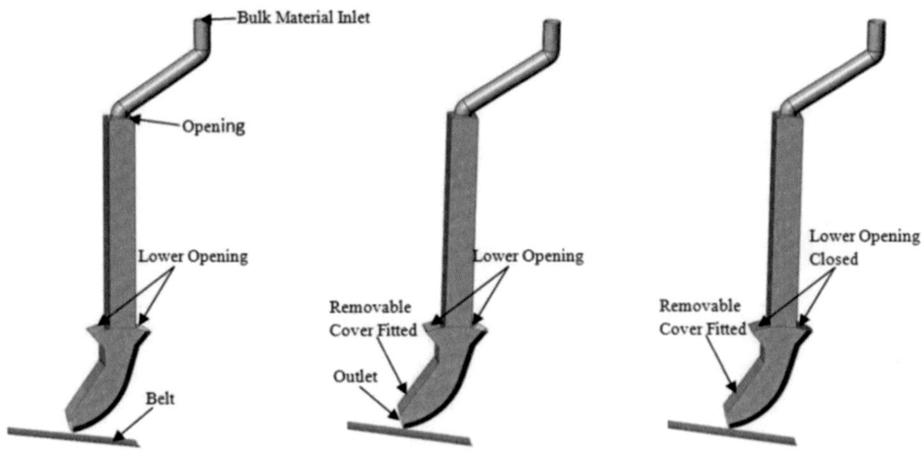

Figure 20. (a) Chute A: Baseline Case; (b) Chute B: Baseline + Removable Cover; (c) Chute C Baseline + Lower opening closed + Removable cover.

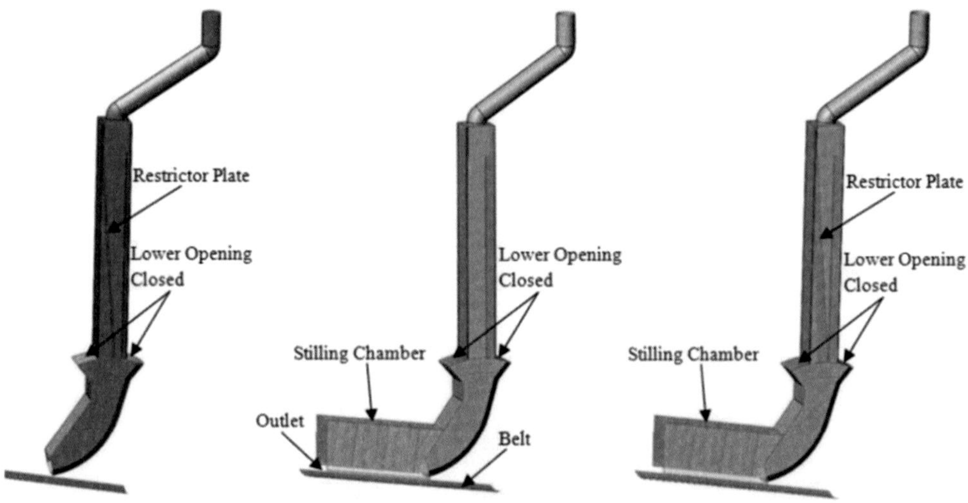

Figure 21. (a) Chute D: Restrictor + Removable Cover + Lower opening closed (b) Chute E: Baseline + Stilling Chamber + Lower opening closed (c) Chute F: Restrictor + Stilling chamber + lower opening closed.

Figure 22. Scale model of Chute E inside enclosure.

The results for the scale model testing can be seen in Figure 23, which expresses the dust generation for each of the six configurations in terms of the baseline case. From the results, it can be concluded that the performance of Chute A, Chute B and Chute C are similar as the results are within the same range of values. Chute F can be seen to provide the most reduction in emitted dust, while Chute D and E both had a similar performance which was also a large improvement on the initial design. The inclusion of the restrictor plate has a large effect on the dust emission, which is evident in a comparison of the results between Chute C and Chute D. This concept of streamlining the flow and directing the air containing the dust back into the granular stream is in keeping with the objective of passive dust control which is discussed in more detail in the following section. The stilling chamber also had a positive effect on reducing the dust emissions. The function of the stilling chamber is to reduce air velocities thereby reducing the maximum particle size of the particles that can be suspended in the air.

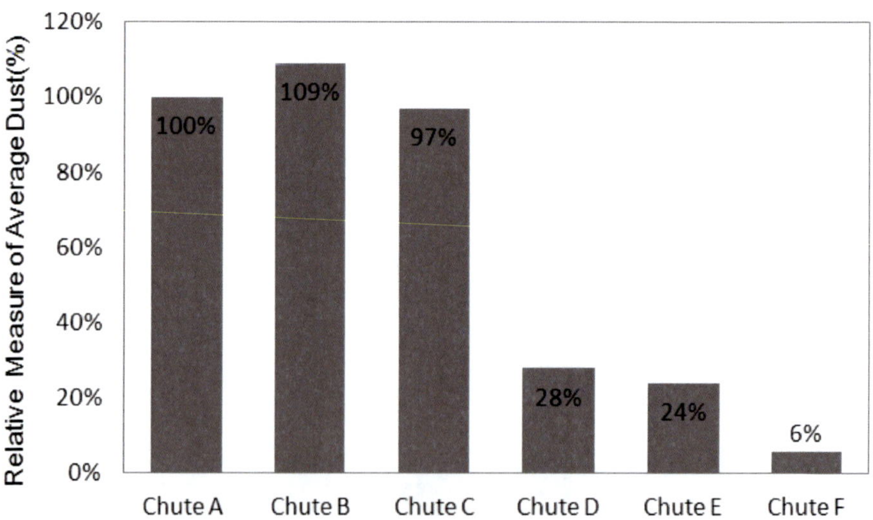

Figure 23. Scale model testing results for average dust emissions.

In addition to the dust measurements taken as described above, a Pitot tube was also used for flow measurements of the air exiting the discharge chute. The Pitot tube was placed at a location of 50mm from the chute outlet, as shown in Figure 24, and seven measurement locations were used for each chute configuration. The location of the Pitot tube varied depending on the chute configuration, as the addition of the stilling chamber meant that the chute discharge location also varied. However, the two measurement locations were comparable in terms of positioning from the moving stream of the bulk solid. Once again, multiple readings for the air velocity were taken at each measurement location to ensure meaningful results.

The results for air flow measurements can be seen in Figure 25. The figure shows Chute A and Chute B have the highest air velocities, followed by Chute C. A comparison of the discharge air velocities with the dust emission results indicates that the higher levels of dust are related to the higher discharge velocities of the air in the chute. In contrast to Chutes A-C, configurations D, E and F have a lower range of measured air velocities which correspond to a lower dust emission for these chutes.

For each of the six chute configurations, a CFD multiphase simulation was performed using two phases, a granular phase and an air phase. The one main advantage that Fluent has over CFX in regards the application of bulk solids flow is its ability to represent friction at the interface between the boundary and the granular phase, which it does so through the specularity coefficient.

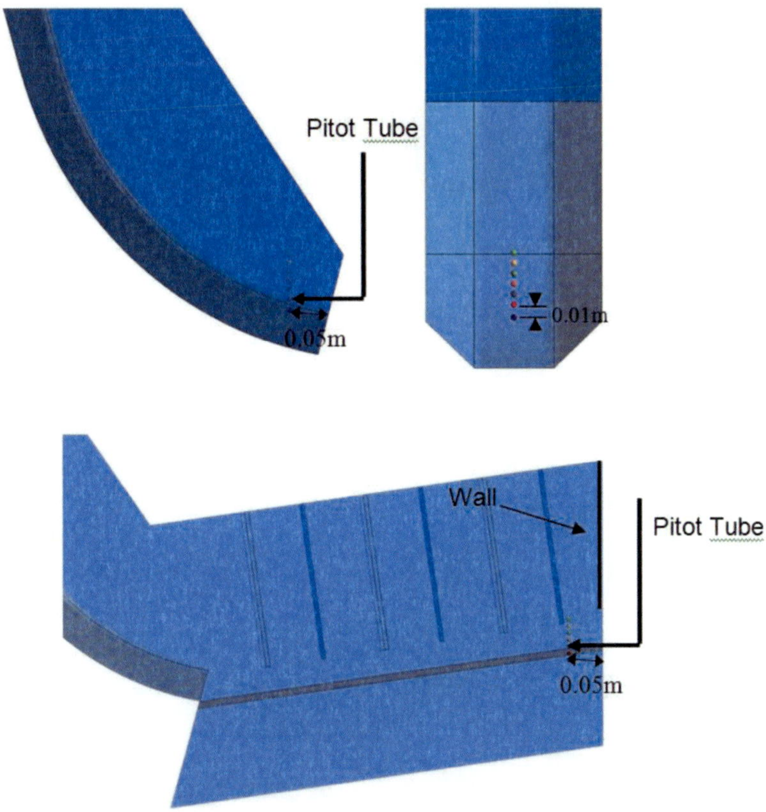

Figure 24. Location of measurement points for the Pitot tube (a) Chutes A-D (b) Chutes E-F.

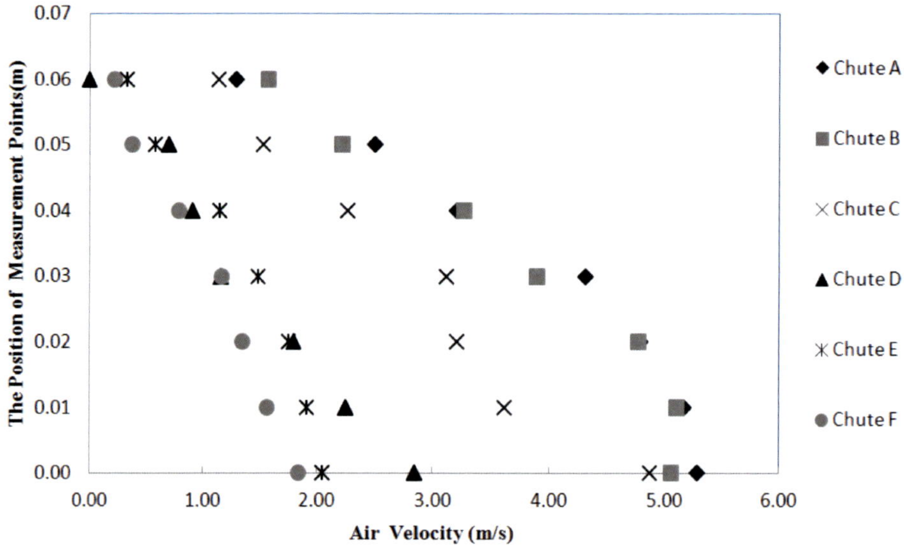

Figure 25. Air velocity measurements using a Pitot tube.

For the granular phase, 4mm particles were used as an average of the Iron Ore particles in the experiment, with the volume fraction of each phase being 60% and 40% for the granular phase and air phase respectively. The inlet velocity was set to be 0.35 m/s, representing the conditions experienced in the laboratory testing. For the full details of the CFD parameters used in this series of simulations the reader is referred to Chen et al [7].

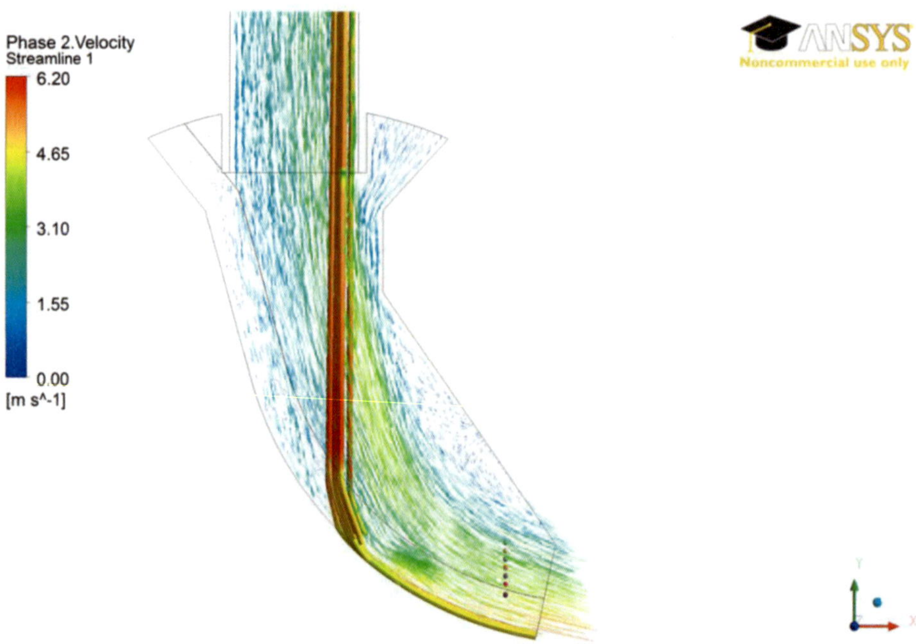

Figure 26. Particle streamlines and air vectors for Chute A.

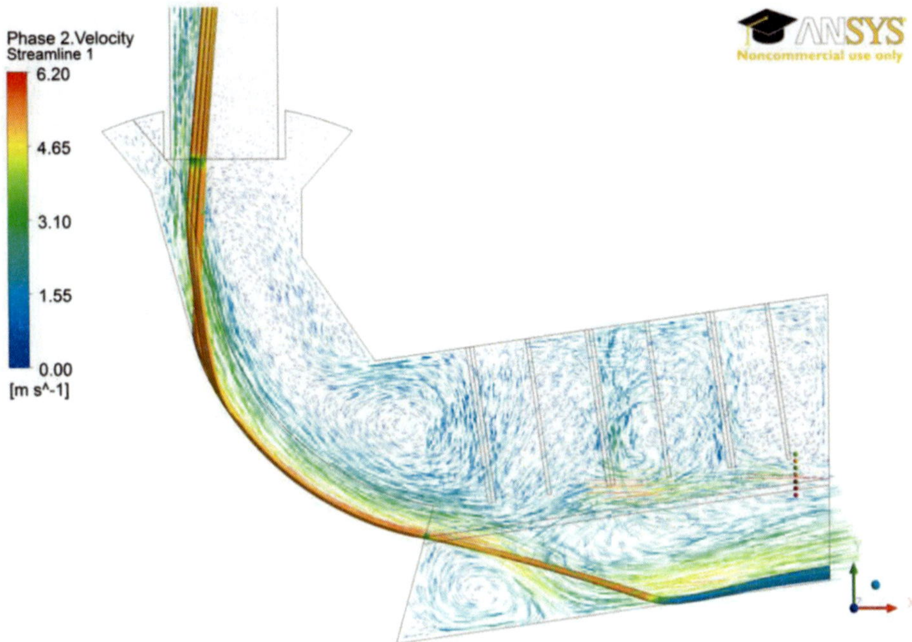

Figure 27. Particle streamlines and air vectors for Chute F.

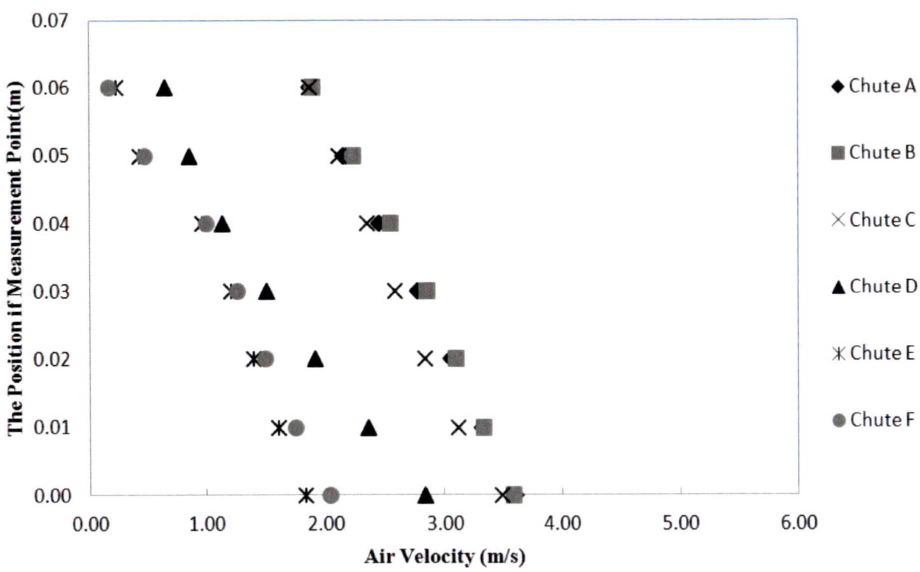

Figure 28. Air velocity results from CFD simulations.

Representative images for two of the CFD simulations are shown in Figure 26 (Chute A) and Figure 27 (Chute F). In these figures the flow of the granular phase is shown by the streamlines and the flow of the air is shown with the air vectors. The simulation does not include a direct representation for the dust phase, so no direct comparison can be made for dust emissions between each configuration. However, both qualitative and quantitative comparisons can still be made. Qualitatively, from comparisons between Figure 26 and Figure 27, the effect the restrictor plate has on the granular stream can be visualised. For the simulation including the restrictor plate, it can be seen that the impact angle between the stream and the chute is much lower than in the simulation without the restrictor plate. Obviously, with a lower impact angle then there will be a lower emission of dust into the surrounding air. It is of interest to compare the air velocities at the outlet of the chute and stilling chamber. The air velocities for each of the chute configurations from the CFD simulations can be seen in Figure 28, with these velocity results taken from approximately the same locations as measured in the laboratory. A comparison of the results presented in Figure 25 and Figure 28 shows a good correlation between the magnitudes of the air velocities, particularly for Chutes D-F. For the configurations of chutes A-C, the difference in the results in most likely due to the flow being more turbulent, which is evident through the higher dust emissions, resulting in larger velocities and more transient air flows. This transient-like effect is reflected in the greater variation in results measured with the Pitot tube in the experiments for Chutes A-C. This two phase study showed the type of results that can be achieved by a multiphase simulation using a granular phase and an air phase. While this type of simulation cannot make direct correlations for the dust emissions, it does allow qualitative and quantitative observations that can lead to increased knowledge in regards to likely dust emissions. The addition of the experimental testing allowed a direct correlation between discharge air velocities and dust emissions, which can then be taken to the CFD model to determine a trend between each of the configurations studied in terms of dust generation.

Three Phase Simulation

A three phase simulation was used to provide analysis on a simple spoon chute as part of the grain loading project as discussed previously. The particular example that is presented here involves a comparison of the constant radius spoon, as presented in a previous section, as well as the same spoon with the inclusion of a tapered chute on entry to the curved spoon. These two chutes can be seen in Figure 29.

As discussed previously, the CFD model employed utilised the Eulerian-Eulerian approach and modelled three phases to represent the experimental testing. The three phases that were modelled were a continuous phase representing air, a 1mm solid dispersed phase representing the grains and also a 0.1mm solid dispersed phase representing dust particles. These sizes of particles were chosen as representative of the grains, with particle density of the 1mm and 0.1mm particles both set to the density of the grains. The exact geometry of the chute tested experimentally was modelled using CFD software, although only a 2D simulation was considered for simplicity. To allow for the 60° cut off angle of the chute used in experimental testing, two separate domains were modelled, as shown in Figure 29, so flow profiles at 60° could be extracted during post-processing.

Boundary conditions for the simulation were set with the inlet velocity specified as 3 m/s and the outlet specified to be at atmospheric pressure, with these boundary conditions being chosen to recreate the experimental testing scenario. One major difference between Ansys CFX and Ansys Fluent, at the time of writing this chapter, is that there is no mechanism within CFX to model friction between the walls and the granular phase. In this case, with CFX, the boundary conditions were set to be no slip for the continuous phase and free slip for the granular phases. In terms of mass flow rates for the individual phases, volume fractions were set to 25% for the 1mm particles, 3% for the 0.1mm particles and 72% air. A full summary with more complete details of the simulations parameters used in this set of simulations can be found in Donohue et al [8].

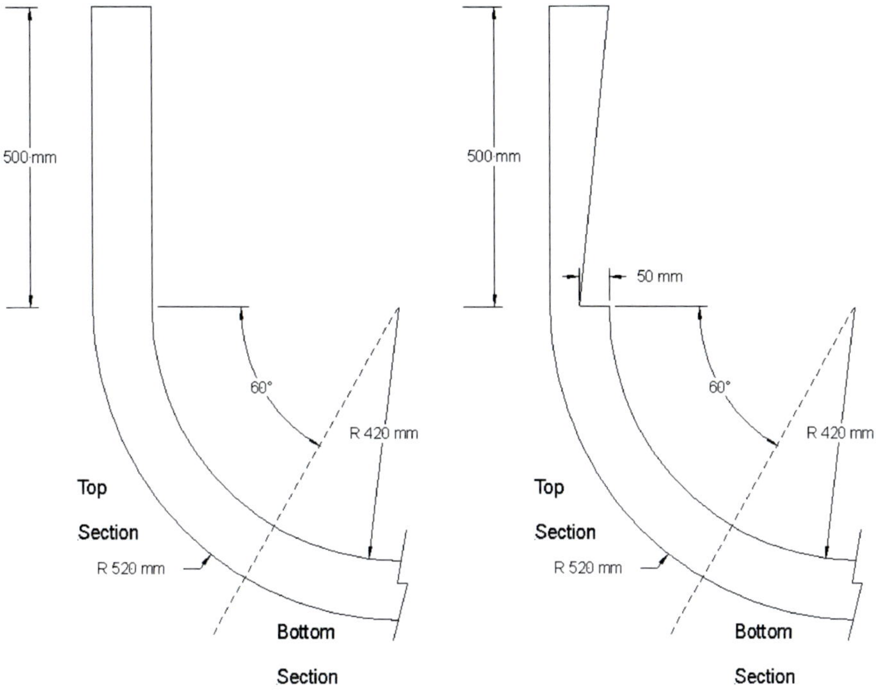

Figure 29. (a) Standard chute geometry (b) Converging chute geometry.

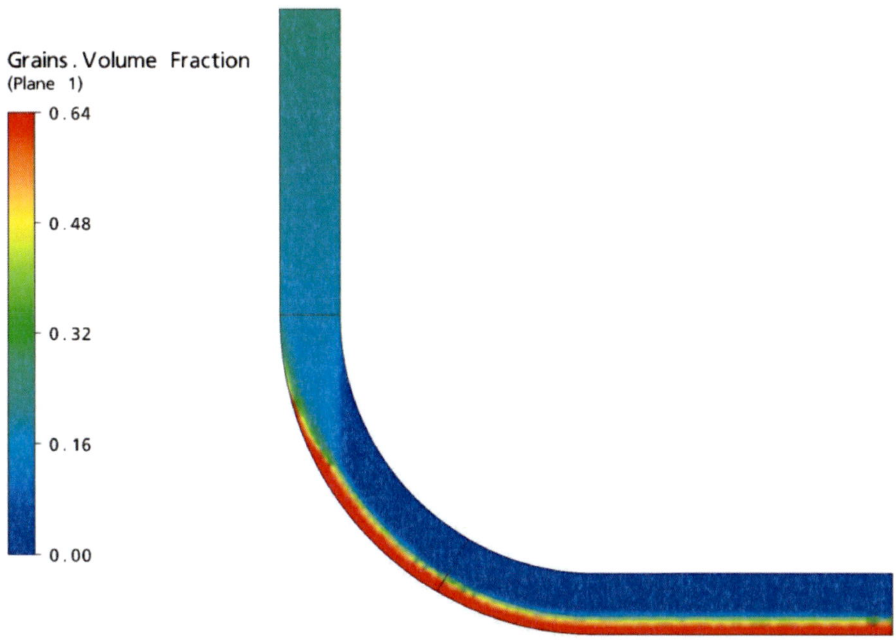

Figure 30. Volume solid fraction of the grains (1mm particles) in the standard chute.

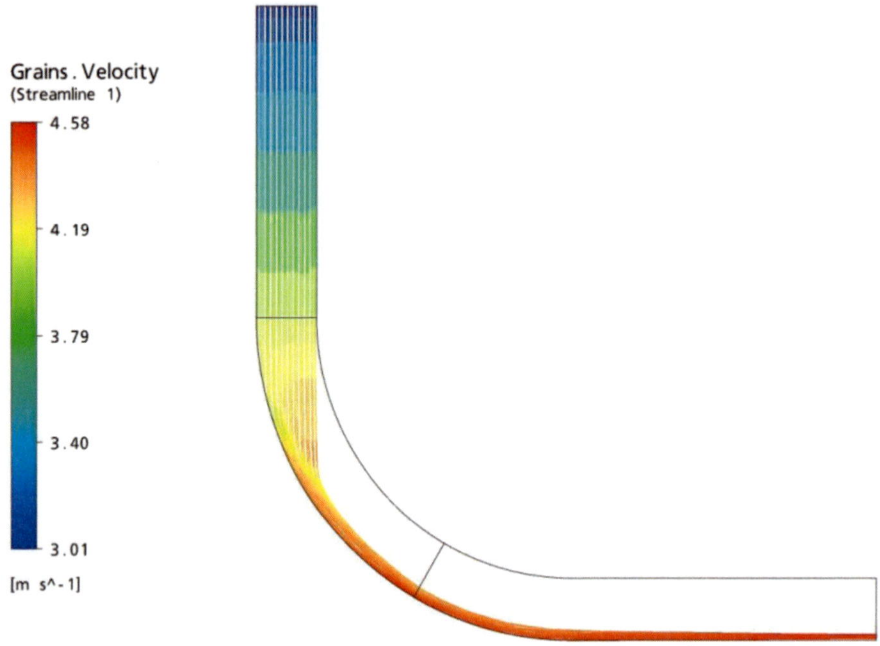

Figure 31. Velocity of the grains (1mm particles) through the standard chute.

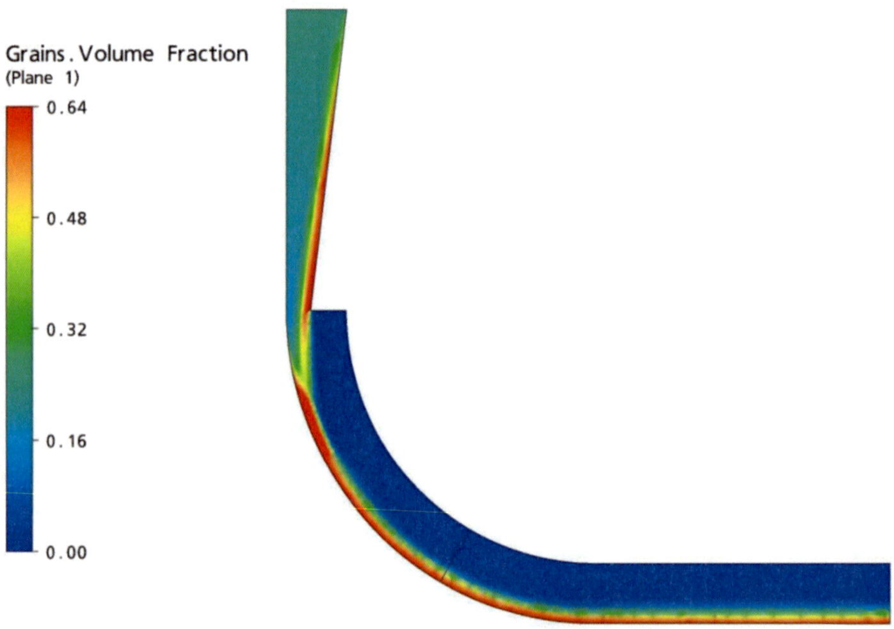

Figure 32. Volume solid fraction of the grains (1mm particles) in the converging chute.

Results for the simulation of the standard chute are presented in Figure 30 (grains volume fraction) and Figure 31 (grains velocity). Figure 30 shows that as the grains fall from the inlet they disperse by a small amount so that the volume solid fraction decreases as the grains fall. When the particles fall onto the chute the volume solid fraction increases as the particles compact. Figure 31 shows the velocity distribution of the grains through the standard chute. It can be seen that the velocity increases as the grains fall and then the velocity continues to increase as the grains slide through the chute. The velocity continues to increase as the wall is modelled as a frictionless surface. This highlights one of the shortcomings of the Eulerian-Eulerian approach in that in practice the velocity of the grains would not actually continue to increase due to wall friction.

The converging chute simulation results can be seen in Figure 32 (grains volume fraction) and Figure 33 (grains velocity). Figure 32 looks much the same as the corresponding figure for the standard chute in that the grains have compacted as they impact with the bottom of the chute. However, as the next figure shows, the velocity of the grains have increased compared with the standard chute, this being due to the converging section of the chute. In Figure 33 it can be seen that the velocity of the grains has increased compared with the standard chute which is due to the convergence of the chute as the grains fall (decreasing the impact angle ensures a lower momentum loss).

As with the previous simulation the grains continue to increase their velocity as they travel through the curved section of the chute. However, it can be seen that the velocity of the grains in the converging chute is of a larger magnitude. As the velocity of the grains increase the bed depth of the material stream decreases, since the throughput remains constant. The following discussion shows how this increased velocity and reduced bed depth have a positive effect on dust control. Velocity profiles are compared for the material as it discharges from the chute for the two varying profile chutes considered. Figure 34 and Figure 35 show the volume solid fraction of the grain phase and the dust phase at the different heights of the outlet.

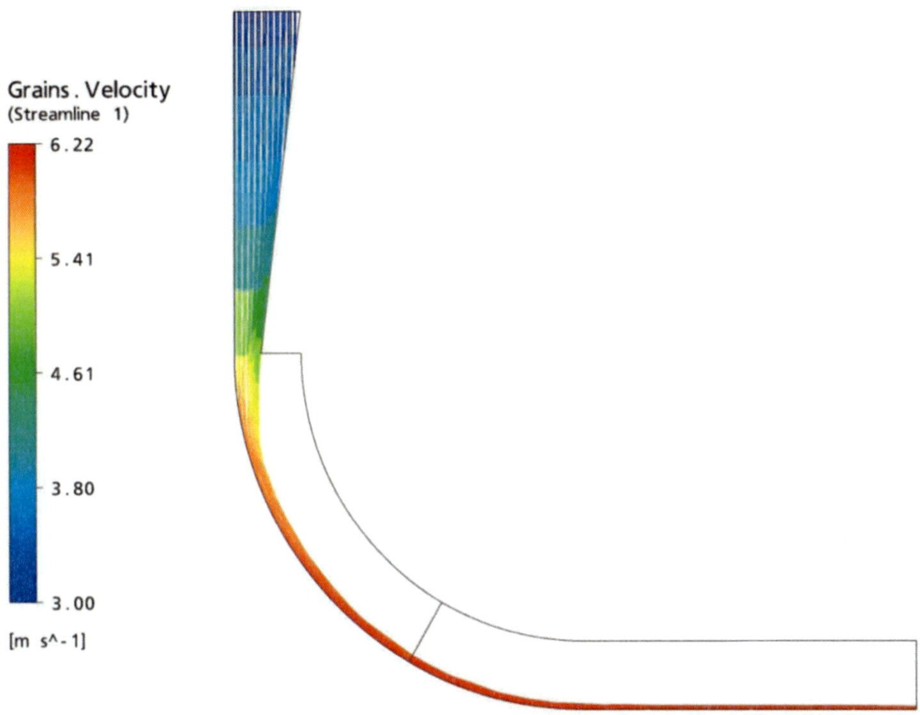

Figure 33. Velocity of the grains (1mm particles) through the converging chute.

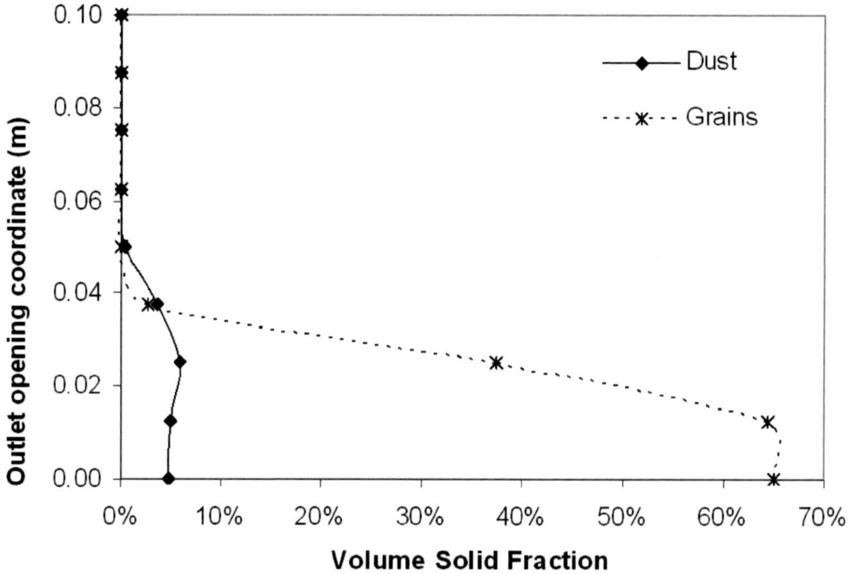

Figure 34. Volume Solid Fraction of solid dispersed phases at outlet of standard chute.

As expected, the volume solid fraction of the grains at the bottom of the outlet is at a maximum, and decreases with increasing height. The stream thickness is approximately 40 mm for the standard chute, while for the converging chute it is slightly smaller than 40 mm. The reduced stream thickness is due to the increased velocity of the grains. Comparisons of the dust profiles can also be made, and Figure 34 shows that at around 40 mm there is an equal amount of dust as there are grains, with the dust located near the top of the stream. As the material discharges from the chute, any dust on top of the material stream is likely to become airborne.

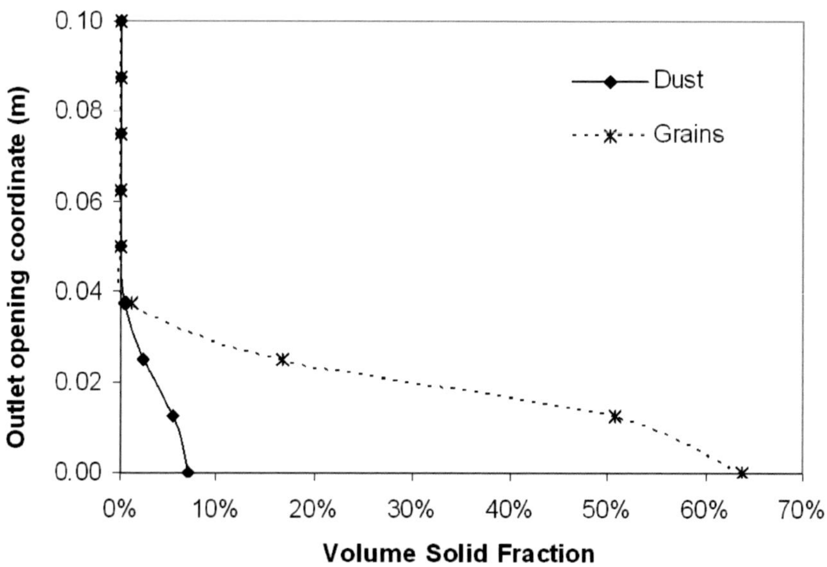

Figure 35. Volume Solid Fraction of solid dispersed phases at outlet converging chute.

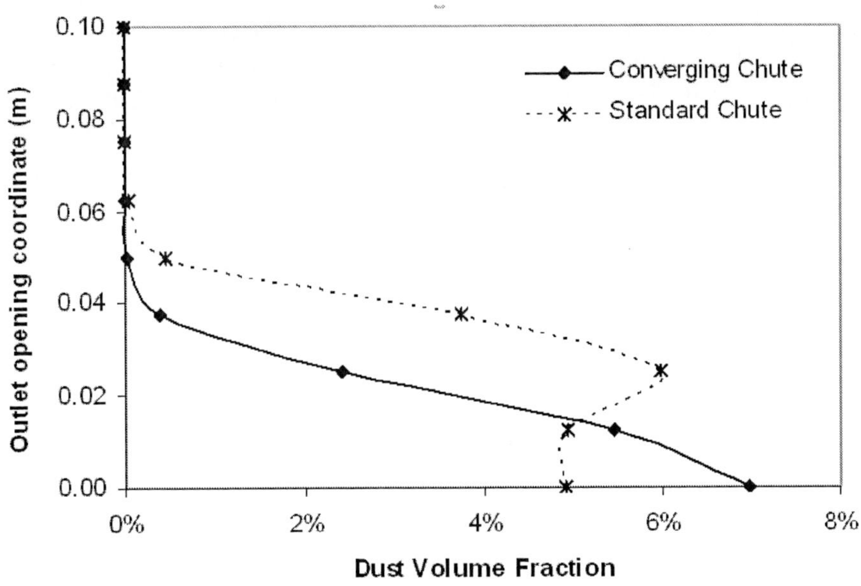

Figure 36. Comparison of dust profiles for the standard and converging chutes.

In comparison, Figure 35 shows that the dust is encapsulated within the material by the grains and so only a minimal amount of dust lies on top of the material stream. This is advantageous as when the material travels through the chute and discharges on to a stockpile the encapsulated dust within the material stream will be embedded into the stockpile and hence dust generation is reduced. Figure 36 shows a further comparison between the two dust profiles which highlights the effectiveness of the converging chute in directing the dust to the bottom of the material stream.

The results show that the addition of a third phase to represent dust in a multiphase simulation can reveal specific information regarding the behaviour of the dust. This example also quantified one of the main themes of this chapter in that converging the flow into a streamlined manner successfully encapsulates the fine particles which reduces dust emission. One of the major drawbacks of three phase modelling is the associated time needed for the simulations and the sensitivity of the solver. Both these factors are currently inhibiting the application of three phase modelling of entire transfer chutes.

Discrete Element Modelling

Discrete Element Modelling (DEM) is a numerical method for simulating the flow of granular bulk solids. The basic principle of DEM is to model each individual particle as a separate entity that can undergo a range of forces as in reality. These forces typically include gravity, contact forces with other particles and walls, with cohesive and adhesive forces being typically applied as well if the bulk solid is cohesive in nature. Calculations for the forces, and resulting displacements, are made for every particle at very small time steps throughout the simulation. As a result, DEM simulations can be very computationally intensive. The reader is directed to [9,10] for more fundamental information about DEM and its uses in bulk solids handling.

Discrete Element simulations are used increasingly for the simulation of material flow in transfer chutes. Because of the complex design of transfer stations, there is an increasing need for the simulation of flow through transfer chutes to adequately analyse the flow. In recent years computer simulations that consider individual particles such as DEM have become increasingly popular in industry. With the help of DEM it is not only possible to simulate and visualise the material flow but it also enables the analysis of forces and moments acting on different parts of the transfer station, including the actual chute and the receiving belt. Combining the analysis of forces and particle velocities it becomes possible to predict the wear of such parts, which can be used to optimise the design with respect wear minimisation and increased life [11].

For a first evaluation of the dust emission it is often sufficient to analyse the material flow in the system, as described previously in the section on the continuum method. Generally, it can be said that dust is generated where there is an abrupt change of the particle speed or direction. These changes in direction cause a change in the porosity of the bulk solid which has a twofold effect;

- A bulk solid dispersing or dilating as, for example during freefall, will tend to induce air into the falling stream of material
- A bulk solid that is being compacted, for example in an impact zone, will cause an exhaustion of air

It has already been established that there will always be a certain amount of dust producing fine particles suspended in the air throughout transfer stations. However, this is particularly important in the impact zones where there will be an exhaustion of air. Therefore, it is necessary to ensure a smooth redirection of the material flow with the minimum loss of momentum. Furthermore, it is necessary to ensure for a compacted material flow, any subsequent dispersion of the flow will increase the porosity, thus leading to an increase in the air flow induced into the stream. This dispersion can potentially create more dust as it will need to be exhausted from the granular stream upon impact. With the help of these two guidelines for smooth redirection and compact material stream, it is possible to evaluate the general dust emission behaviour of many transfer chutes. This approach is similar to the continuum method presented previously, and in fact DEM can be used as a complimentary tool in conjunction with the continuum method to design transfer chutes that can reduce dust emission by passive control.

As an example of this type of approach, consider the two designs of a transfer chute presented in Figure 37. The figure shows a comparison of the steady state material flow in a transfer chute with original design (left) and optimised design (right). The design criteria discussed above to improve the dust emission characteristics can be easily evaluated via the simulated particle flow and the particle velocity (particle colour).

The above example shows the benefits of implementing the basic guidelines of ensuring a smooth redirection of flow and a compact material stream.

The main advantage of this type of DEM approach over a method such as the continuum method is the ability to visualise the effect that the change in geometry has on the flowing granular stream. While the use of DEM on its own, or in conjunction with the continuum method, has its benefits, even more information can be gleaned from the use of the DEM coupled with CFD. This analysis technique is discussed in the next section.

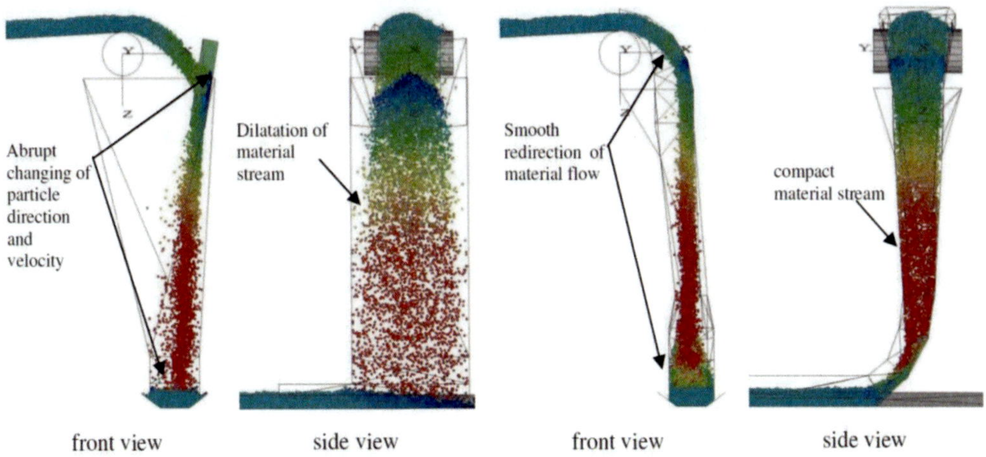

Figure 37. (a) Original Design (b) Optimised Design.

Computational Fluid Dynamics Coupled with Discrete Element Modelling

Recent advances in computing power has meant that the direct simultaneous coupling of CFD with DEM has become more popular and as a result there are a number of commercial codes available. However, in many instances the focus is on applications where the fluid/gas is the driving mechanism of flow for the particles (for example, see [12,13]) where there is high fluid/gas velocities or small particles. In the experience of the Authors there are very few options available for dealing with the CFD-DEM direct coupling on the scale of transfer chutes where the particle-fluid interaction needs to be resolved accurately. In addition, simultaneous coupled CFD-DEM simulations show an even higher numerical effort than either DEM or CFD when used separately.

In the following sections, a discussion is first presented on an alternative to the direct coupling of CFD to DEM where the results are translated manually between the two software platforms, and secondly, a brief review of the Authors' experience with the direct CFD-DEM coupled analysis technique.

Indirect Coupling

Due to the high mass flow rates present in transfer chutes, it is usually necessary in DEM simulations to neglect the mass fraction of the fine bulk material as the inclusion of these fine particles would make the number of particles too large within the simulation. The effect of this is that these fine particles are the most likely to be influenced by the air flow in a transfer chute, with the flow of the large particles typically unaffected by the surrounding air. Given this is the case, a more applicable and efficient approach than direct CFD-DEM coupling is the use of a serial combination of DEM and CFD simulations which can be applied with the well known continuum method already discussed in this chapter.

For the combined serial analysis of DEM and CFD, it is necessary to run a two stage CFD analysis. Firstly the air flow inside the granular stream is simulated using a single phase flow through a porous medium, and secondly the air flow surrounding the flow of the bulk solid is simulated, for example inside the transfer chute enclosure. For the CFD calculation of the air flow in the material stream the following information from the DEM simulation, or alternatively the continuum method can be used, is essential:

- the idealized geometry profile of the bulk material flow
- the porosity distribution inside the material flow

In addition, it is necessary to assume several boundary conditions in the CFD simulation along the bulk material profile to define the interaction of the air stream inside and outside of the bulk material stream.

To illustrate the application of this serial CFD and DEM approach, an inline, slow speed conveyor transfer has been used in which there is an enclosure surrounding the transfer (see Figure 38).

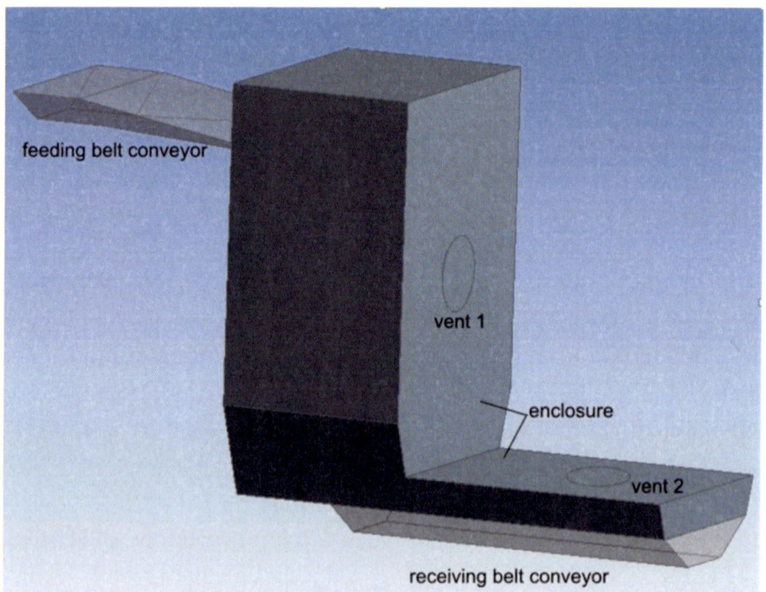

Figure 38. 3D CAD model of the combined CFD and DEM serial approach.

In this particular example the height difference between the two conveyors is 2 m, the throughput of the system is approximately 750 tph with a belt width of 1100 mm. Furthermore, two circular vents with 400mm diameter, having different positions and a variable pressure drop are considered in the model. The inclusion of the two vents is to replicate dust extraction fans that can be used to reduce dust emissions. The highest amount of dust will be generated during the impact on the receiving conveyor. Hence, it can be easily estimated, that vent 2 - in opposition to vent 1 - will most likely exhaust the most dust polluted air. In this case, the general bulk material flow is not affected by the enclosure, and hence only the feeding and receiving belt was considered in the DEM simulation, as shown in Figure 39 (a). The air flow inside the material stream was calculated based on a porosity map out of the DEM simulation, as shown in Figure 39 (b), and is visualized by the streamline profile in Figure 40. The boundary conditions for the simulation of the bulk material stream were chosen to reflect the movement of air around the moving bulk material stream.

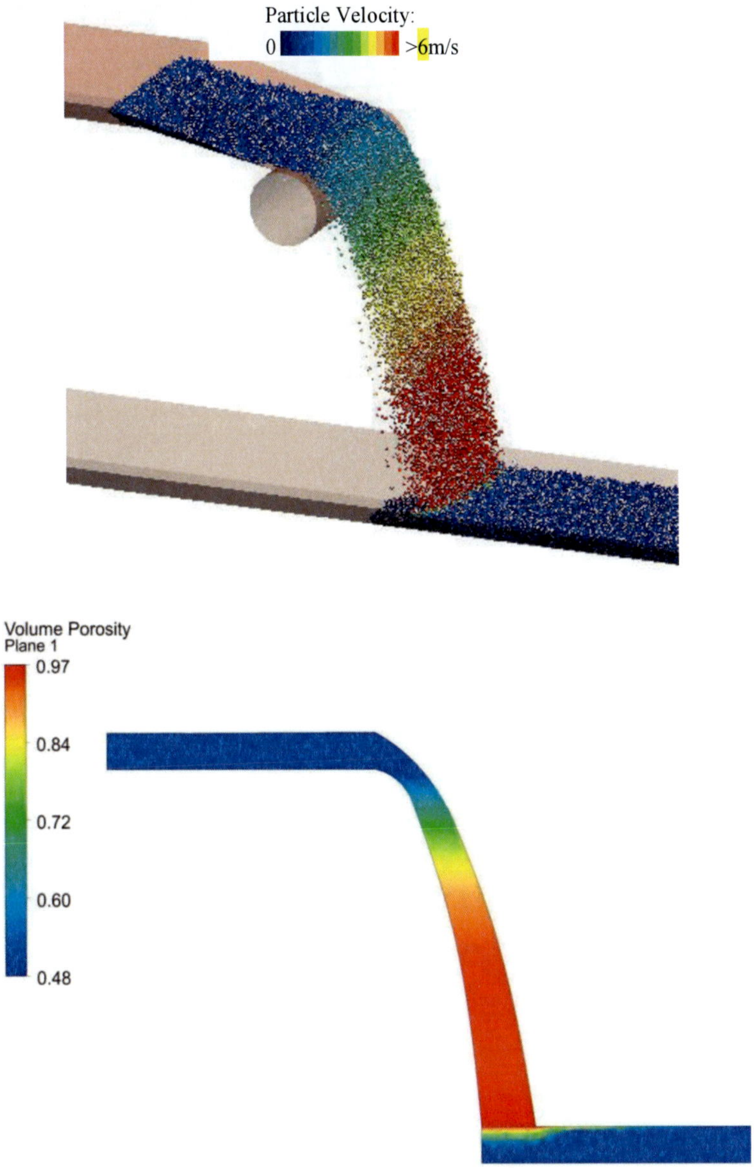

Figure 39. (a) Bulk material flow in the DEM simulation; (b) Porosity distribution in the idealised bulk material flow profile for the CFD simulation.

These air movements include air being induced into the material stream [14] as it falls and also air being exhausted as the stream impacts with the receiving belt. For this simulation, the porosity in the impact zone on the receiving belt was variable but still had a relatively ordered structure. This ordered structure caused the air flow within the material stream to be much more ordered than what is expected in reality. Hence it was necessary to apply an external pressure boundary condition to the top surface of the receiving conveyor belt. This ensured the airflow in the region of the impact zone more closely resembled what we would expect in real circumstances with large amounts of air exhausting from the material stream.

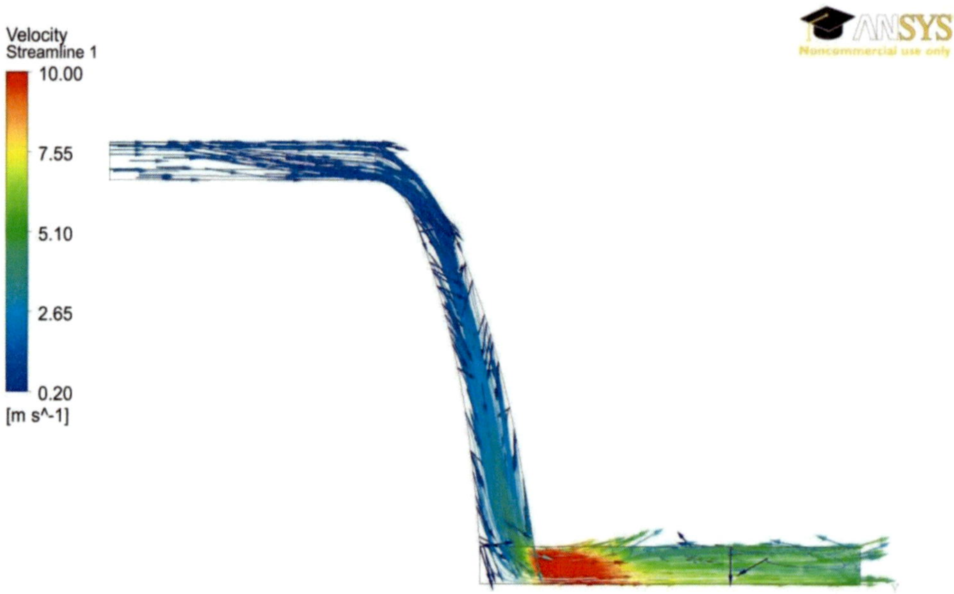

Figure 40. Velocity stream lines of air which is induced in falling material stream.

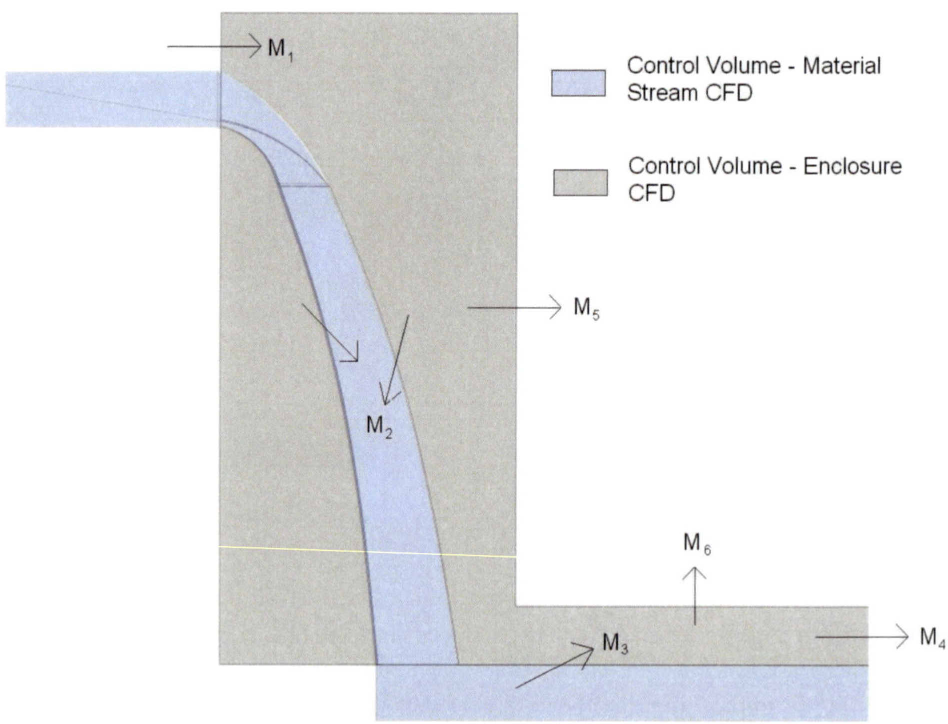

Figure 41. Mass flow rates measured at different positions in the CFD model.

Some of the important boundary conditions for the calculation of the air flow in the enclosure, outside the bulk material stream, are shown in Figure 41.

The air flows designated can be explained as follows:

- M_1 – This is a zero relative pressure opening which allows air to flow into or out of the enclosure, depending on the solution obtained from the other specified boundary conditions
- M_2 – This is the flow of air induced by the falling stream of bulk solid. This boundary condition is specified from the results of the CFD simulation for the bulk material stream
- M_3 – This is the flow of air being exhausted due to the rapid compaction of the falling stream upon impact. This boundary condition is specified from the results of the CFD simulation for the bulk material stream
- M_4 – This is an zero relative pressure opening which allows air to flow into or out of the enclosure, depending on the solution obtained from the other specified boundary conditions
- M_5 – This is a relative pressure specified opening, where the pressure was either 0 or -50 Pa
- M_6 – This is a relative pressure specified opening, where the pressure was either 0, -50 Pa or -100 Pa

The results of the CFD simulations are shown in Figure 42 and Table 3. Figure 42 (a) shows the air flow inside the enclosure while using Vent 1 (Vent 2 closed), and Figure 42 (b) shows the air flow inside the enclosure while using Vent 2 (Vent 1 closed). The results shown in Figure 42 were for the case of -50 Pa relative pressure drop. In Figure 42 (a) it can be clearly seen that the air, exhausted by Vent 1, does not come from the impact zone of the material stream on the receiving belt conveyor, rather it comes from the opening above the incoming conveyor belt. Hence, the dust content of the exhausted air will be very little.

Figure 42. (a) Velocity streamlines for Vent 1 open.

Figure 43. Velocity streamlines for Vent 2 open.

The image in Figure 42 (b) shows that almost the whole amount of the exhausted air of Vent 2 comes from the impact zone. Hence, it can be estimated that a large amount of the dust polluted air is flowing into Vent 2. Table 3 shows the results for the mass flow rates and volumetric flow rates of the air inside the enclosure at different positions and for the different test cases. Positive values represent air flowing into the enclosure. For this transfer chute geometry, it can be easily assumed that the air with the highest dust content can be found on top of the receiving conveyor (region M_3 in Figure 41 (b), an average volumetric flow of 1.14 m^3/s exhausts from the material stream on the receiving belt).

Table 3. Air mass flow rates and volumetric flow rates at different positions in the CFD model

Condition	M_1 (kg/s) Q_1 (m^3/s)	M_2 (kg/s) Q_2 (m^3/s)	M_3 (kg/s) Q_3 (m^3/s)	M_4 (kg/s) Q_4 (m^3/s)	M_5 (kg/s) Q_5 (m^3/s)	M_6 (kg/s) Q_6 (m^3/s)
Vent 1 (no pressure)	+0.8 +0.96	-0.87 -1.05	+0.95 +1.14	-0.98 -1.18	+0.1 +0.12	N/A
Vent 1 (-50 Pa)	+2.03 +2.44	-0.87 -1.05	+0.95 +1.14	-0.98 -1.18	-1.12	N/A
Vent 2 (no pressure)	+0.97 +1.17	-0.87 -1.05	+0.95 +1.14	-0.85 -1.02	N/A	-0.2 -0.24
Vent 2 (-50 Pa)	+1.13 +1.36	-0.87 -1.05	+0.95 +1.14	-0.11 -0.13	N/A	-1.09 -1.31
Vent 2 (-100 Pa)	+1.23 +1.48	-0.87 -1.05	+0.95 +1.14	+0.26 +0.31	N/A	-1.56 -1.88

Vent 2 exhausts the air from this region, and, depending on the pressure drop of the vent, from the opening above the receiving belt (M_4). Putting the results from Table 1 into context, a pressure drop of 50 Pa for Vent 2 shows a volumetric flow of 1.31 m^3/s, and while this is more than the 1.14 m^3/s coming from region M_3, there is still an outflow of air at the opening above the receiving belt (M_4) of 0.13 m^3/s. This means that there is a possibility of dust escaping the enclosure. If the case of a 100 Pa pressure drop at Vent 2 is then considered, the volumetric flow through Vent 2 increases to 1.88 m^3/s out of the enclosure, which results in the air flow at the opening to be reversed so that air is drawn in through this opening. This indicates that no dust polluted air escapes from the enclosure.

This example shows how a serial combination of CFD and DEM can be used to analyse the flow of air in a transfer chute. The advantage of this method is that simulation time is saved, thus avoiding the computationally expensive direct CFD-DEM interaction, while still using results from both software platforms. Also shown was the ability of CFD to model dust extraction fans which is a form of active dust suppression.

Direct Coupling

The Authors of this chapter do not have any direct experience with the CFD-DEM coupling, so this brief discussion will include only a basic introduction to the method and some of the most common applications. Direct coupling takes place when there is a simultaneous solving of the Navier Stokes equations from computational fluid dynamics and the equations of motion from the discrete element method. The timescale of this simultaneous solving of equations is generally such that a number of DEM time step calculations are carried out for every CFD time step calculation. An important consideration in these simulations is the fluid cell size, and the relationship of the fluid cell size to the particle diameter. In terms of the flow in transfer chutes, another key factor is to accurately model the momentum transfer between the fluid and the discrete particles. At the time of writing this chapter, the technology linking the direct coupling of CFD to DEM is not sufficiently developed for ease of application and remains an of active area of research. This is particularly evident in the field of materials handling where the simulations often require regions of dense granular flows encompassing relatively large control volumes.

As highlighted in a recent review paper [15] there have been two main applications of CFD-DEM research, with those applications being fluidization and pneumatic conveying. As this review paper illustrates, there have been numerous research studies carried out on these two particular areas.

Another application of coupled CFD-DEM includes the study of cyclones which are used to separate solids from fluids, with the work of Chu et al [16] demonstrating that CFD-DEM can capture some key features in a cyclone separator. The common trend in each of these research topics is that it is the fluid that is the driving force in the simulation, as without the injection of a moving fluid into the control volume there would be no movement of the discrete phase. In the particular application of transfer chutes, the opposite is true. The movement of the bulk solid due to mechanical conveying is the reason for generating the fluid velocity field. To date, there have been very few instances of using a combined CFD-DEM method to predict the emission of dust in transfer chutes, with the work of Goniva et al [17] illustrating one approach to this problem. It is expected that in the future this area will receive more attention as it is potentially an excellent method for analyzing the flow of granular material through transfer chutes and the resulting affect on the surrounding air.

CONCLUSION

The focus of this chapter has been on methods for passive dust control in conveyor transfer chutes. Before an effective design of plant equipment that minimises dust generation can progress, it is important to characterise the material being handled in terms the "dustiness" of the material. A number of experimental and analytical techniques are available for this type of analysis, with the Australian Standards outlining a rotating drum test that seeks to find a moisture content of the material that eliminates the dust produced to a certain level. Another useful test for characterisation was also presented in which a wind tunnel is used to simulate air flowing over a stockpile to investigate the dust lift-off under certain environmental conditions. This testing can be expanded to also include active dust control methods such as the application of surface veneer treatments.

A range of analysis techniques were also presented which can be used to study individual components of a transfer chute or the entire transfer. Each of the methods presented have relative advantages and disadvantages in comparison with each other and can be summarised as follows;

- The continuum method for predicting the flow of the bulk solid is well established and is based on actual flow properties measured in the laboratory. This method is dependent only on solving a single set of equations and so can be carried out quite fast. There is no direct correlation to predict the air flow of the surrounding material.
- A CFD single phase simulation can be used to look at the flow of air only. This is a good technique for those applications that allow it to be applied. In the materials handling field, unfortunately there are not many applications for which an air only simulation will be of any benefit.
- A CFD two phase simulation can be carried out by simulating a granular phase and air phase and uses some properties from the material sample. However, some of the parameters within the software are difficult to measure and correlate to the material sample. A direct representation of the resulting air flow from the simulation is possible.
- A CFD three phase simulation can be used to add in a third phase of 'dust' particles so that a direct representation of the dust phase is possible. In comparison with the two phase simulation, simulation times are much longer and the solver (in CFX) is less robust and more sensitive to initial conditions. A direct representation of not only the airflow is possible but also the dust as well.
- The DEM method on its own can be used to visualise the granular flow and view the behaviour of individual particles. Like CFD, some of the parameters used in DEM are difficult to measure and correlate to the material sample. Solve times for DEM are dependent on many factors such as particle size, stiffness, software code. However, the computational time is generally faster than a two phase CFD simulation. As with the continuum method, there is no direct correlation to predict the flow of air surrounding the material stream.
- An indirect coupling of DEM or continuum modelling with a single phase CFD simulation can be used to reveal velocity and porosity information for the granular stream to input into a CFD simulation. In this method some assumptions need to be

made for the induced air into the material stream. Similar comments apply regarding the parameters in the software as with previous methods. This method can deliver relatively fast qualitative results.

- The direct coupling of DEM and CFD is the newest analysis technique. For application to bulk materials handling the method is still in the early stage of development. Simulation times would be expected to be in the range bounded by the time needed for a DEM simulation and the time needed for a three phase CFD simulation. As the development continues for this technique it should prove a valuable method to analyse the dust emissions in transfer chutes.

REFERENCES

[1] U.S. Department of Labor, Occupational Safety and Health Administration, *Dust Control Handbook for Minerals Processing*, 1987.

[2] Roberts, A. W.; Ooms, M.; Bennett, D. J.; Bulk Solid Conveyor Belt Interaction in Relation to Belt Cleaning, *Bulk Solids Handling*, Vol. 7 (3), 1987, pp 355-362.

[3] Roberts, A. W.; An Investigation of the Gravity Flow of Non-cohesive Granular Materials through Discharge Chutes, *Transactions of A.S.M.E., Jnl. of Eng. in Industry*, Vol. 91 (2), Series B, 1969, pp 373-381.

[4] Roberts, A. W.; Chute Performance and Design for Rapid Flow Conditions, *Chemical Engineering Technology*, Vol. 26 (2), 2003, pp 163-170.

[5] Djukic, M.; Planner, J.; Evaluation of Dust Emission from Coal Wagons, *Bulk Europe 2010 International Conference*, Glasgow, Scotland, 9-10th September 2010.

[6] Wheeler, C.A.; Krull, T.; Roberts, A.W.; Wiche, S.J.; Design of Ship Loading Chutes to Reduce Dust Emissions, *Process Safety Progress*, Vol. 26 (2) 2007, pp 229-234.

[7] Chen, X.L.; Wheeler, C.A.; Donohue, T.J.; McLean, R.; Roberts, A.W.; Evaluation of Dust Emissions from Conveyor Transfer Chutes using Experiments and CFD Simulation, *International Journal of Mineral Processing* (imprint).

[8] Donohue, T.J.; Roberts, A.W.; Wheeler, C.A.; McBride, W.; Computer Simulations as a Tool for Investigating Dust Generation in Bulk Solids Handling Operations, *Particle and Particle Systems Characterisation*, Vol 26, 2009, pp 265-274.

[9] Gröger, T.; Katterfeld, A.; Application of the Discrete Element Method in Materials Handling: Basics and Calibration; *Bulk Solids Handling* Vol. 27 (1), 2007, pp 17-23.

[10] Katterfeld, A; Gröger, T.; Application of the Discrete Element Method in Materials Handling: Transfer Stations; *Bulk Solids Handling* Vol. 27 (3), 2007, pp 158-167.

[11] Katterfeld, A.; Gröger, T.; Simulation Based Wear Prediction of Transfer Chutes, *Bulk Europe 2008 International Conference*, Prague, Czech Republic, 11-12th September 2008.

[12] Feng, Y.; Zhang, S.; Xu, B.; Yu, A. B.; Zulli, P.; Discrete Particle Simulation of Gas Fluidization of Particle Mixtures, *AIChE Journal*, Vol. 50, 2004, pp 1713-1728.

[13] Feng, Y.; Yu, A. B.; Microdynamic Modelling and Analysis of the Mixing and Segregation of Binary Mixtures of Particles in Gas Fluidization, *Chemical Engineering Science*, Vol. 62, 2007, pp 256-268.

[14] Hemeon, W.; Plant and Process Ventilation, Industrial Press, New York, USA, 1963.

[15] Zhu, H.P.; Zhou, Z.Y.; Yang, R.Y.; Yu, A.B.; Discrete particle simulation of particulate systems: A review of major applications and findings, *Chemical Engineering Science*, Vol. 63, 2008, pp 5728-5770.

[16] Chu, K.W.; Wang, B.; Yu, A.B.; Vince, A.; Modelling of the multiphase flow in dense medium cyclone. *The Fourth International Conference of Discrete Element Method*, Brisbane, Australia, 27–29th August, 2007.

[17] Goniva, C.; Katterfeld, A.; Kloss, C.; Simulation einer Übergabestelle hinsichtlich Emission und Ausbreitung von Staub sowie Materialabnutzung, Fachtagung Schuttgutfirdertechnik 2011, Magdeburg, Germany. (Title translates as "Simulation of a Transfer Chute focussing on dust emissions and wear").

In: Dust: Sources, Environmental Concerns and Control
Editors: Laurent B. Wouters and Michel Pauwels
ISBN: 978-1-61942-547-7

Chapter 4

BLOWING DUST AND DUST STORMS IN THE ARABIAN PENINSULA WITH PARTICULAR REFERENCE TO THE ARABIAN GULF

M. P. de Villiers
Weather Services International, Europe,
Birmingham, UK

ABSTRACT

Dust lifted into the atmosphere can be carried far afield and, apart from disruption to human activities, can have serious and life threatening implications. The Arabian Peninsula is a major dust producing region. This chapter briefly discusses the health and environmental effects of dust storms and, in more detail, the dynamics and the two atmospheric mechanisms that cause dust storms and allow dust to be raised in suspension for prolonged periods in the atmosphere. Particular reference is made to blowing dust and dust storms during the summer and winter Shamals, as well as those caused by the Kaus wind, the less frequent Nashi wind and the usually short lived thunderstorm associated dust storm. Seasonal and diurnal distribution and frequency is discussed. Attention is drawn to the importance of numerical weather prediction models and a simple dust forecast index proposed. The latest advances in satellite imagery for dust detection are discussed and a dust storm forecasting methodology is presented.

1. INTRODUCTION

Mention dust and most people think of the dust that collects on furniture and ornaments in the home, but there is more to dust than meets the eye. Apart form affecting human activities, dust carried aloft in blowing dust and dust storms can be a health hazard. The attendant poor visibility and gusty winds affect commercial activities and can be a danger to road transport and can lead to the closure of airports and seaports. Dust storms are responsible for most of the estimated two billion metric tons of dust that are carried up into the atmosphere annually and one dust storm can lift more than 200 metric tons of dust (Griffin et al 2002). The Arabian Peninsula (Figure 1), with its frequent dust storms, is one of five major dust producers (Idso 1976, Goudie 1983).

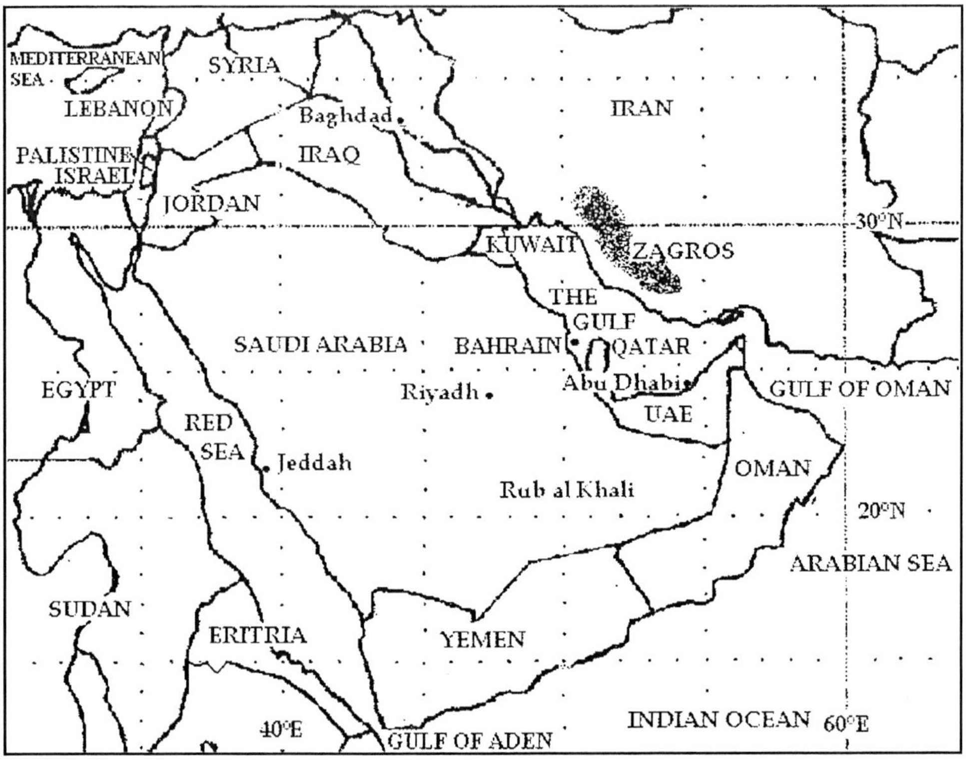

Figure 1. Map of the Arabian Peninsula with places names mentioned in the text.

A dust storm, or sand storm, can be defined as particles of dust, or sand, lifted to great height by a strong and turbulent wind in sufficient concentration that the visibility is reduced to less than 1000 metres (m) and a severe storm when the visibility is below 200m (UKMO 1991, Safar 1985). A dust storm is also known by the Arabic word Haboob. Initially it was a name used in the Sudan for a dust storm caused by a thunderstorm, but the name is being increasingly used worldwide to describe any type of dust storm (Miller et al 2008, Wilkerson 1991, Idso et al 1972).

This chapter briefly discusses the health and environmental effects of dust storms. The two atmospheric mechanisms that cause dust storms are presented and the dynamics that cause dust to be raised in suspension in the atmosphere for prolonged periods (and the short-lived saltation process) are discussed. The types of dust storms that occur in the Arabian Peninsula area, with particular reference to the Arabian Gulf, are discussed in detail within the context of the atmospheric mechanisms. Particular reference is made to the summer and winter Shamal dust storms, as well as those caused by the Kaus wind, the less frequent Nashi wind and the usually short lived thunderstorm associated dust storm. Their seasonal and diurnal distribution and frequency is discussed. A numerical weather prediction (NWP) model derived dust prediction index is suggested, as well as a short section on the latest advances in satellite imagery for dust detection is presented. Finally, blowing dust and dust storm prediction guidelines are given.

2. Health and Environmental Effects

Among the material of varying size, carried aloft and described as dust, sand and less often clay and silt, there are other materials carried aloft, such as soot, salt, oxides and phosphates (Galvin et al 2002, Wilkerson 2009, Perkins 2001, WMO 1983). Probably more important, are "micro-organisms, such as bacteria, viruses and fungi and soil pollutants, such as herbicides and pesticides" (Griffin et al 2002), which have an effect on public health and sensitive ecosystems. Air samples taken in the Caribbean, by Griffin et al (2002), during Africa originating dust events reveal that of the microbes present, 25% are species of plant pathogenic bacteria and fungi, while 10% "are opportunistic human pathogens (organisms that can infect people who have a lowered resistance)." Originally it was thought that bacteria, viruses and fungi could not survive for long in the atmosphere without being destroyed by ultraviolet radiation, lack of nutrients and desiccation (Griffin et al 2002). However, Imshenetsky et al (1977) discovered the opposite. Samples, taken by meteorological rockets at altitudes of 49 to 77 km and brought back to the surface, survived tests such as exposure to gamma-radiation, high vacuum and ultraviolet radiation levels and freeze-drying. They also found that the concentration of these micro-organisms were greater after dust events.

High concentrations of mineral dust smaller than 2.5 microns (μm) can be found in dust storms. Due to their small size they are of particular concern to human health as they can easily penetrate breathing systems, cannot be exhaled and are deposited in the lung (Prospero 2006). Studies in the United States and Europe have proven that high concentrations in the lungs are associated with increased death rates and hospital admissions with severe respiratory ailments such as aggravated asthma and chronic bronchitis (Pope 2000). The day after a severe sand storm occurred in Saudi Arabia's Eastern Province, people reported at various hospitals complaining about breathing problems such as choked throats and breathlessness, many of these had a history of asthma (Arabianbusiness.com 2009). The news report added that, according to Arab News, "the air was so thick with sand that most patients actually complained of experiencing an iron-like taste on their tongues." On another occasion it was reported that "unprecedented sandstorms in Kuwait, the likes of which have not been seen for 30 years, are making the public physically and mentally ill" (Arabianbusiness.com 2008a). Furthermore, apart from reported physical illness, especially among residents with allergies, psychologists also said the sand storms caused "seasonal depression" and that "environmental pollution and changes play a great role in creating psychological and behavioural tension and anxiety."

Further evidence of health related effects of dust storms comes from the Aral Sea. Diversion of water sources for agricultural purposes, has reduced the sea to less than half its original 60 000 square kilometres, leaving a large barren and dry surface. The consequence is that storm raised dust with high concentrations of pesticides are now common. O'Hara et al (2000) found that "dust deposition rates across eastern Turkmenistan are among the highest in the world and that the dust is contaminated with pesticide. In similar vein, due to increased water use and poor rains, Lake Chad, on the edge of the Sahara Desert, has shrunk to 5% of its original size since the 1960s, making this a fertile area for dust storms (Perkins 2001). A more positive note is that dust storms are an important source of iron and nitrogen species, which are a boon when they are carried to nutrient deficient areas (Trapp et al 2010).

Particulate matter can be carried great distances. Depending on the prevailing winds and the season, Sahara dust can be carried to Europe, the Middle East, Caribbean, southern United States and South America (Perkins 2001, Perry et al 1997). This is borne out by the fact that iron in red soil of the Caribbean, which is naturally composed of limestone, came from African dust. Another example is that at regular intervals wind borne aerosols from the Asian continent reaches North America, such as in a huge dust storm in China in April 1998 that may already be affecting ecosystems in North America. A study of the Fraser River watershed, in British Columbia, concluded that wind borne pollutants from Asia could be contaminating lake fish and sediments (Prospero 1999, Wilkening et al 2000).

Dust storms present a serious disruption in day to day life, especially with respect to the operation, maintenance and deterioration of mechanical and electrical equipment (Hoidale and Hinds (1975) in Safar 1985). In the Middle East a sand storm during 13 April 2011 at Kuwait, halted oil shipments, closed Kuwait International Airport, closed all schools and colleges and caused minor traffic accidents (ArabianBusiness.com 2011a and 2011b). This was the third time in as many weeks, the previous two being on 4 April and 26 March (Arabianbusiness.com 2011c and 2011d). In another sand storm in April 2008 a strong north-westerly Shamal wind that gusted up to 48 kilometres per hour "battered much of the Gulf," in particular Kuwait, Bahrain and Qatar and caused more than 500 traffic accidents in Kuwait in 48 hours while in Bahrain a further 20 accidents were reported. Hundreds of people went to hospitals suffering from asthma and allergic reactions (Arabianbusiness.com 2008b). Obviously, if one is out in the open in a dust storm, it can be very uncomfortable. Thesiger (1994) in November, or December, 1949, during a primitively equipped camel trip into Rub al Khali (The Empty Quarter) of western Oman describes camping in a bitterly cold north-easterly dust storm. "The wind had been southerly, but the next day it went back to the north-east and blew another gale. Bitterly cold off the frozen uplands of central Asia, it was thick with sand whipped from the dunes around us. We sat throughout the day, without shelter, in a reddish obscurity, half-smothered by the flying grains which, reaching a height of about eight feet above the ground, rasped our skins, filled our eyes, noses, and ears, and were gritty between our teeth. Sunset brought relief, for the wind dropped and the stars came out."

3. Atmospheric Mechanisms and Dynamics

For dust to be raised into the atmosphere and held in suspension for any length of time the soil must be dry, a strong wind is needed, as is a well-mixed and deep boundary layer that is inherently unstable, that is, it has a nearly dry adiabatic lapse rate (Jauregui 1989, Safar 1985 and Blair 1957). There are two atmospheric processes that meet these requirements. One is a strong pressure gradient between high pressure and low pressure cells; the other is the downdraft outflow beneath a thunderstorm (Wheaton and Chakravarti 1990, WMO 1983). Although the air is usually thermally stable in the area of the high pressure cell, the dust is lifted due to turbulence and vertical motion caused by the steep pressure gradient induced wind (de Villiers and van Heerden 2007). However, increased instability, when the environmental lapse rate (ELR) approaches the dry adiabatic temperature lapse rate (DALR), assists upward transport of dust particles (Safar 1985 and Blair 1943), such as occurs during strong surface heating during the day, or the influx of colder air aloft (Coles 1938 in

Wilkerson 1991, WMO 1983). Dust raised in this way can cover over a million square kilometres (WMO 1983). This synoptic scale process can be predicted by NWP models. With respect to thunderstorms, a strong downdraft outflow induced wind creates a gust front and severe turbulence, which raises dust. This mesoscale event is more difficult to detect on NWP models (Miller et al 2008). However, areas with parameters, or indicators, associated with thunderstorm development are available as a guide to possible occurrence. For example, vertical velocity, atmospheric moisture, convective available potential energy (CAPE) and instability indices such as, lifted index, K index and total totals (de Villiers 2010).

In general discussion little distinction is usually made whether the cause of the storm is dust, or sand. The public media prefers to refer to sand storms, as evidenced in the news reports in the previous section, although strictly speaking dust storm is correct. Sand tends to creep along the surface in a strengthening wind. The near surface wind (measured at 10m above ground) needs to reach about $10ms^{-1}$ before surface turbulence is strong enough to raise sand grains to about a metre above the ground. The lifted sand then "bounces" across the surface and disturbs other surface grains of sand, which in turn lift and fall in a process known as saltation, but the sand very soon settles when the wind drops. Smaller dust particles can be more easily lifted to a great height and held in suspension for days before slowly settling. From its source dust carried aloft is carried downwind and tends to spread in the shape of an ever increasing teardrop, unless it is lifted across a broad front.

Grains of desert sand are usually accepted as having a diameter, or longer axis, greater than 0.07 to 0.08 mm (70 to 80 μm) and dust less than this. Dust can be further sub-divided into silt, which is as small 0.002 mm (2 μm) and clay even smaller (UKMO 1994, Hoock 1984 in Wilkerson 1991). Dust less than 0.02 mm (20 μm) can be carried long distances (Gillette 1979, in Wilkerson 1991).

The stronger the wind the stronger will be turbulent vertical currents and the greater likelihood of dust carried aloft. Bagnold (2005) suggests that $^1/_5$ of the mean wind speed gives a rough guide to the vertical speed, the implication being that particles with a terminal velocity less than the upward vertical velocity will be carried upward and remain in suspension.

Researchers, such as Chepil and Woodruff (1957) have established that there is a direct relationship between suspended dust concentration and visibility. However, from a practical point of view wind speed observations are more widely and readily available to investigate this correlation. It is generally accepted that dust begins to be lifted at $5\text{-}8ms^{-1}$ (Bagnold 2005, Morales 1979 and Coles 1938 in Wilkerson 1991). From dust storm data in Kuwait, Safar (1985) concluded that dust storms occur at $\geq 9ms^{-1}$. However, full scale blowing dust storms require speeds exceeding $15ms^{-1}$ (WMO 1983). A study at Abu Dhabi International Airport (ADIA) in the United Arab Emirates (UAE) found that $>7.5ms^{-1}$ the visibility is reduced to 5000-8000m, $>10ms^{-1}$ the visibility will be 1000-2000m and $>15ms^{-1}$ the visibility will be <1000m (de Villiers and van Heerden 2007). Membery (1983) in a study at Bahrain noted that the presence of a northerly nocturnal jet of $\geq 15ms^{-1}$ between 300m and 900m altitude is a forerunner of a strong surface wind during the day and blowing dust, or a dust storm.

As the wind decreases so does the upward vertical velocity. When the upward vertical velocity of the air no longer exceeds the dust's terminal vertical velocity, the dust will begin to settle. Larger particles will settle first with smaller particles tending to drift further downwind before settling. An effective rule of thumb that appears to work is to use an average settling rate of 300m per hour (Lawson 1971).

4. Dust Storm Types

Dust storms in the Middle East occur in three basic wind regimes. These are; the summer and winter north-westerly Shamal, which Rao, et al (2001) have defined as lasting at least 3 hours in a day at $\geq 8.5 ms^{-1}$; the south-easterly to south-westerly Kaus, and the north-easterly Nashi (UAE Climate 1996, UKMO 1991, Middleton 1986).

In addition there is the much shorter duration dust storm that occurs with thunderstorms, which are usually associated with the passing of cold front troughs (de Villiers 2010, UAE Climate 1996, Safar 1985, Membery 1983,).

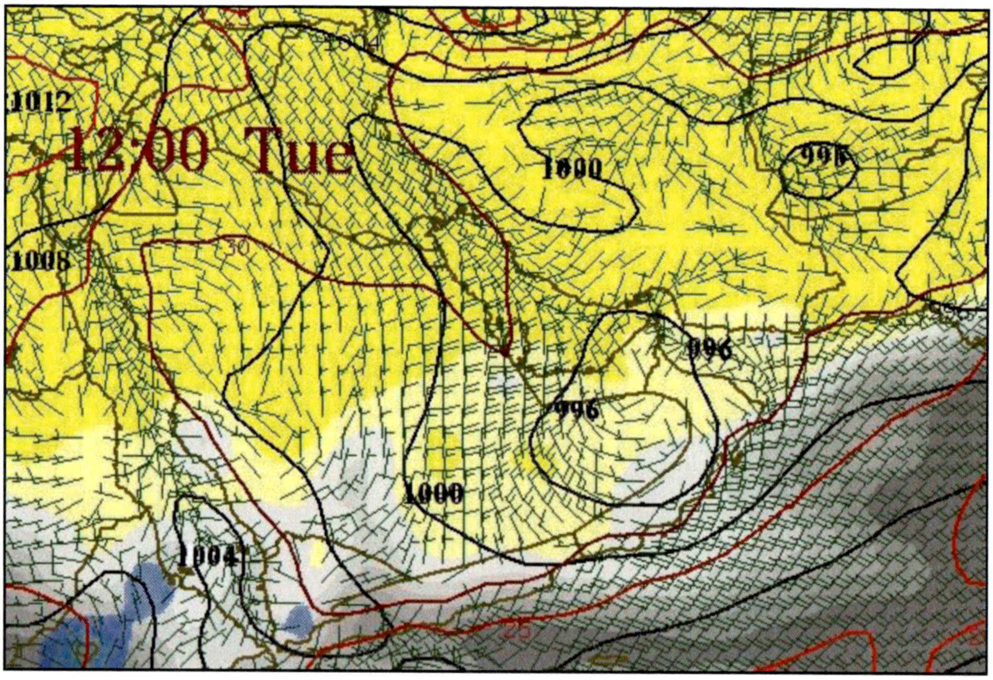

Figure 2. Example of a summer Shamal dust event circulation. The GFS NWP model prognosis valid at T+12 hours 21 June 2011 1200 UTC (1600 LT) showing wind arrows, mean sea level hPa pressure (black lines), 850 hPa level air temperature (red lines), cloud cover (shades of yellow (clear), white and dark grey (cloudy)) and precipitation (blue) (Courtesy J. Griffiths, Weather Services International).

The persistent summer Shamal occurs due to the deep monsoon low pressure over Asia and a semi-permanent high pressure cell over northern Saudi Arabia. The pressure gradient between the two pressure systems results in an intensified and persistent north-westerly wind over the Arabian Gulf and surrounding Gulf States. The Shamal is also enhanced by a lee low effect west of the Zagros Mountains, which are situated immediately inland of the Arabian Gulf over south-western Iran (Membery 1983). However, during the summer months, an approximately north-south trough from the Iran/Asian low to the southern Arabian Peninsula heat low often lies across the UAE and Rub al Khali (Empty Quarter) of eastern Saudi Arabia (Rao et al 2003), so that the Shamal often does not reach the eastern parts of the Arabian Gulf and Arabian Peninsula. The typical surface pressure and wind pattern can be seen in the NWP Global Forecast System (GFS) model prognosis in Figure 2, valid at 1600 Local Time (LT = UTC + 4 hours) 21 June 2011. It predicted $10\text{-}13 ms^{-1}$ north-westerly winds from southern

Iraq across the Arabian Gulf to low pressure of 996 hectoPascal (hPa) over Rub al Khali. Blowing dust was subsequently observed over south-eastern Iran, adjacent north-western Iran, Kuwait and southward over the eastern part of Saudi Arabia to Rub al Khali. Some examples of the weather at 1900 LT were visibility 1600-2500m due to blowing dust in a 12ms^{-1} gusting 17ms^{-1} north-westerly wind in Kuwait and 2500m in an 8ms^{-1} northerly wind at Al Ahsa inland of Qatar.

Much less frequent winter Shamals occur in association with mid-latitude upper air troughs that pass from west to east through the region, mainly during the months of November to March. In character with these baroclinic systems, a surface low pressure cell exists below the upper air south-westerly flow ahead of the trough, while a surface high pressure cell follows to the west of the low pressure cell. The trough usually passes across Syria, Iraq and northern Iran, the Shamal being in the north-westerly air flow between the low and the high pressure cells. In association with the trough, a secondary low often develops and moves from the Red Sea across Saudi Arabia to southern Iran (Rao, et al 2001), followed by a southern extension of the Shamal. A typical synoptic situation can be seen in Figure 3. The NWP Eta model T+30 hour prognosis valid at 0600 UTC 13 March 2003 indicates high pressure intensifying from the west in the wake of the low pressure/trough system in the north over Iran with a 10 ms^{-1} Shamal in the northern Arabian Gulf into to a secondary low of 1008 hPa over the southern part of the Arabian Gulf.

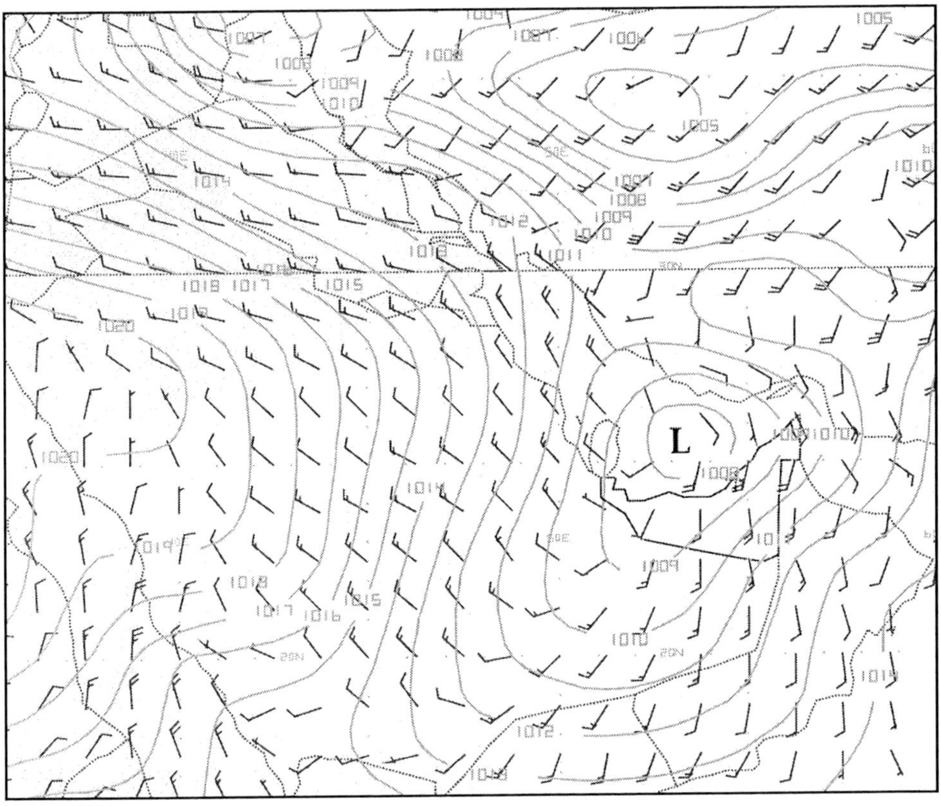

Figure 3. Example of a Kaus wind and winter Shamal dust event circulation. The GFS NWP model prognosis valid at T+30 hours 13 March 2003 0600 UTC (1000 LT) showing wind arrows and mean sea level hPa pressure (grey lines).

The Kaus wind is regarded as a moderate to gale-force southerly wind, accompanied by cloudy weather and rain that is associated with the passage of a winter depression (AMS 1989, Middleton 1986). However, in the south in the UAE it is a hot, dry and sometimes strong southerly, or south-easterly wind that occurs during the day and sometimes part of the night. It is most likely to produce dust storms when it occurs east of winter secondary lows and are most prevalent in this region during March and April (UAE Climate 1996). Referring to the example in Figure 3, a $10ms^{-1}$ Kaus wind is predicted over the UAE east of the secondary low. At ADIA on the 12th the visibility deteriorated to 1000m due to blowing dust in $8\text{-}10ms^{-1}$ Kaus winds off the desert to the south and on the 13th a 300-600m visibility dust storm lasted nearly 7 hours in 10-16 ms^{-1} Kaus winds. During this time the daylight air temperatures rose to 40-41°C with the relative humidity down to 5%. The Kaus wind also briefly produces dust storms when occasionally easterly troughs, or easterly waves move from east to west through the southern part of the Arabia Peninsula during summer (Taha, et al 1981).

During the winter months the strong anticyclone over Asia occasionally extends southward to Iran and Pakistan in the wake of a cold front, or tough and, over the UAE, Sultanate of Oman and Rub al Khali, produces "strong north-easterly cold and dry surges" (U.A.E. Climate 1996). This Nashi wind can be the bearer of dust storms. A typical example on 12 and 13 December 2003 (de Villiers and van Heerden 2011) can be seen in the model prognosis in Figure 4.

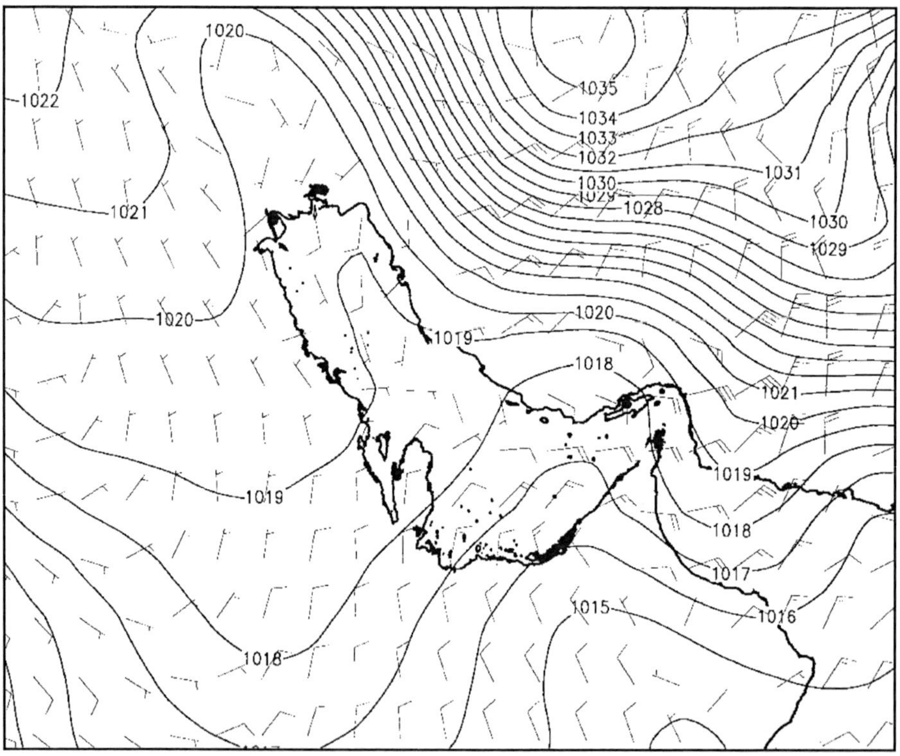

Figure 4. Example of a winter Nashi dust event circulation. The GFS NWP model prognosis valid at T+24 hours 12 December 2003 1200 UTC (1600 LT) showing wind arrows and mean sea level hPa pressure (grey lines).

The model predicted winds of $13ms^{-1}$ to $20ms^{-1}$ over dust source regions in south-eastern Iran reaching Oman and the UAE as $10ms^{-1}$ easterlies. On this day dust brought from source regions over south-eastern Iran reduced the visibility to 100m along the Gulf of Oman coast and 500m at Dubai in the UAE late in the evening. Early the following morning a considerably weakened system reduced the visibility to 1500m at ADIA.

In winter the passage of easterly migrating upper air troughs are the main producers of cloudy weather and rain. However, embedded thunderstorm line squalls develop in the south-westerly flow ahead of the trough (Taha, et al 1981) and although heavy and persistent rain can occur (Membery 1997) there are more likely to be strong wind gusts with temporary dust storms (Tantawy 1961). The dust storms form in a horizontal vortex at the leading edge of the horizontal outward rush of air, the outward flow being caused by the strong downdraft, or microburst out through the bottom of the thunderstorm cell (Fujita 1986). A particularly severe thunderstorm in a passing trough occurred at ADIA 1730 LT 18 March 2002 that caused structural damage in the city centre. The visibility fell to 1000m in blowing dust in a south-westerly gust front wind of $15\text{-}17ms^{-1}$ that gusted to $22ms^{-1}$ and lasted about 30 minutes. This was closely followed by 9.5 mm of downburst rain that fell in about 40 minutes, a significant amount in this part of the world. Incidentally, during the night and morning prior to the passing of the trough a $10\text{-}13ms^{-1}$ Kaus wind reduced the visibility to 800m in a dust storm, but the wind dropped and the visibility improved to 3000m prior to the thunderstorm.

In the pre-monsoon spring period of May and June and the post-monsoon autumn period of October and November, the Arabian Sea is invaded by tropical disturbances, some of which may intensify into tropical storms, but development into a full blown tropical cyclone is a rare event. The majority of tropical cyclones occur in the post-monsoon period, but their influence is usually limited to the southern-eastern part of the Arabian Peninsula, where apart from the rain they can, as a precursor to the rain, produce dust storms in the strong wind. The most frequent disturbances are much weaker easterly troughs, or easterly waves, moving to the west across the southern part of the Arabian Peninsula (Taha, et al 1981). The migration temporarily deepens the Arabian Peninsula heat low (Al-Brashdi 2007) and cause wind blown dust and dust storms from southerly Kaus winds (de Villiers and van Heerden 2007).

With respect to the depth of dust storms, Wilkerson (1991), referring to summer Shamal dust storms in Iraq and adjoining countries states that the average depth is 900-2400m, but, based on satellite derived temperatures, they can reach 4600-5500m, while post-frontal dust storms are between 2450-4600m and thunderstorm induced dust storms range from 1500-2400m. Safar (1985), commenting on dust storms in Kuwait says they reach about 3000m, but can reach 5500m.

5. Frequency and Duration

Discussion of the dust storm producing wind directions in section 4 draws attention to the fact that summer Shamal dust is caused by the Asian Monsoon low, while in the winter months the emphasis is on the southward shift of mid-latitude troughs, which affect mainly the northern part of the region, but can extend to the southern-most edge of the Arabian Peninsula (Taha, et al 1981) and associated secondary depressions that migrate eastward from the Mediterranean (Kendrew 1961). The greater persistence of the summer Shamal is supported by a study at Doha, in Qatar, It was established that 51% of Shamal days occurred in the summer months of May to July against 26% during the winter months of November to March (Rao et al 2001).

The seasonal difference, at five sites, is reflected in the average number of days with blowing, or suspended dust in Figure 5 and the percentage wind direction frequency in Figure 6. Data for four of the sites were obtained from Wilkerson (1991).

Data for Abu Dhabi were extracted by the author from ADIA observations 1994 to 2003, inclusive. There is a distinct increase to maxima during the summer months at Baghdad, Kuwait and Riyadh, which can be attributed to the summer Shamal, while the high percentage frequency of Kaus winds, in Figure 6, suggests that in winter the dominant feature is prefrontal dust brought from the desert of Saudi Arabia and to a lesser extent a postfrontal winter Shamal. Although, the winter Shamal is possibly dust free due to a dampened surface after the passing of frontal rain.

Abu Dhabi, to the east of Riyadh at about the same latitude south, has distinctly fewer days with blowing, or suspended, dust throughout the year (Figure 5). In summer this is due to it being, for the most part, under the trough from the Asian low to the eastern Arabian Peninsula low and therefore protected from the full force of the summer Shamal. Summer blowing dust and dust storms do occur when there is occasional deepening of the heat low due to easterly waves passing to the west over Rub al Khali. In winter Abu Dhabi is usually too far south to be greatly influenced by the troughs passing mainly to the north. Nevertheless, most of the dust events that do occur can be attributed to Kaus winds (Figure 6) ahead of the passage of troughs and secondary depressions. Incidentally, during the peak late winter months of February and March and the summer month August in the 1994-2003 ADIA data (Figure 5), 68% of the dust events were due to Kaus winds.

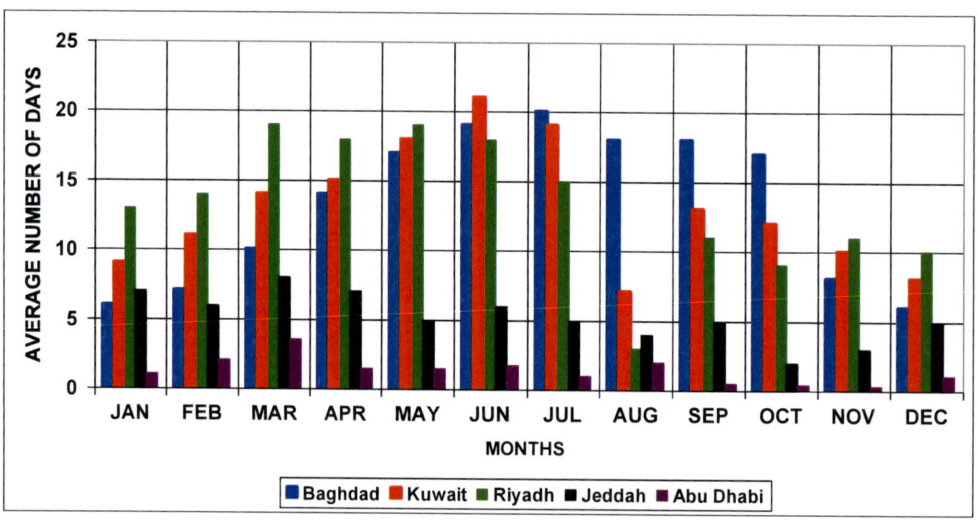

Figure 5. The average number of days with blowing, or suspended, dust at Baghdad, Kuwait, Riyadh and Jeddah of uncertain years, possibly 1973-1989 from Wilkerson (1991) and ADIA observations 1994 to 2003 (inclusive) when the visibility was ≤5000m.

Winter Shamals tend to be dust free due to frontal rain dampened conditions to the north. Less frequent are the cold Nashi winds from the north-east (Figure 6). Further confirmation of the difference in dust event frequency at ADIA in the south and Baghdad in the north can be seen by the monthly average number of days with dust storms in Figure 7.

In the west, Jeddah, on the Red Sea and slightly further south than Riyadh, not only experiences fewer dust events than in the north (figure 5), but it also has a distinct lack of Shamal winds reaching this far to the south-west and a distinct maximum of Kaus winds (Figure 6). As the extent of the Kaus wind extends to 180 degrees it does mask a variety of wind directions. These can vary from a summer south-westerly wind carrying dust from the Sudan to low pressure over the Arabian Peninsula, to an easterly wind carrying dust from Rub al Khali if there is lower pressure to the west over Egypt (Wilkerson 1991). The easterly wind tends to prevail in winter when there is high pressure to the north over north-western Saudi Arabia (as mentioned in the previous section).

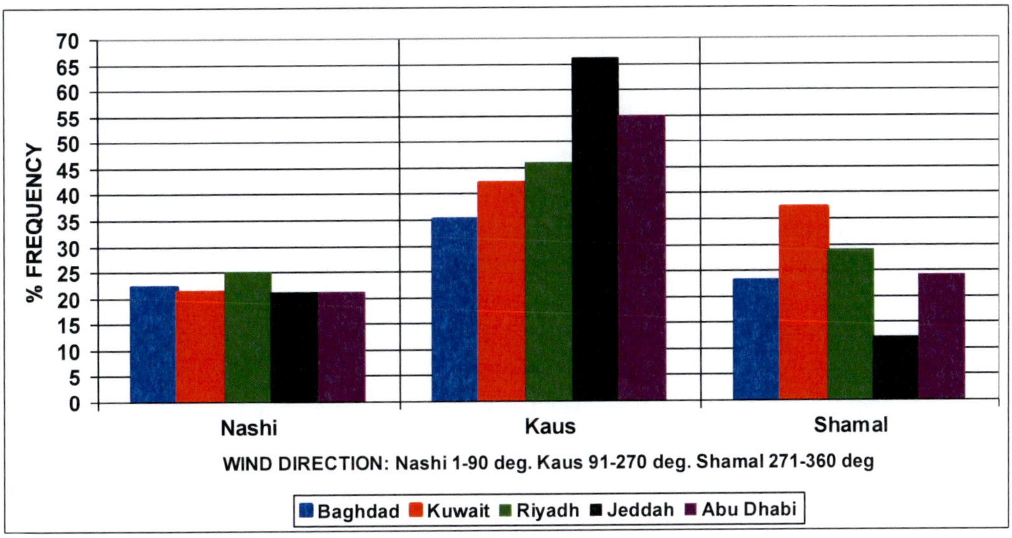

Figure 6. The percentage wind direction frequency of blowing, or suspended dust at Baghdad, Kuwait, Riyadh and Jeddah of uncertain years, possibly 1973-1989 from Wilkerson (1991) and ADIA observations 1994 to 2003 (inclusive).

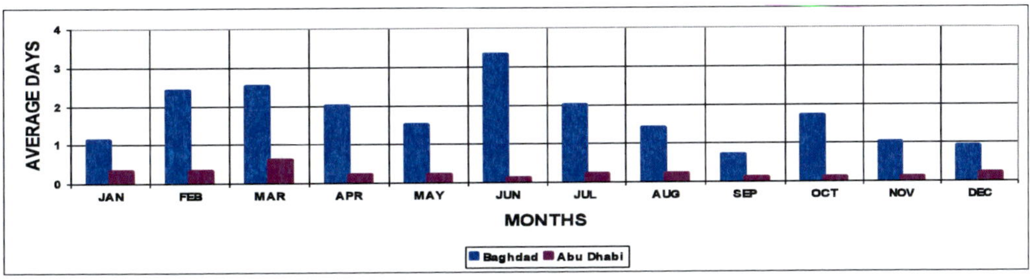

Figure 7. The monthly average number of days with dust storms at ADIA (purple) from observations 1982 to 2003 (inclusive) extracted by the author and Baghdad (blue, years unknown) from Awad in Wilkerson (1991).

Little data are available with respect to the cold Nashi wind and no documented research could be found by de Villiers and van Heerden (2007). Their study, describing a Nashi dust storm over the UAE and Oman, established that although Nashi wind dust events at ADIA occurred less often than Kaus events (Figure 6), the Nashi wind was more likely to deteriorate to a dust storm with the visibility <1000m than a Kaus wind, the percentage of dust storms being 31% and 13%, respectively. Figure 6 shows that the wind is similarly prevalent at the other 4 sites, although, at Jeddah, as mentioned in the previous paragraph, it is more likely to be a modified warm to hot easterly wind after passing across the desert.

The Shamal is also strongest and most frequent during the later part of the afternoon and early evening and least frequent overnight (Rao et al 2001) and the visibility is most likely to be at its worst during daylight hours when the wind is at its strongest (UKMO 1994). This diurnal variation is due to the earth's daily radiation balance. Increased heating during the day causes turbulent eddies, which reduce stability, but increases vertical mixing thereby drawing stronger winds above the ground to the surface. At night the ground cools and cools the air immediately above it by radiation, resulting in increased stability, which decouples the surface wind from the stronger wind aloft and due to friction becomes weaker. As a result dust is more readily lifted aloft by day and causes poorer visibility. By night the dust is able to settle which allows the visibility to improve (Kessler 1985).

Land and sea breezes are common throughout the year in the tropics (Miller 1966). This is especially true of the Arabian Gulf coast where land and sea breezes are virtually a daily occurrence. In a Kaus wind at ADIA reduction of visibility to below 5000m, was found to be most frequent during the morning when it was coincident with the land breeze. Conversely, during the late afternoon and into the early night the opposing sea breeze reduces the Kaus wind, more often than not to the point where it would cease. Dust is therefore able to settle and the visibility improves. Conversely, dust reduced visibility in a Shamal was most likely during the middle and late afternoon when it would be strengthened by the sea breeze and diurnal deepening of the heat low over Rub al Khali (de Villiers and van Heerden 2007, Rao et al 2003).

Dust in suspension due to the persistent synoptic situation, such as during the summer Shamal, can exist from one day to possibly more than a week, but usually about 3 days, whereas the winter post-frontal Shamal usually lasts1-1½ days as a trough moves through from west to east, but it can last 2-3 days when the front stalls (Wilkerson 1991 and Perrone 1979 in Wilkerson 1991). Much further to the south at ADIA, de Villiers and van Heerden (2007) state the longest duration of dust in suspension during their study was 21 hours. However, dust storms when the visibility is below 1000m are usually of much shorter duration. A press report in section 2 mentioned a dust storm in Kuwait that lasted as much as 2 days, but Safar states that they occur in Kuwait, on average, 27 days in the year, or a total of 128 hours, that is, a more likely average duration of 4.7 hours. De Villiers and van Heerden (2007) found that at ADIA the average was 1¼ hours and the longest 4½ hours. Dust storms associated with thunderstorms tend to be of shorter duration. Idso et al 1972 quote an average duration of 3 hours and a maximum of 6½ hours, while de Villiers and van Heerden (2007) found an average duration of 37 minutes and the longest nearly 60 minutes, but these figures are subjective, because they exclude visibility below 1000m when there was dust and rain mixed. The two can co-exist until the rain "washes" the dust from the atmosphere.

An important consideration is whether the surface is wet, or dry. Blowing dust usually does not occur within 24 hours of being properly wetted (Wilkerson 1991). In the north, in

Iraq for example, many of the dust source areas during winter and spring (November to April), are too wet due to the passage of rain bearing troughs and low pressure systems (Wilkerson 1991). To the south, in the much drier desert climate of ADIA with an annual mean of 84 mm precipitation, this is not so relevant. However, looking at the number of dust days with visibility below 5000m at ADIA, in Figure 5, a somewhat peak period of dust events in February and March coincides with the two highest rainfall months at ADIA. The explanation is that although rain does occur with the passage of toughs they more often do not produce rain, nor do the passing of secondary depressions

6. NWP Model Sand Storm Forecast Index

The preceding sections have already highlighted the dynamical and synoptic conditions that indicate potential dust storms and prolonged dust in suspension and attention has been drawn to the value of NWP models in section 4 and Figures 2 to 4. However, a tool that would be beneficial is a NWP model derived dust storm forecast index that combines the characteristics for dust storms and presents the index values on a map. A simple index would need to incorporate the following:

1. *Wind speed*: Strong winds are required throughout the boundary layer. Membery (1983) found that a nocturnal jet between 300-900m altitude was a good indicator of strong winds developing at the surface, although de Villiers (2011) found that the jet at Abu Dhabi during Shamal conditions was sometimes as high as 1200m. The wind speed to be used by the index programme could be averaged between the surface and 1200m. This would have the disadvantage that overnight lighter surface winds would decrease the average wind speed. Alternatively the wind at the mid-point of 600m could be used, or the programme could use the detected maximum wind speed between 300-1200m. Higher wind speed values being more favourable for dust storms.
2. *Deep well mixed boundary layer*: This implies a uniform wind speed and velocity throughout the boundary layer, that is, little or no wind shear. Marked wind shear inhibits the dispersion of sand and dust particles throughout the boundary layer. This can be determined by using the vector difference between the forecast wind at the top of the boundary layer and the surface. In NWP model terms, this could be the difference between 1000 and 850 hPa. The closer the two values, the lower the wind shear. Therefore, the lower the value indicated the better conditions are for development.
3. *Nearly adiabatic lapse rate*: An environmental lapse rate (ELR) that tends to the dry adiabatic lapse (DALR = ±1°C/300m) rate indicates weak static stability. That is, the air is more unstable. This facilitates upward motion of sand and dust particles. Apart from prevailing synoptic weather conditions, this is most likely in the early afternoon when surface heating is at its maximum and there is unlikely to be a boundary layer, temperature inversion. This can be determined by using the difference between the potential temperature at the top of a boundary layer and the surface, the suggested top being about 850 hPa, but preferably 700 hPa, to enable deeper upward

percolation of the dust. The closer the two values are to each other, the nearer the ELR is to the DALR. Therefore, the lower the value indicated the better conditions are for development.

These three values can be added together to provide a simple dimensionless index. But, there is a problem. A maximum wind speed is a favourable result while the others require a minimum value to be favourable. The solution would be to use the inverse of the wind to produce a minimum favourable result, or the inverse of the others to produce a maximum value. Such an index was used in a trial at ADIA during two years from 2003. The results were inconclusive due to the limited number of dust storm events.

Another parameter that could be incorporated is the NWP model predicted upward (negative) vertical velocity of the air in, for example, the surface to 850 hPa layer. The inverse value would need to be used to produce a maximum value.

The index does not take into consideration the all important dryness of the surface and its composition, or texture. Whether the surface is dry or not, needs to be determined according to locally derived criteria. Most of the Arabian Peninsula is almost perpetually dry, so soil moisture is not a problem. Similar knowledge of the suitability of the sand, or dust, is also necessary. The index would not be suitable for predicting dust storms due to thunderstorm downbursts.

7. Satellite Imagery

The advent of more sophisticated methods have made satellite imagery an increasingly important tool for the detection of blowing dust and dust storms, particularly at the development stage and then monitoring its spread and track with time. This was not always the case. When using early generation black and white visual and infrared images it was difficult to detect dust. On a visual image, dust lifted over a desert has the same texture and colouring as the surface, which makes it very difficult, if not impossible, to detect unless it is transported over a markedly different surface that has a much darker appearance in the image, such as water, or vegetation. Low sun angle complicates identification, because the dust in suspension creates a shadow and therefore resembles a cloud. Thermally, the lifted dust is at a temperature that is very little different from the surface and it is therefore difficult to detect using infrared imagery. Very good knowledge of the surface terrain is required to be able to have any chance of dust detection in areas devoid of surface observations to corroborate its presence. A method of detection is to increase the thermal sensitivity of the infrared image. Thermal images indicate a gradual change in temperature from hot to cold. By making the change more rapid, or steeper, the contrast is increased, thereby making detection easier. It is also important to know the surface temperature, either based on surface observations or seasonal differences (Wilkerson 1991).

Another method is to create a visual and infrared composite image. This type of image is effective for indicating dust over a land/water surface, but it is not usually readily available as it requires computer software to delete the infrared image over the water along the coastline and delete the visual image over land along the coastline, before joining them together as a single image (Lee 1989).

A more effective and recent innovation is to use previously unavailable "multiple narrow-band channels in the visible spectrum to enhance dust" (Miller 2003) and distinguish it from cloud whether over land or water. The principle spectral channels used are red (IR12.0–10.8), blue (10.8) and green (IR 10.8 to 8.7). Depression of the cloud signal in the red colour produces a cyan (blue/green mix) appearance, "while the enhanced dust component assumes shades of red and bright pink" (Miller 2003). Two examples of platforms that can be used to obtain dust enhanced images are the Moderate Resolution Imaging Spectroradiometer (MODIS) on the Earth Observing System (EOS) Terra and Aqua satellites and the EUMETSAT Meteosat Second Generation satellite. The EUMETSAT Spinning Enhanced Visible and InfraRed Imager (SEVIRI), observes the Earth's full disk every 15 minutes. Dust is a RGB (Red, Green and Blue) composite composed from data from a combination of the SEVIRI IR8.7, IR10.8 and IR12.0 channels. It is designed to monitor the evolution of dust storms during both day and night (EUMETSAT 2011). An example can be seen in Figure 8.

Figure 8. A dust storm (shown bright pink) over Kuwait, at the northern end of the Arabian Gulf observed by Meteosat/SEVIRI at 1500 UTC (1900 LT) 21 June 2011, with less intense pink colouring of blowing dust southward along the western coast to more marked pink blowing dust over south-eastern Saudi Arabia. Notice the swirl into low pressure over Rub al Khali (© EUMETSAT 2011).

Compare this figure with the NWP model surface wind and pressure prognosis in Figure 2. The "pink" dust in the image is being carried southward by the Shamal from Iraq, over Kuwait, inland of Bahrain and in a cyclonic swirl into the low pressure over Rub al Khali. At

Kuwait the wind was north-westerly at $12ms^{-1}$ gusting $17ms^{-1}$ with the visibility down to 1600-200m in blowing dust. Inland of Bahrain at Al Ahsa (Saudi Arabia) the wind was $8ms^{-1}$ with 2500m in blowing dust. Unfortunately, due to the absence of observations, the wind speed and visibility in the pink area in Rub al Khali cannot be verified. However, the image does provide verification of the model prognosis, as well as confirmation of blowing dust and a dust storm in the bright pink area at Kuwait.

8. Prediction Guidelines

Taking into consideration the preceding sections and the author's experience, a set of prediction guidelines are proposed. The NWP models give ample long term (>24 hours) warning of synoptic patterns that have the potential to generate blowing dust and/or dust storms.

- In summer, identify an increased surface pressure gradient between the semi-permanent high pressure over northern Saudi Arabia and the Asian low pressure. Dust raised in this manner in source areas in Iraq can be carried south to the heat low over Rub al Khali.
- In summer, seek evidence of deepening of the heat low over the Rub al Khali, such as due to an easterly wave moving to the west across the southern part of the Arabian Peninsula.
- In winter, identify mid-latitude upper air troughs that could pass through the region from the 700, 500 and 300 hPa wind and height prognoses. In particular those that indicate a surface low pressure cell below the south-westerly flow ahead of the trough and a following well developed surface high pressure.
- In winter, identify a secondary low that may be predicted to move from the Red Sea across Saudi Arabia to southern Iran. This can cause a Kaus wind as far south as the UAE and a following Shamal.
- Identify suitable surface wind directions and speeds $\geq 8\ ms^{-1}$ at dust source regions and $\geq 15ms^{-1}$ at 1000m.
- Is there little, or no, directional wind shear in the boundary layer?
- Instability. Look for an ELR that tends to the DALR.
- Is there upward (negative) vertical velocity up to 700 hPa.
- Are conditions favourable for thunderstorm development?
- Identify upper air high humidity areas and precipitation fields.

Precipitation will dampen the surface and prevent the lifting of dust.
Additionally, in the short term (<24 hours)

- Use satellite dust enhanced imagery to detect blowing dust and dust in suspension.
- Use atmospheric soundings to identify an ELR that equals the DALR (instability).

The top of the dust layer will be where the ELR becomes less than the DALR.

General rules:

- Monitor surface observations at source regions to determine the projected track based on NWP model wind velocity and the pressure pattern.
- Take into consideration the dryness of the surface dust/sand at source regions. Blowing dust usually does not occur within 24 hours of being properly wetted.
- Significant dust starts to be lifted when the surface wind $\geq 8ms^{-1}$ and begins to settle when it drops below this speed.
- On average, dust settles at 300m/hour.
- A nocturnal wind $\geq 15ms^{-1}$ at 300-900m is a good indicator of strong winds to come at the surface.
- Visibility improves with distance from the source region.
- Visibility improves at night when the wind becomes lighter.
- Within range of the land/sea breeze effect, blowing dust winds can be strengthened, or weakened depending on whether the effect is co-incident, or not.
- Summer Shamal borne dust can last for one day to possibly more than a week, but more usually about 3 days.
- Winter Shamal borne dust will last for a day to 1½ days, but can last for 2 to 3 days.
- Dust storms (visibility < 1000m) last < 2 hours, at the most < 5 hours.
- Thunderstorm generated dust storms (visibility <1000m) last about 30 minutes and not more than 60 minutes (at ADIA).
- In general, the wind speed and visibility relationship (at ADIA) in Table 1 below can be applied (de Villiers 2010).

Table 1. Wind speed and visibility relationship at Abu Dhabi

Wind speed	Visibility		
$>7.5ms^{-1}$	<8000m	often <5000m	
$>10ms^{-1}$	<5000m	often <2000m	most likely <1000m
$>12.5ms^{-1}$	<2000m		most likely <1000m
$>15ms^{-1}$	<1000m		

CONCLUSION

Dust lifted into the atmosphere contains more than people usually realise and carried further afield from source regions than realised. Harmful bacteria, viruses and fungi, as well as soil pollutants can also be present. The dust can cause respiratory problems and frequent exposure to blowing dust and dust storms can cause mental problems. So, apart from affecting human activities, it can also be a health hazard. The Arabian Peninsula is one of the important dust produces regions of the world. It is therefore important to know and understand when the conditions are ripe for blowing dust, or dust in suspension so that suitable precautions can be taken. This chapter has discussed the conditions that favour blowing dust and dust storms, the synoptic weather patterns that cause them and provided a suggested forecast methodology using NWP models and satellite imagery and suggested a dust storm forecast index. Satellite imagery as a monitoring and short term forecast tool has become particularly important in view of the availability of new generation enhanced imagery that allows dust in suspension to be identified by day, or night.

Dust situations fall into four causal groups. There is the frequent and persistent summer Shamal that occurs due to the deep monsoon low over Asia and higher pressure over northern Saudi Arabia and the less frequent winter Shamal that is associated with the passing of mid-latitude troughs. A secondary effect of passing troughs is blowing dust and storms that occur in the cold winter Nashi wind, which arrives from the north-east around the eastern edge of a strong anticyclone that can develop over Asia in the wake of the trough. Dust brought by the Kaus wind also occurs due to winter migrations of mid-latitude troughs and secondary lows, as well as deepening of the summer heat low when easterly waves pass through. In addition there is less frequent and random dust raised by thunderstorms. The winter Shamal and Kaus can be dustless, due to source regions having been previously dampened by rain bearing troughs. This is particularly true of the source areas in south-eastern Iraq from November to April. However, in the UAE the frequent Kaus wind is much more likely to be a dry and therefore dust laden wind, especially in association with secondary lows during March and April and summer deepening of the heat low. A rare occurrence is dust brought by pre and post monsoon tropical disturbances. The depth of the dust storms can vary from 900m to as much as 5000m.

Dust begins to be raised and visibility impaired when the wind exceeds about 7-8ms^{-1} and depends on surface soil composition, but research at ADIA shows that blowing dust with the visibility below 5000m occurs at 8-10ms^{-1} and dust storm proportions with the visibility below 1000m when the wind speed exceeds 10ms^{-1}. Dust tends to settle and the visibility improve overnight when nocturnal cooling results in increased boundary layer stability and the surface wind decreases. Research at ADIA showed that the frequent land and sea breezes in the region, when operating in tandem with the dust generating wind, have the effect of strengthening the wind, lifting more dust and decreasing the visibility.

The duration of dust events can vary from one day to more than a week with respect to the summer Shamal, but usually average about 3 days, whereas the winter dust Shamals last about 1½ days, but can be 2 to 3 days in a slow moving trough. At Abu Dhabi Shamal dust does do not last more than 2 days. It also has distinctly fewer Shamal dust days than in the north due to it often being under a trough that extends from Asia to the eastern Arabian Peninsula that keeps the Shamal at bay to the west. Dust storms, when the visibility is less than 1000m, can last about 2 days in Kuwait, but average 4.7 hours, while at Abu Dhabi the average is 1¼ hours and the longest 4½ hours. Dust storms due to thunderstorms average 3 hours with a maximum of 6½, but at Abu Dhabi the average was found to be 37 minutes and the maximum nearly one hour.

REFERENCES

Al-Brashdi, H. A. S. 2007. *Forecasting techniques over the Western Hajar Mountains in the Sultanate of Oman.* MSc dissertation. University of Pretoria.

American Meteorological Society (AMS). 1989. *Glossary of Meteorology.* Boston: American Meteorological Society.

Arabianbusiness.com. 2011a. Kuwait halts oil exports due to sandstorm; suspends flights. Quoting Bloomberg. Wednesday, 13 April 2011 12:37 PM. *http://www.arabianbusiness.com/kuwait-halts-oil-exports-due-sandstorm-suspends-flights-393650.html?service=printer&page*. Accessed 2011-06-19.

Arabianbusiness.com. 2011b. Kuwait resumes oil exports after sandstorm ends. Quoting Reuters, Thursday, 14 April 2011 4:19 PM. *http://www.arabianbusiness.com/kuwait-resumes-oil-exports-after-sandstorm-ends-393874.html*. Accessed 2011-06-19.

Arabianbusiness.com. 2011c. Kuwait halts oil exports due to sandstorm, KNPC spokesman says. Quoting Bloomberg, Monday, 4 April 2011 5:07 PM. *http://www.arabianbusiness.com/kuwait-halts-oil-exports-due-sandstorm-knpc-spokesman-says--391923.html*. Accessed 2011-06-19.

Arabianbusiness.com. 2011d. Kuwait shuts airport, halts oil activities in dust storm. Andy Sambridge, Saturday, 26 March 2011 10:10 AM. *http://www.arabianbusiness.com/kuwait-shuts-airport-halts-oil-activities-in-dust-storm-389992.html?service=printer&page*. Accessed 2011-06-19.

Arabianbusiness.com. 2009. Asthma cases follow sandstorm in Saudi – paper. Neeraj Gangal, Saturday, 1 August 2009 3:04 PM. *http://www.arabianbusiness.com/asthma-cases-follow-sandstorm-in-saudi-paper-15310.html*. Accessed 2011-06-19

ArabianBusiness.com. 2008a. Sandstorms making people mentally ill. Dylan Bowman, Sunday, 29 June 2008 5:41 PM. *http://www.arabianbusiness.com/sandstorms-making-people-mentally-ill-47486.html*. Accessed: 2011-06-19.

Arabianbusiness.com. 2008b. Severe sandstorms batter Gulf. Dylan Bowman, Saturday, 19 April 2008 1:26 PM. *http://www.arabianbusiness.com/severe-sandstorms-batter-gulf-187552.html?service=printer&page*. Accessed 2011-06-19.

Bagnold, R. A. 2005. *The Physics of Blown Sand and Desert Dunes*. Dover Publications, Inc.: Mineola, New York, pp 5-6.

Blair, T. A. 1957. *Weather elements*. Prentice Hall, Inc.: New York, pp 228 – 229.

Chepil, W. S.; Woodruff, N. P. 1957. Sedimentary characteristics of dust storms. II: Visibility and dust concentration. *American Journal of Science: New Haven,* vol 255, pp 104 – 114.

Coles, F. E. 1938. *Dust-Storms in Iraq, Meteorological Office, Professional Notes* No. 84, HMSO: London.

de Villiers, M.; van Heerden, J. 2011. Nashi dust storm over the United Arab Emirates. *Weather*. Vol 66, pp 79-81.

de Villiers, M. . 2010. *Predicting the development of weather phenomena that influence aviation at Abu Dhabi International Airport*. Thesis, University of Pretoria.

de Villiers, M. P.; van Heerden, J. 2007. Dust storms at Abu Dhabi International Airport. *Weather*. Vol 62, pp 339-343.

EUMETSAT. 2011. *http://oiswww.eumetsat.org/IPPS/html/MSG/RGB/index.htm*. accessed 2011-07-20).

Fujita, T. T. 1985: The Downburst, Microburst and Macroburst. Satellite and Mesometeorology Research Project. *Research Paper* 210 (NTIS PB-148880, February), Department of Geophysical Sciences, University of Chicago.

Galvin, J.; Black, I.; Priestley, D. 2011. Mesoscale weather features over the Mediterranean: Part 1. *Weather*. Vol 66, pp 72-78.

Gillette, D. A. 1979. Environmental Factors Affecting Dust Emission by Wind Erosion. *Saharan Dust, Mobilization, Transport, Deposition*, SCOPE.

Goudie, A. S. 1983. Dust storms in space and time. *Progress in Physical Geography: London*, Vol 7, pp 502 - 530.

Griffin, D. W.; Kellogg, C. A.; Garrison, V. H.; Shinn, E. A. 2002. The global transport of dust. *American Scientist*. Vol 90, pp 220 – 235.

Hoidale, G. B.; Hinds, B. D. 1975. *Boundary layer dust occurrence II: Atmospheric dust over the Middle East, Near East and North Africa*. Atmospheric Science Laboratory: New Mexico.

Hoock, D. 1984. A survey of available data characterizing the battlefield dust environment. *Battlefield Dust Environment, Symposium I, US Corps of Engineers,* Washington, DC, December.

Idso, S. B. 1976. Dust storms. *Scientific American*. Vol 235, pp 108-114.

Idso, S. B.; Ingram, R. S.; Pritchard, J. M. 1972. An American Haboob. *Bulletin American Meteorological Society*. Vol 53, pp 930-935.

Imshenetsky, A. A.; Lysenko, S. V.; Kasakov, G. A.; Ramkova, N. V. 1977. Resistance of stratospheric and mesospheric micro-organisms to extreme factors. Life Science *Space Research*. Vol 15, pp 37-39.

Jauregui, E. 1989. The dust storms of Mexico City. *International Journal Climatology*. Vol 9, pp 169–180.

Kendrew, W. G. 1961. *The climate of the continents*. Oxford University Press, New York, 5th edition.

Kessler, E. 1985. Severe weather. *Handbook of applied Meteorology*. Ed: Houghton, D. New York: John Wiley & Sons.

Lee, T. F. 1989. Dust tracking using composite visible/IR images: A case study. Weather and Forecasting. *American Meteorological Society*. Vol 4, pp 258-263.

Lawson, T. J. 1971."Haboob Structure at Khartoum," *Weather*. Vol 26, pp 105-112.

Membery, D. 1997. Unusually wet weather across Arabia. *Weather*. Vol 52, pp 166- 174.

Membery, D. A. 1983. Low level wind profiles during the Gulf Shamal. *Weather*. Vol38, pp 18-24.

Middleton, N. J. 1986. Dust Storms in the Middle East. *Journal Arid Environments*, Vol 10, pp. 83-96.

Miller, S. D.; Kuciauskas, A. P.; Liu, M.; Ji, Q.; Reid, J. S.; Breed, D. W.; Walker, A. L.; Mandoos, A. A. 2008. Haboob dust storms of the southern Arabian Peninsula, *Journal Geophysical Research – Atmospheres*. Vol 113, pp 26.

Miller, S. D. 2003. An improved land/ocean dust enhancement applicable to MODIS. *American eteorological Society 12th conference on satellite meteorology and oceanography*. Paper 3.2. 9-13 February.

Morales, C. 1979. The Use of Meteorological Observations for Studies of the Mobilization, Transport, and Deposition of Saharan Soil Dust, *Saharan Dust, Mobilization, Transport, Deposition, SCOPE*, 1979.

O'Hara, S.; Wiggs, G. F. S.; Mamedov, B.; Davidson, G.; Hubbard, R. B. 2000. Exposure to airborne dust contaminated with pesticide in the Aral Sa region. *The Lancet*. Vol 355, pp 627-628.

Perkins, S. 2001. Dust, the thermostat. *Science News*. Vol 160, pp 200-201, September 29.

Perrone, T. J. 1979. Winter Shamal in the Persian Gulf. *NAVENVPREDRSCHFAC Technical Report* TR 79-06, Monterey, CA, August.

Perry, K. D.; Cahill, T. A.; Eldred, R. A.; Dutcher, D. A.; Gill, T. E. 1997. Long-range transport of North African dust to the eastern United States. *Journal Geophysical Research*. Vol 102, pp 11225-11238.

Pope, C. A. III. 2000. Epidemiology of fine particulate air pollution and human health: Biologic mechanisms and who's at risk? *Environmental Health Perspectives*. Vol 108 (supplement 4), pp 713-723.

Prospero, J. M. 2006. Saharan dust impacts and climate change. *Oceanography*. Vol 19, pp 60-61.

Prospero, J. M. 1999. Long-term measurements of the transport of African mineral dust to the south-eastern United States: Implications for regional air quality. *Journal Geophysical Research*. Vol 104, pp 15917-15927.

Rao, P. G.; Hatwar, H. R.; AI-Sulaiti, M. H.; AI-Mulla, A. H. 2003. Summer shamals over the Arabian Gulf. *Weather*. Vol 58, pp 472-477.

Rao, P. G., Al-Sulaiti, M. H., Al-Mulla, A. H. 2001. Winter shamals in Quatar, Arabian Gulf. *Weather*. Vol 56, pp 444 - 451.

Safar. M. I. 1985. *Dust and dust storms in Kuwait*. Directorate of General Civil Aviation, Meteorological Department. State of Kuwait, pp. 15.

Taha, M. F.; Harb, S. A.; Nagib, M. K.; Tantawy, A. H. 1981. *The Climate of the Near East*. World Survey of Climatology. Chief editor: Landsberg, H. E. Amsterdam, Oxford, New York: Elsevier Scientific Publishing Company, pp 183-192, 215-228, 253-255.

Tantawy, A. H. I. 1961. The role of the jet stream in the formation of desert depressions in the Middle East. In High level forecasting for turbine engined aircraft operations over Africa and the Middle East. *World Meteorological Organization. Technical Note*, No 64, pp 159-171.

Thesiger, W. 1994. *Arabian Sands*. Motivate Publishing: Dubai, pp 241, 280.

Trapp, J. M.; Millero, F. J.; Prospero, J. M. 2010. Temporal variability of the elemental composition of African dust measured in trade wind aerosols at Barbados and Miami. *Aerosol Chemistry and Impacts on the Ocean: Elsevier*. Vol 120, Issues 1-4, pp 71-92.

U.A.E. Climate. 1996. First Edition. *Ministry of communications*. Abu Dhabi: Cultural Foundation Publications.

United Kingdom Meteorological Office (UKMO). 1994. Handbook of Aviation Meteorology. *Her Majesty's Stationary Office: London*.

United Kingdom Meteorological Office (UKMO). 1991. Meteorological Glossary, 6th edition. *Her Majesty's Stationary Office: London*.

Wheaton, E. E.; Chakravarti, A. K. 1990. Dust storms in the Canadian Prairies. *International Journal Climatology*. Vol 10, pp 829 – 837.

Wilkening, K. E.; Barrie, L. A.; Engle, M. 2000. Trans-Pacific air pollution. *Science*. Vol. 290, No. 5489, pp 65-67.

Wilkerson, W. D. 1991. Dust and sand forecasting in Iraq and adjoining countries. AWS/TN-91/001. Air weather Service, Scott Air Force Base, Illinois. *AWS/USAFETAC publications*.

World Meteorological Organization (WMO). 1983. Meteorological Aspects of Certain Processes Affecting Soil Degradation – Especially Erosion. Technical Note No.178. WMO No. 591. *World Meteorological Organization: Geneva*.

In: Dust: Sources, Environmental Concerns and Control
Editors: Laurent B. Wouters and Michel Pauwels
ISBN: 978-1-61942-547-7

Chapter 5

Sources of Dust: Their Control and Hazard in Kuwait

Adeeba Al-Hurban[*]
Earth & Environmental Sciences Dept.,
Faculty of Science, Kuwait University, State of Kuwait

Abstract

Dust and dust storms are always associated with widespread consequences, hazards and complications on the environment, public health, soil quality, economy … etc. Rates of dust fallout deposition on multiple large scale areas throughout the world significantly indicates the adverse effect of dust storms on the deposition of soil particles in arid and semi-arid areas. Soil erosion, sediment suspension and sediment deposition are considered to be the major processes that are involved in a dust storm and could contribute in a diversity of environmental hazards. Soil quality, particularly its fertility, can be directly affected by soil erosion due to the deterioration of its nutrient contents. Deflation of clay and silt particles by the wind action is the major contributor to the high sand content of soil. Suspended particles during a dust storm directly affect the atmospheric environmental quality and greatly reduce the visibility in some severe cases to less than 0.5meter, which results in traffic accidents and their consequent complications. Sediment deposition and sand approach results in considerable economics losses in the urban areas, different utilities' plants, field crops and cultivated areas, and disruption of communications. The factors playing a significant role in dust storms and dust fallout could be the environment type, geomorphology and relief differences.

1. Introduction

Duststorms are natural events by variant factors of which the most significant is severe and widespread drought in the dry land areas, which, in the extreme, takes place during summer months. Duststorms directly affect the air quality by reducing it, as well as the

[*] Tel: 0096524985951 / 7143; Fax: 0096524987143; E-mail: q8geo@hotmail.com or Adeeba@kuc01.kuniv.edu.kw

visibility extent and the major hazards it has on the public health, especially with individuals suffering from breathing-related problems.

A duststorm is a meteorological phenomenon which commonly takes place in arid and semi-arid regions. This phenomenon occurs when these areas are subjected to strong winds at the leading edge of a cold front allowing small, light dust particles to be blown into the air and transported by saltation or suspension, forming duststorms or sandstorms. These sandstorms can rapidly reshape the appearance of the subjected area through the shifting and re-forming of dunes by wind movement and soil erosion from one place and deposition in another.

The Arabian peninsula and its surroundings in general, and its northeastern part in particular, is considered one of the major terrestrial sources of moving sand in the world, with minor contributions from Iran, Pakistan and India into the Arabian sea, and China's important storms deposit dust in the Pacific. Different names are used in different areas to describe dust bearing winds. It is known as Khamasine in Egypt, Ghibli in Libya, Haboob in Sudan, Sahel in Morocco, Chili in Tunisia, Loo in India and Shamal or Toze in the Gulf region (Olbruck, 1973).

The State of Kuwait as part of the Arabian Peninsula, is considered to be one of the major regions of dust and duststorms in the world. These phenomena are common weather events in the arid and semi-arid regions because much of the surface is covered by loose, very fine, mobile sediments. According to Thomas (1989), in non-mountainous arid lands, a large proportion (up to 40% in the Sahara and 50% in Arabia) are composed of depositional landforms with wind as an important geomorphological agent. The atmospheric dust is composed of foreign moving particles of minute size that are carried by wind. Fine dust is smaller than 0.001 mm in size, which is so light as to remain within calm air as suspended particles. Coarse dust may be larger than 0.005 mm, which is raised from the ground by wind-storms and can be carried for thousands of miles.

It is worth mentioning that poor management of the dry lands significantly contributed in increasing duststorms from desert margins and changing both the local and global climate and even affecting the local economies.

2. Dust and Duststorms: Definitions and Types

2.1. Definitions

Duststorms and sandstorms have been defined in various ways. The term *"sandstorm"* is usually used when approaching desert sandstorms in general, and the Sahara desert in particular, as sand is the predominant particle size than dirt or rock and larger sand particles are blown closer to the surface. But the term *"duststorm"* is often used when finer airborne particles are blown long distances, particularly in cases of the urban areas affected by the blowing duststorm.

The Weather Service Convention defines the duststorm at the time of observation only if the prevailing visibility at the station is less than 1 km (Nickling, 1984; and Nasralla, 1986). Orgill and Sehmel (1976) employed a visibility limit of less than 11.3 km, while Oliver (1945) utilized a visibility limit of 700 m. Goudie (1978), reported the annual frequency of duststorms, reducing visibility to less than 1000 m in different locations in the world. The highest frequency (63%) was reported for Mexico City (Jauregul, 1973) and (60%) for the Kazakhstan (Lydolph, 1977). The area of southern Iraq and Kuwait comes next, where duststorms range in frequency between 27% and 37% days/year. Nasralla (1986) defined the duststorm as blowing dust or sand events associated with visibility less than 11.3 km.

As an example, these phenomena occur almost through all seasons in the State of Kuwait. They are more frequent during the spring and summer. This is mainly due to firstly: the dry fresh (15-24 m/s) northwesterly winds blowing from Iraq and local lands (Safar, 1984). Secondly: Kuwait is surrounded by Iraqi desert from the north and the northwest and the desert of Saudi Arabia from the west and South. Thirdly: most of the surface of the area is covered by loose sediments (Khalaf *et. al.*, 1984).

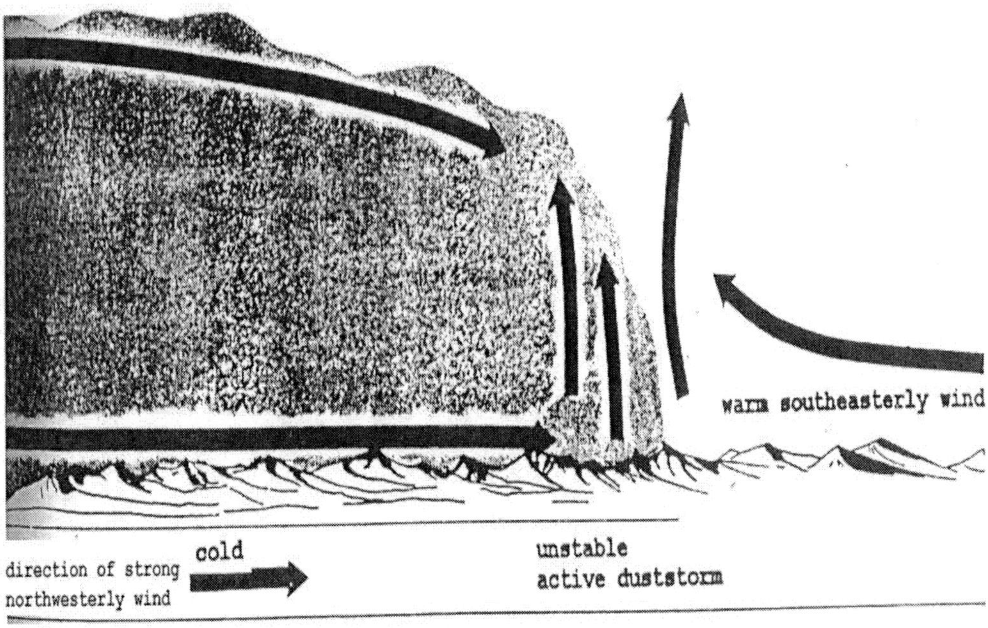

Figure 1. A sketch of the formation of a duststorm (Modified after Al-Kulaib, 1985).

2.2. Dust Types

There are three main different types of dust in Kuwait if taken as an example of arid and semi-arid regions suffering frequently from duststorms phenomena. These three types are namely: duststorms, rising and suspended dust. Dust may take one of these forms depending on the density, particle size, effect on visibility and wind speed. Duststorms and suspended dust are the most important types of dust weather phenomena, within these two types, the visibility may reduce to less than 200 meters.

There are relationships between the wind speed and direction on one hand and the formation of duststorms on the other hand. From May to August, approximately 40% of dust movement is associated with wind speeds between 10 to 12 m/s and in April about 33% is associated with wind speed between 7.5 to 9.5 m/s. From May to August, up to 78% of the dust is associated with northwesterly winds. During June and July, up to 91% of the dust is associated with wind blowing from the northwest. During April, 44% of dust is associated with wind blowing from the northeast, 29% from the southeast and 13% from the northeast. In terms of sediment budget, as much as 1680 tons of dust may be deposited in one year on one square kilometer of land. This will have an impact on the movement of the coarser fractions.

Bagnold (1941) stated that dust particles have a diameter of less than 0.08 mm. Others define them according to the silt/sand boundary (less than 0.0625 mm). Another subdivision can be made according to the distance that the particles were transported: dust carried from few kilometers to less than 100 km is of diameter between 0.005-0.05 mm. According to Jackson, et. al., (1973), material that can be transported over greater distances has a diameter between 0.002-0.01 mm. Such material can form suspended dust and be kept suspended in the troposphere as an aerosol for many days and sometimes more than a week.

Duststorms Rising dust

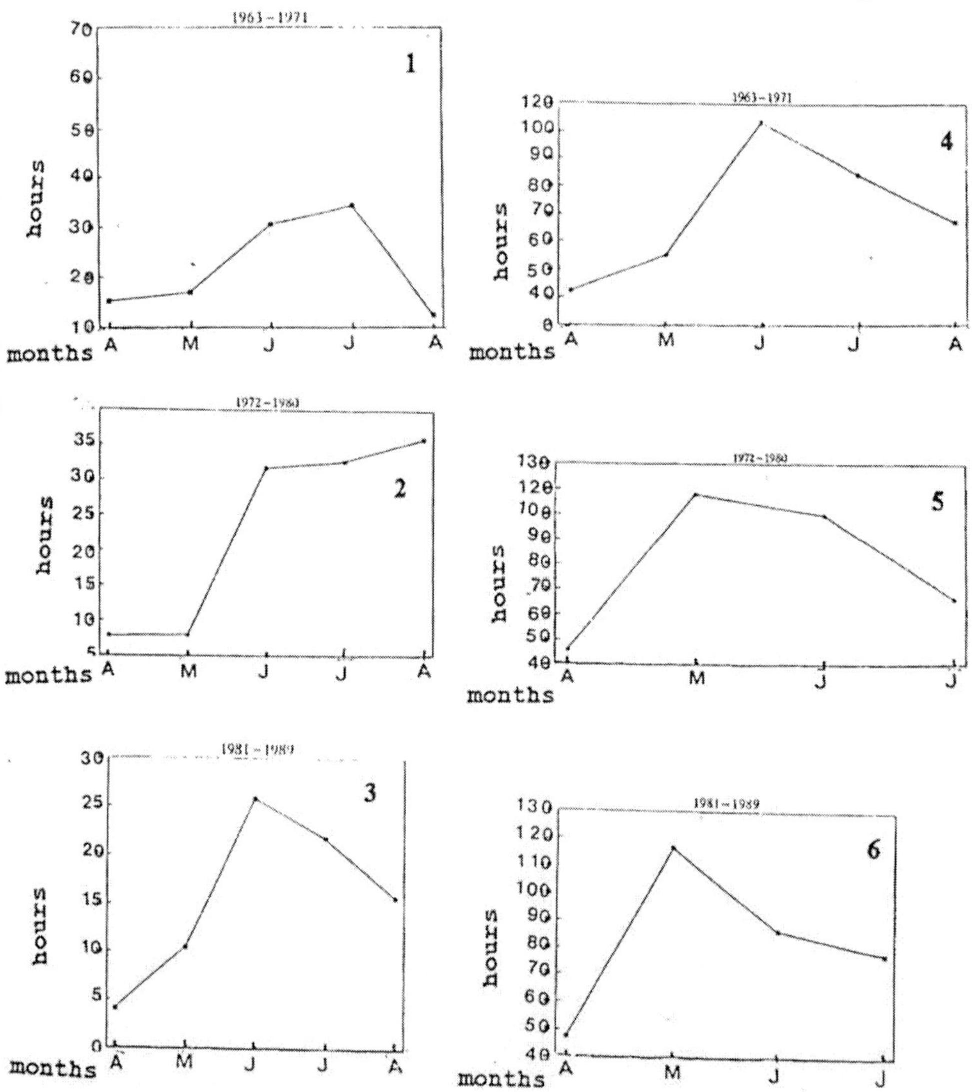

Suspended dust

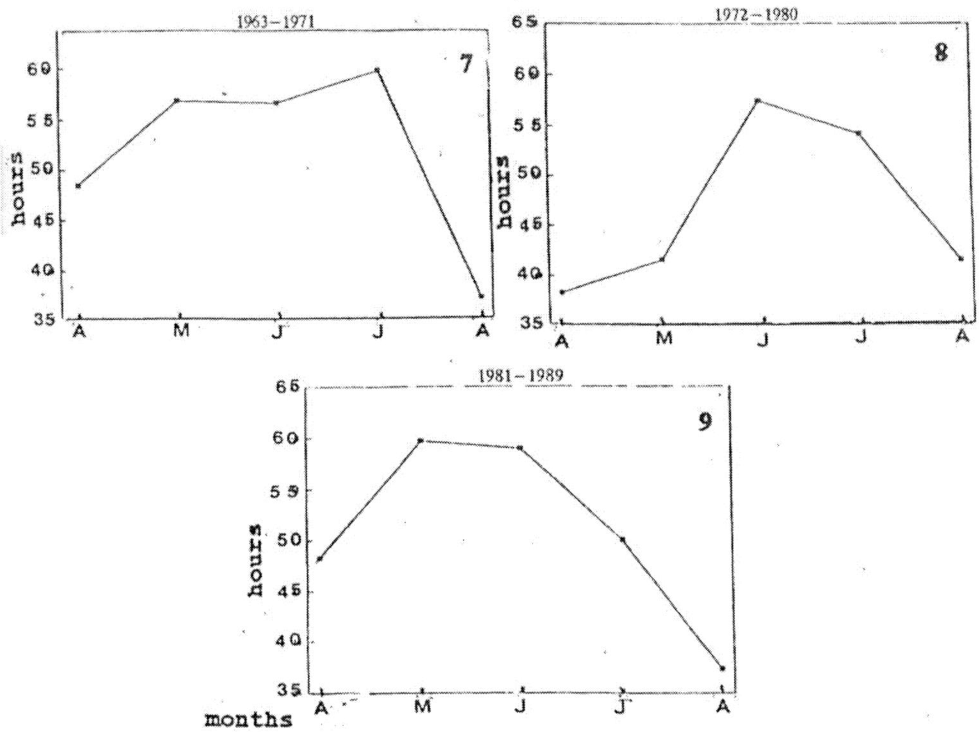

Figure 2. The mean number of hours of duststorms, rising dust and suspended dust for a period: per month 1963-1989 (After Al-Ostad, 1994).

The atmospheric dust may take one of various forms depending on the density, particle size, and wind speed. Rising and suspended dust occurs when the wind speed exceeds (5.78-8.44 m/s). Duststorms occur when the wind speed approaches storm level (14.22-16.89 m/s). During the movement of the westerly low pressure systems in spring and early summer seasons, occasional dust weather occurs during unstable situations such as thunderstorms and the passage of cold fronts.

Wind updraft ahead of the cold front, with fresh warm and humid southeasterly winds coming from the Gulf, and strong to fresh cold northwesterly winds behind the cold fronts tends to bring unstable situations and can increase dust fallout (Figure 1).

Al-Ostad (1994, 1998) has studied the hourly rate of the three types of dust in the State of Kuwait for 27 years (1963-1989) during the spring and summer (Figure 2). From Table 1 it can be noticed that duststorms during the months of April to August of the first period (1963-1971) averaged a total of 110.6 hr. per year, while those during the second period (1972-1980) averaged 116.5 hr. per year and values for the third period (1981-1989) averaged 78.5 hr. per year. This means that dust storm hours increased during the second period (1972-1980) and decreased during the third period (1981-1989). This decline can be explained partially by lower wind speeds and partially by lack of available dust.

Turning next to rising dust, from Table 1, the number of rising dust hours during the first period (1963-1971) reached 403.4 hr. while those in the second period (1972-1980) reached

365.3 hr. and in the third period (1981-1989) 357.1 hr. In other words, the total number of hours of rising dust decreased during the entire period.

As indicated from Table 1, the total hours of suspended dust during the first period (1963-1971) reached 212.5 hr., while during the second period (1972-1980) and the third period (1981-1989) they reached 232.8 hr. and 254.3 hr., respectively.

This means that suspended dust increased during the third period (1981-1989), while duststorm and rising dust hours decreased during the last period (1981-1989). This data suggests that duststorms and rising dust are decreasing, while suspended dust is increasing.

This data indicates that the months of June and July have the most dust activity. This was also shown by Safar (1985) who studied the percentages of hours of the three types of dust for the period of 1962-1984 in Kuwait. He found that the average percentages of hours per month equaled 13% of the time. This percentage increases to 27% in June and 24% in July. November and December experience the least dust activity with only a 5% incidence. Safar (1985) also studied the frequency of dust with wind speed and direction during the daytime in the summer for a period of 10 years in Kuwait. He concluded that 66% of the dust during May to August was accompanied by wind speeds in the range 7.6 to 12.4 m/s, and 78% of the dust in May to August was associated with wind blowing from the northwesterly direction. During June and July, 89% of the dust was blowing from the northwesterly direction. Although experiments have shown the wind speed necessary to lift dust, this does not mean that the wind speed and wind direction are the only factors controlling a duststorms occurrence. There are other factors, such as, the nature of the land surface and convective currents.

Table 1. The mean number of hours of duststorms, rising dust a suspended dust. Period: 1963-1989. After Al-Ostad, 1994

Duststorms						
	Month					Total (hr.)
Period	A	M	J	J	A	
1963-1971	15.3	17.1	30.8	34.7	12.7	110.6
1972-1980	7.9	8.0	31.8	32.8	36.0	116.5
1981-1989	4.1	10.5	25.9	21.9	15.8	78.5
Rising dust						
	Month					
Period	A	M	J	J	A	Total (hr.)
1963-1971	52.3	65.4	113.9	94.6	77.3	403.4
1972-1980	45.1	45.9	108.0	99.8	66.5	356.3
1981-1989	28.0	47.6	117.0	86.8	77.7	357.1
Suspended dust						
	Month					Total (hr.)
Period	A	M	J	J	A	
1963-1971	39.0	47.8	47.6	50.9	27.2	212.5
1972-1980	38.3	41.6	57.4	54.1	41.4	232.8
1981-1989	48.3	59.7	58.9	50.1	37.3	254.3

Table 2. Mean monthly dustfall in Kuwait (gm/m^2). Source: Ministry of Public Health, Environmental Protection Department, Annual Reports, Kuwait

Month \ Year	1981	1982	1983
January	22.2	48.0	46.6
February	18.5	75.9	35.4
March	14.2	68.7	69.3
April	26.6	86.5	48.9
May	15.4	151.3	127.9
June	29.2	228.7	197.8
July	57.6	187.2	239.9
August	84.9	457.0	201.4
September	54.7	92.7	65.0
October	45.9	149.9	52.3
November	24.7	55.7	58.2
December	38.0	135.0	54.1
Mean	36.0	144.7	100.0

The significance of dust movement can be seen in data concerning the amount of dust fallout. Table 2 shows mean monthly dustfall in Kuwait based on observations by the Meteorological Department of Kuwait International Airport and other observation sites of dust fallout in Kuwait, which are being fed to the Ministry of Public Health, Environmental Protection Department. The yearly pattern is generally consistent but there are fluctuations from year to year and December of 1982 was clearly out of line with the other data. The important fact is that quite considerable amounts of dust are being transported and deposited.

2.2.1. Duststorms

Duststorms are accompanied by fresh or strong winds (about 18 knots (33.336 km/h) or more). The horizontal visibility is less than one km but when the visibility is deteriorated below 200 meters, the storm is called a severe duststorm.

2.2.2. Rising Dust

Winds are generally moderate. Horizontal visibility is equal to or more than one km.

2.2.3. Suspended Dust

With suspended dust, horizontal visibility is less than 1 km, but when the wind speed is moderate (6-14 m/s), visibility may range between 1 km and 5 km due to dust not raised by local winds.

3. The Main Processes of Dust-Fallout

The main processes that take place during duststorms are: soil erosion, transport and deposition, as well as, dust abrasion. It has been suggested that flutings on ventifacts and some yardangs are the result of abrasion by fine dust. Soil erosion has a direct impact on the soil quality, especially its fertility as a result of the deterioration of soil nutrient contents. This phenomenon is usually responsible for the relatively high sand content of soil due to the deflation of clay and silt particles. The second process is soil deposition. Several areas in the world indicate the significant effect of duststorms on the deposition of soil particles in arid and semi-arid areas. Morales (1977) has reviewed the north African climate with particular emphasis on the production of dust formation and transport in the Sahara. He estimated that transportation may consist annually of $60x10^{12}$ gm to $200x10^{12}$ gm of dust. Rates of deposition in individual duststorms expressed in tons of sediments per square kilometer can reach as high as 500 gm/m^2 (Idaho (USA)), 455 gm/m^2 (north Africa), 300 gm/m^2 (Nebraska (USA)), 162 gm/m^2 (Colorado (USA)) and 126 gm/m^2 (Caspian Sea area) on desert margins (Goudie, 1978). The third process is sediment suspension, which represents the transportation phase and is probably the most familiar of all the stages. It is the phase that is seen by the public and reported by various organizations and disciplines.

3.1. Causes of Duststorms

As the wind blows with increasing force on loosely held particles, they, in the first place, begin to vibrate then saltate, (i.e. leep), and hence start to repeatedly strike the ground, loosen and break into smaller particles of airborne particles starting to travel in suspension. As the wind speed increases, aggregation of dust grains will take place and start moving by a range of mechanisms: suspension, saltation and creep.

Recently, it was found that as the sand particles start saltating, they result in a static electric by friction. This acquires a negative charge in relation to the ground, which results in loosening more sand particle which undergo saltating, by which the previously predicted number of particle is doubled.

It is known that drought, arid conditions, and wind nature induce particles to be loosely held. An intense thunderstorm results in an outflow of rain-cooled air, which, in turn, produces a gust front that may represent a dry cold front moving into a dry air mass with no precipitation, which caused the Dustbowl years or Dirty Thirties in the US (Figure 3(a-d)) (1930-1936 and in some areas, up to 1940).

As the dry cold front passes over heated ground, a convective instability is produced keeping the duststorm initiated at the front. In desert areas, such dust and sand storms are usually produced either by thunderstorms outflows or by strong pressure gradients that increase the wind velocity over a wide area. The weight of the airborne particles and the stability of the atmosphere above the ground play a significant role in determining the vertical extent of the raised dust or sand particles.

Not only do the drought and wind contribute to the occurrence of duststorms, but also poor farming and grazing practices play similar roles by exposing the dust and sand to the blowing wind. The practices of the farmers in dry land farming are considerably regarded as a major cause of duststorms. Such practices involve relying on rainfall to water the crops and in order to maintain the moisture in the soil, the farmer leaves the field uncultivated for a year upon harvesting for water buildup in the soil and covers the field with dry earth in order to prevent leaching in the underlying soil. These practices will directly affect dry land agriculture and enforce the susceptibility of the land to duststorms (Figure 4).

The problem of dust and duststorms has been investigated by many researchers and writers all over the world. According to Al-Kulaib's study on Kuwait (1984-1985), winter duststorms are due to the eastward passage of the western depressions.

(a)

(b)

Figure 3. (Continued).

(c)

(d)

Figure 3(a-d). Dustbowl years or the Dirty Thirties in the US (1930-1936 and in some areas, up to 1940). (a) A farmer and his two sons during a duststorm in Cimarron County, Oklahoma, 1936. Photo: Arthur Rothstein. (b) A duststorm approaches Stratford, Texas, in 1935. (c) A duststorm; Spearman, Texas, on April 14,1935. (d) Buried machinery in a barn lot; Dallas, South Dakota, May 1936.

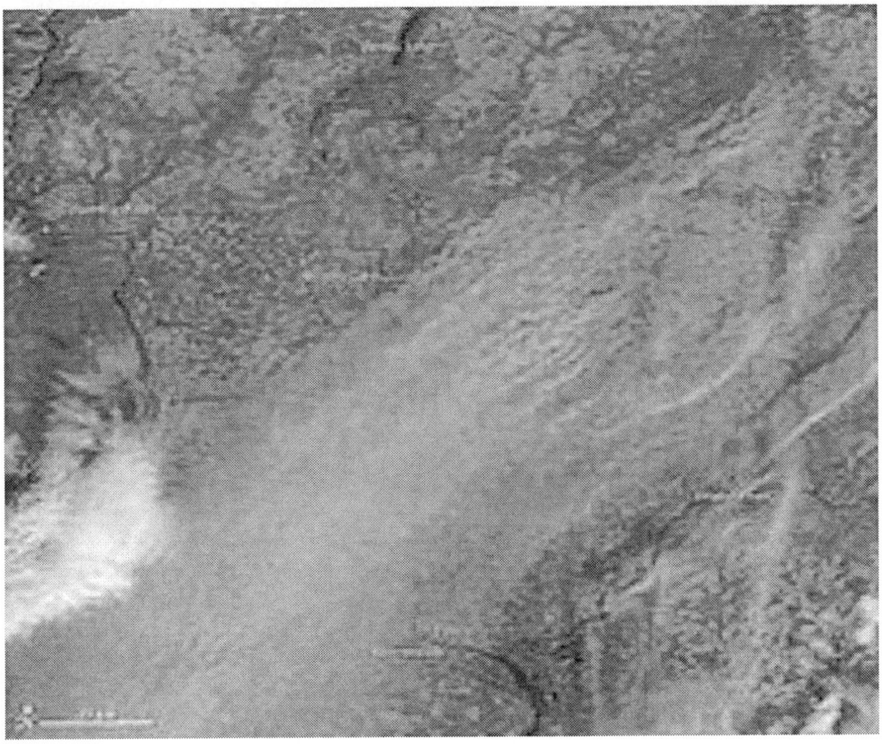

Figure 4. Dry land farming caused a large duststorm in parts of eastern Washington on October 4, 2009. Courtesy: NASA/GSFC, MODIS Rapid Response.

The average number of duststorms per month is about one to two storms. Rising dust ranges from two days in December to four days in February and suspended dust ranges from five days in January to eight days in February. During the spring (March-May), duststorms are developed due to thunderstorms which are produced by active cold fronts. This refers to winter depressions, which cross the country giving stormy weather especially during March, April and May. During late May, duststorms are due to the prevailing northwesterly wind, which is locally called "Shamal"; a result of the active Indian Monsoon low pressure system. The average number of duststorms per month is three to four days, rising dust occurs on about five to six days and suspended dust from 11 days during March and April to 14 days in May.

In the summer (June-September), dust phenomena reaches its peak, particularly during June and July. This is due to the strong wind prevailing from the northwest. This wind is produced by the effects of the Indian Monsoon to the east and Mediterranean ridge of high pressure to the west.

During dusty days, the visibility may decrease to about 500 meters. The average number of duststorms per month during this season, ranges from 5 days in June and July to 2 days in August and to about one day in September.

Rising dust occurs on about 8 days in June and July, 6 days in August and 3 days in September. Suspended dust occurs on about 11 days in June and July and 12 days in August and September. Autumn (October-November) is the quietest season of the year. The average number of duststorms ranges from one day in October to 3 days in November while rising dust occurs from 2 days in October to one day in November and the suspended dust occurs on about 10 days in October and about 6 days in November.

Figure 5. Vehicle use is one of the human activities that destabilizes the land and generates dust.

4. Sources of Duststorms

Factors controlling the occurrence of duststorms included: availability and nature of source sediments, vegetation cover density, prevailing climatic conditions, and the textural characteristics of the surface deposits.

Figure 6. A satellite image showing glacial river dust.

In addition to volcanic eruption and pollution, soil dust from arid and dry regions lifted by prevailing high velocity winds, are regarded among the various sources for the silt-sized particles in the atmosphere of which dust is consisted. Such processes would deflate susceptible surfaces including areas where grazing, ploughing, vehicle use (Figure 5) and other human activities are further destabilizing the land. Dust also exists in local environments, such as homes, offices and other human environments. It consists of differently sourced materials comprising small amounts of plant pollens, human and animal hairs, textile fibers, paper fibers, and minerals from outdoor soil, human skin cells, and many other materials which may be found in the local environment.

Dust is also windblown from glaciers (Figure 6), dry river banks, exposed riverbeds or shorelines (Figure 7), as well as our dirt roads and traffic (vehicle exhausts four wheelers, trucks, cars) (Figure 8), trampled ground, un-vegetated lots (Figure 9), unpaved airfields (Figure 10), paved roads or potholes, gravel pits and stockpiles (Figure 11), construction and mining (Figure 12), agriculture, and vacant lands are major producers of dust in the air of our communities.

Atmospheric dust is produced by saltation and sandblasting of sand-sized grains, which is then transported through the troposphere. Such wind-lifted dust is regarded, as it is in the atmosphere, and may affect air temperature, cause ocean cooling and alter rainfall amounts.

Dust lifting from the ground is controlled by three major factors: nature of the blowing wind, nature of the top soil or even the surface sediments, and the existence of any obstacles to windflow. Other secondary factors are comprised of: the topography of area that the wind will pass over and how it will directly affect the windflow or indirectly affect the formation of certain meteorological systems; and the human activities which mostly contribute in the earth surfaces disturbance and consequently affect the vegetation cover.

Figure 7. Dust from exposed riverbeds or shorelines.

Figure 8. Dust from dirt roads and traffic.

Figure 9. Dust from un-vegetated lots.

4.1. Global Distribution of Duststorm Frequency

The major contemporary sources of duststorms (desert dust) were identified through satellite images interpretation and standard meteorological observation data analysis; using the 1000-m visibility limit as the international standard, as when the visibility is reduced below this limit by the atmospheric dust and raised within the sight of the observer, then a duststorm is logged. Figure 13 shows the world map with the global distribution of major duststorms with main seasonal dust trajectories and Table 3 shows the areas of major global dust sources with key station annual duststorm day frequencies (Thomas, 1997).

Figure 10. Dust from unpaved airfields.

Figure 11. Dust from gravel pits and stockpiles.

Figure 12. Dust from construction and mining.

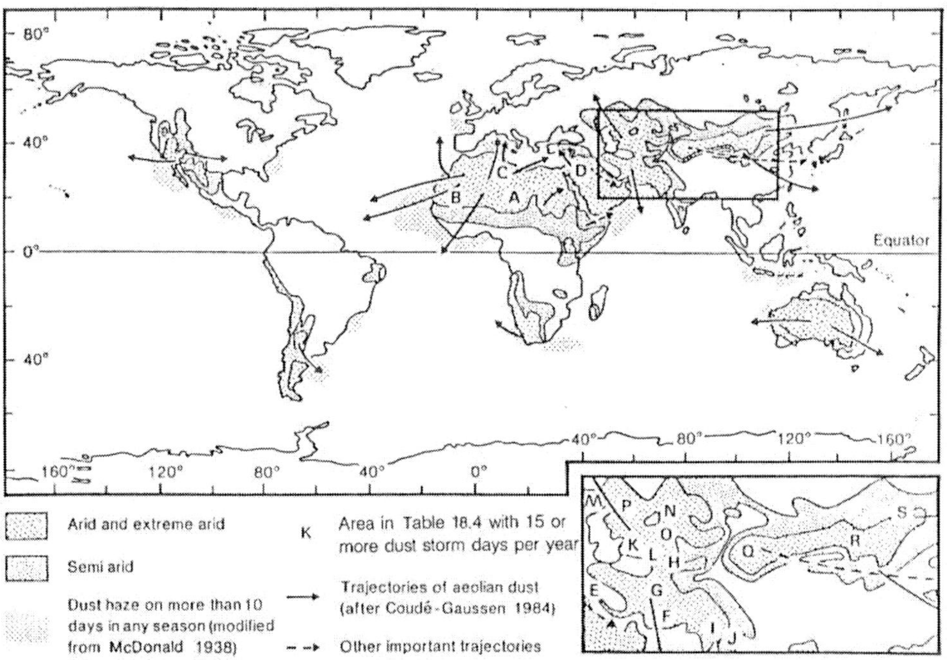

Figure 13. Global distribution of major duststorm areas with main seasonal dust trajectories (modified after Middleton, et. al., 1986b).

Table 3. Major global dust source areas with key station annual duststorm day frequencies (D) (modified after Middleton, 1989)

Source area	Ref. to Fig. 13	Station	*D*	Data period
N and W Africa				
Bodele Depression	A	Maiduguri (Nigeria)	22.5	1955–79
S Mauritania/N Mali/C Algeria	B	Nouakchott (Mauritania)	27.4	1960–84
Libya/E Algeria	C	Sirte (Libya)	17.8	1956–77
Middle East				
N Saudi/Jordan/Syria	D	Abou Kamal	14.9	1959–79
Lower Mesopotamia	E	Kuwait International Airport	27.0	1962–84
SW Asia				
Makran coast	F	Jask (Iran)	27.3	1970–73
Seistan Basin	G	Zabol (Iran)	80.7	1967–73
Afghan Turkestan plains	H	Chardarrah (Afghanistan)	46.7	1974–80
Upper Indus plains	I	Jhelum (Pakistan)	18.9	1951–58
Thar Desert	J	Fort Abbas (Pakistan)	17.8	1951–59
Central Asia				
Turkmenistan	K	Repetek (Turkmenistan)	65.5	1936–60
Karakum Desert	L	Nebit Dag (Turkmenistan)	60.0	1936–60
Rostov	M	Zavetnoe (Russia)	23.3	1936–60
Altai	N	Rubtsovsk (Russia)	25.1	1936–60
Alma Alta	O	Bakanas (Kazakhstan)	47.7	1936–60
Kazakhstan	P	Dzhambeiti (Kazakhstan)	45.9	1936–60
NW Asia				
Taklamakan Desert	Q	Hotien (China)	32.9	1953–80
Gansu Corridor	R	Minqin (China)	37.3	1953–80
Gobi Desert	S	Zamiin Uud (Mongolia)	34.4	1956–86

In Kuwait, the preliminary source for sand and duststorms is the recent surface deposits (Khalaf and Al-Ajmi, 1993). Those deposits are of six categories depending on their textural characteristics and field occurrence: Aeolian, residual deposits, playa, desert plain, slope and alluvial fans, and coastal areas of Kuwait. Figure 14 shows an interpretation of a satellite image of the Arabian Gulf during a severe duststorm sourced from southern Mesopotamian Desert in Iraq, which covers the southern coastal areas of Iraq and the coastal areas of Kuwait where the area of high intensive duststorms extends >50 km into the sea and lighter clouds extend to 150-200 km offshore (Foda, et. al., 1985).

Kuwait's surface is mostly covered by loose transported deposits acting as airborne particles carried by suspension, saltation, and creep across the surface and re-deposited around buildings, roads, farms, utilities plants and other man-made structures leading to acute environmental problems. The transportation scheme of the deposits depends on the particles' grain size, as suspended particles are <0.05 mm in diameter, saltation from 0.08 and 0.3 mm in diameter and creep are of larger sizes (Anwar, et. al., 1986b). Aeolian sand transported as bed-load reaches Kuwait in the form of mobile sand dunes and sand sheets, which are of positive sand budget in the northern and western areas in the shape of sand dunes, wadi fill deposits and rugged vegetated sheets, whereas they are of negative sand budget in the shape of very active sand sheets (Khalaf, 1988). Such an Aeolian sand is originated from the inland areas and entrained as airborne to the coastal environments as a bed-load forming the mobile sand phenomena and as a suspended load forming the dust fallout (Foda, et. al., 1985). It is worth mentioning that the annual rate of sediment transportation to the southern coastal area is about 138,000 m^3/year, while the annual rate of sediment transportation for the desert to the coastline is almost 293,000 m^3/year (Khalaf, et. al., 1982).

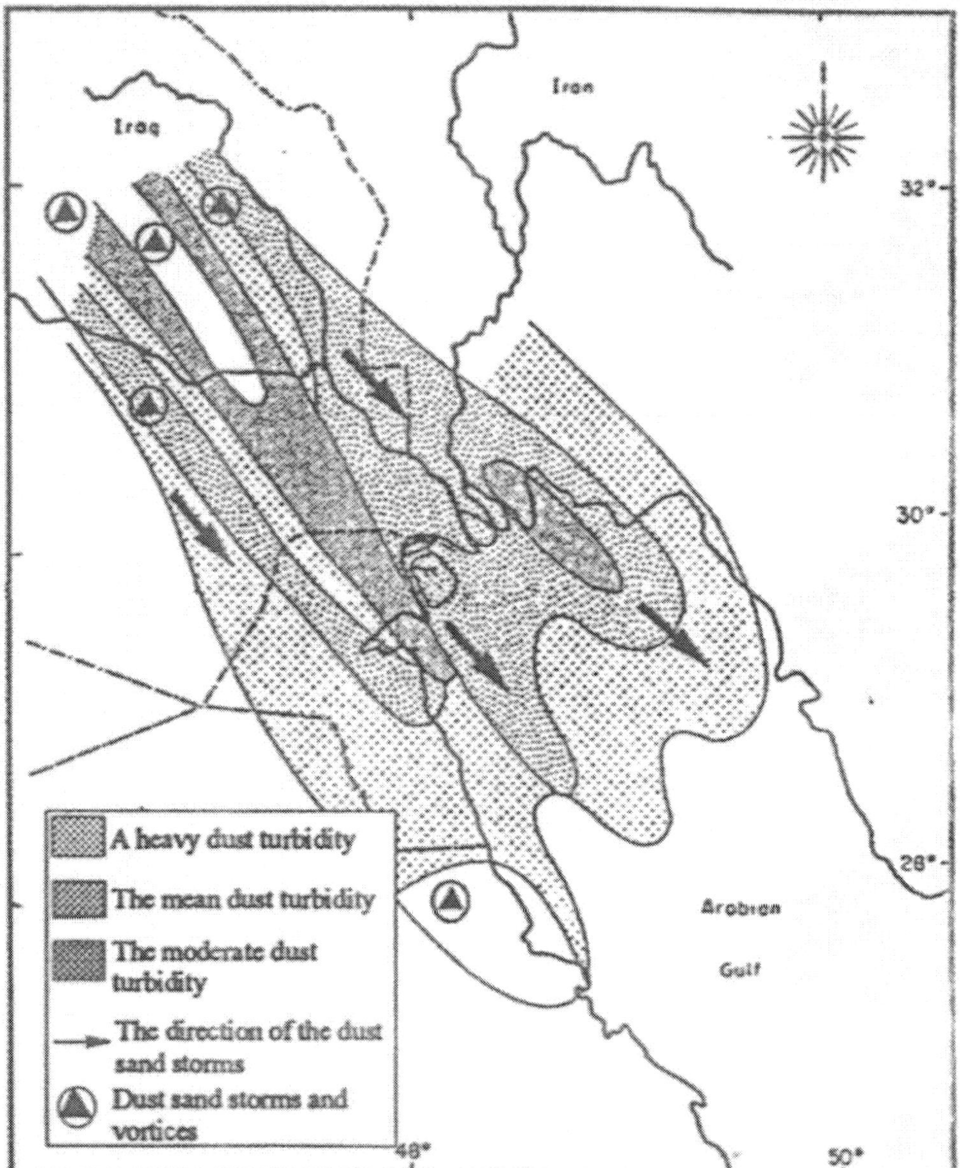

Figure 14. A satellite image interpretation of the Arabian Gulf during a severe dust storm originating from the southern Mesopotamian Desert in Iraq. After Vinogradov, Grigoryev and Lipatov, 1973.

The recent sand deposits of Kuwait were originated from muddy or gravelly muddy older sediments, of which sand was transported in the form of a bed load, whereas silt and clay-sized finer particles were transported in suspension and most probably deposited in the northern Arabian Gulf (Khalaf, et. al., 1980).

According to Khalaf, et. al., (1980), the potential source sediments of such dust fallout in Kuwait include:

1. The muddy sediments of the dry sabkhas in the lower Mesopotamian flood plain.
2. Old sediments of sandstone, limestone and dolostone exposed in the western desert of Iraq.
3. Paraconglomeratic sediments of Dibdibba Formation exposed in southern Iraq and northern Kuwait.
4. Local air picked up particles from the recent surface sediments from playa, sabkhas, and fine-grained mobile sand.

In Kuwait there are human activities in the desert that increase the susceptibility of recent surface deposits to duststorms including:

1. Extensive motor car movements in the desert areas resulting in the damage of the vegetation cover and destroying the semi-consolidated sediments and hence enforcing the deflation ability of the winds.
2. Extensive urban development which involve enormous excavation activities leading to the piling of loose prolonged uncovered materials.
3. Environmentally uncontrolled quarrying activities during which sand and gravel are exploited from the northwestern areas leading to the formation of quarry pits that expose tremendous amounts of fine materials to wind action.
4. Exploitation of gravel from the northern area by which the naturally developed log gravels are removed and consequently fresh fine sediments are exposed to wind action.
5. Overgrazing by cattle throughout the year that would destroy the vegetation cover and weakening the soil's binding forces and consequently resulting in the soil dust deflation by wind action.

In the middle east, Idso (1976) recognized Saudi Arabia as one of five world regions of intense dust generation. Middleton (1986a) stated that the Lower Mesopotamian plains witnessed the highest number of duststorms days per year, as well as Central Saudi Arabia witnessed a moderate level of duststorm activity (Figure 15), where Riyadh city witnessed an average duststorm of 7.6 days per year and an average of 76 days of reduced visibility to <11 km.

In southwest Asia, the desert area extending from Iran through Afghanistan and Pakistan into northwest India is the source of soil dust (Washington, *et. al.*, 2003).

In China, the major source areas for duststorms are The Taklamakan and the Badain Juran (Zhang, et. al., 1998), where about 800 Teragram (Tg) of dust is blown into the atmosphere annually, which is probably half of the global production of dust (Zhang, Arimoto, and An, 1997).

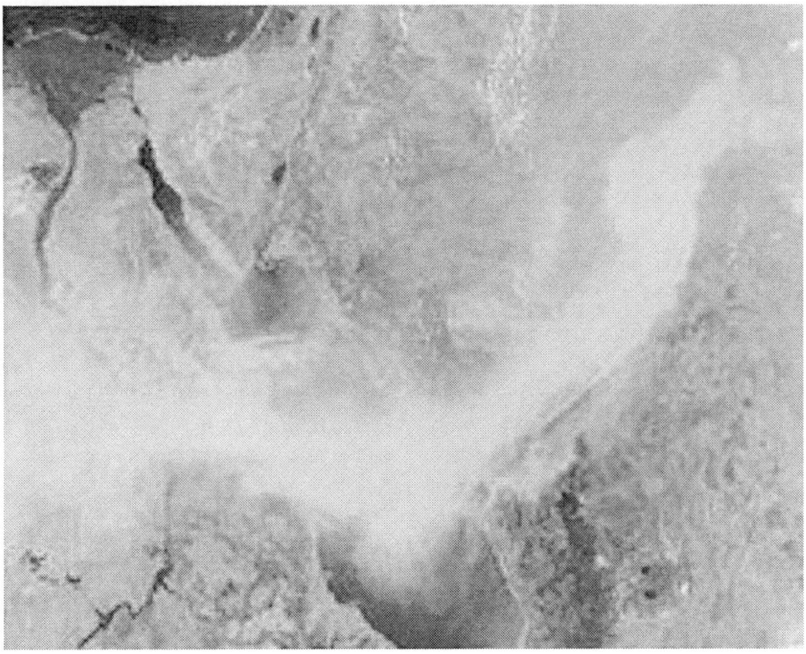

Figure 15. A duststorm crossing the Red sea from Egypt to Saudi Arabia on May 13, 2005.

5. Hazardous Impacts of Dust and Sand Storms

Dust and sandstorms are affecting the environment in variant disciplines. They could have physical, environmental, economic, social, human health ... etc. effects (Table 4) (Thomas, 1997).

5.1. Physical and Environmental Impacts

A sandstorm can possess a huge power that it can move whole sand dunes, as well as the duststorms which their leading edges can reach extreme heights that may reach more than a kilometer due to the large amounts of dust that duststorms can carry (Figure 16). The Sahara desert is considered the main source of duststorms, especially the Bodélé Depression and the area covering the confluence of Mauritania, Mali, and Algeria (Figure 17).

It has been noticed that the occurrence of the Saharan desert had been considerably increased by 10-folds during the second half of the previous century since the 1950s. This caused lost topsoil in Niger, Chad, northern Nigeria, and Burkina Faso. In the same context, Mauritania was subjected to duststorms twice annually, but nowadays, Mauritania witnesses about 80 duststorms annually. Duststorms and suspended dust are the most important types of weather phenomena, and with these two types, visibility may be reduced to less than 200 meters. Dust may block roads, damage materials and equipment and affect transportation and severely pollute the air. Very poor visibility (less than 500 meters) occurs more frequent during the summer, especially in June, July and August. Visibility classes and their incidence are shown in Table 5.

Table 4. Some environmental consequences and hazards to human populations caused by sandstorms (After Thomas, 1997)

Consequence	Example
Environmental	
Soil erosion	Kalma *et al.* (1988)
Soil nutrient gain	Swap *et al.* (1992)
Plant nutrient gain	Das (1988)
Ocean nutrient gain	Young *et al.* (1991)
Salt deposition and groundwater salinisation	Logan (1974)
Playa formation and relief inversion	Khalaf *et al.* (1982)
Sediment input to streams	Goudie (1978)
Case-hardening of rock	Conca and Rossman (1982)
Stone pavement formation	McFadden *et al.* (1987)
Desert varnish formation	Dorn (1986)
Loess formation	Zhang *et al.* (1991)
Silcrete development	Summerfield (1983)
Calcrete development	Coudé-Gaussen and Rognon (1988)
Bauxite metal enrichment	Brimhall *et al.* (1988)
Climatic change	Maley (1982)
Ocean sedimentation	Rea (1994)
Glacier mass budget alteration	Wake and Mayewski (1993)
Radiation budget alteration	Coakley and Cess (1985)
Ventifact sculpture	Whitney and Dietrich (1973)
Yardang sculpture	Breed *et al.* (1989)
Rock polish	Lancaster (1984)
Human-related	
Transport disruption	Burritt and Hyers (1981)
Disease transmission (human)	Leathers (1981)
Disease transmission (plants)	Clafin *et al.* (1973)
Radioactive dust transport	Bacon and Sarma (1991)
DDT transport	Riesborough *et al.* (1968)
Air pollution	Hagen and Woodruff (1973)
Radio communication problems	Martin (1937)
Animal suffocation	Choun (1936)
Animal madness	Choun (1936)
Rainfall acid neutralisation	Lôye-Pilot *et al.* (1986)
Machinery problems	Jones *et al.* (1986)
Warfare disruption	Agence France Press (1985)
Closing of business	Gillette (1981)
Reduction of property values	Gillette (1981)
Reduction of solar power potential	Reinking *et al.* (1975)
Car-ignition failure	Clements *et al.* (1975)
Drinking water contamination	Clements *et al.* (1975)
Electrical insulator failure	Kes (1983)

Table 5. Mean number of hours visibility lies within specified ranges. Period: 1976-1980. After Safar (1985) Kuwait International Airport

Month	Limits of visibility (meters)					Mean monthly total number of hours visibility
	< 400	400-<800	800-<1500	1500-<5000	5000-<8000	
January	7	2	5	104	111	229
February	2	2	4	62	105	175
March	4	4	5	66	116	195
April	4	6	10	79	125	224
May	2	4	5	65	125	201
June	7	21	25	124	159	336
July	16	17	25	131	152	341
August	0.4	4	7	86	133	230.4
September	1	1	3	66	103	174
October	7	2	4	82	128	223
November	6	1	3	35	84	129
December	10	5	1	69	101	186
ANNUAL TOTAL	66.4	69	97	969	1442	2643.4
MEAN	5.5	5.7	8.1	80.1	120.2	219.6

Figure 16. A sandstorm approaching Al Asad, Iraq, just before nightfall on April 27, 2005 where huge amounts of dust and sand are carried by this storm.

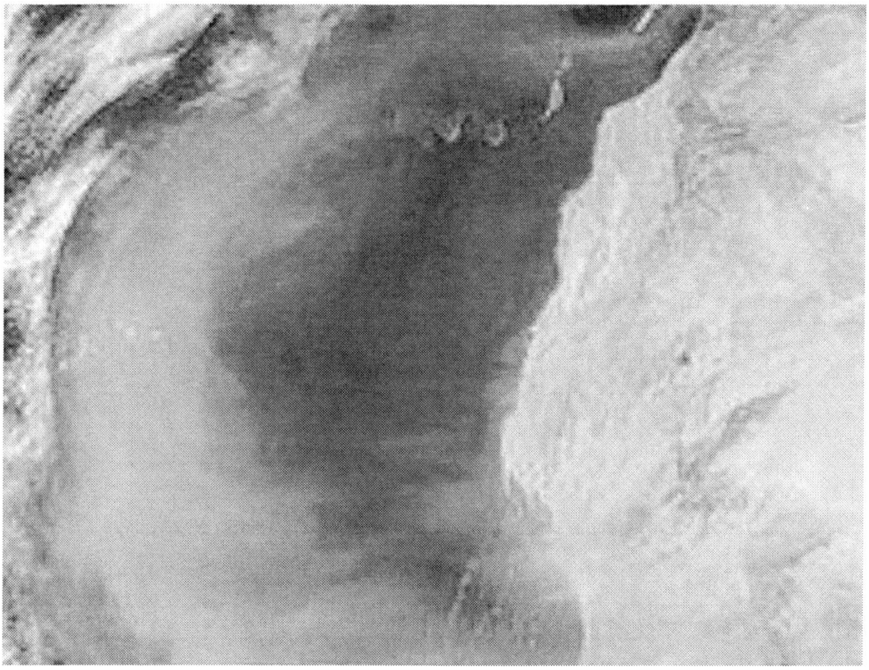

Figure 17. An intense Saharan duststorm sent a massive dust plume northwestward over the Atlantic Ocean on March 2, 2003.

Dust particles are of radioactive effect on the climate through reflection and absorption of solar radiation, which will result in severe cooling days or hot days depending on the dust particles' properties and their position in the atmosphere.

Dust particles as tropospheric aerosols are one of the significant component of the earth's climatic system and can modify climate through their direct radiative scattering and absorption (Tegen, *et. al.*, 1996) and indirect radiative effects by affecting on clouds microphysics (Rosenfeld, *et. al.*, 1997). They can also affect the processes of atmospheric chemistry. Dust can also influence the nutrient dynamics and biogeochemical cycling of ecosystems and can considerably impact the soil characteristics, oceanic productivity, and air chemistry. In addition, due to their transportation over large distances, they can influence great distances from their sources.

5.2. Economic Impact

Duststorms result in soil loss in the dry lands, which, in turn, will remove the organic matter and nutrient-rich particles. This will reduce the soil fertility, adversely affect and damage the young crop plants by their abrasive effect while transportation tremendously reduces the crop productivity. Moreover, reduced visibility will affect aircrafts and road transportation, which subsequently causes financial and human loss. Duststorms will reduce the sunlight reaching the surface, which, in turn, will have vital complications on plants photosynthesis and productivity and reduction of the livestock forage. Increased clouds of dust and sandstorms will increase the heat blanket effect, which will, in turn, affect the ecosystem stability.

5.3. Social Impact

Among the complications of land degradation caused by duststorms are the reduction of livestock forage due to reduced plant productivity, as well as the reduction ecosystem biodiversity, water availability and farmland yields. As a result of the loss of land resources, the spread of poverty and hunger will increase. The land deterioration and reduction of natural resources provision will push the wheel of migration in search of food and relief, which, in turn, will increase the number of environmental refugees that poses pressure on neighborhood areas and results in tremendous social problems.

5.4. Health Impact

When duststorms pass over habituated lands and approach urban and populated areas, this will contribute to urban air pollution and threatens public health, which will make people with respiratory systems problems and disorders severely suffer while breathing in dust particles. Duststorms can be a major contributor to diseases spread all over the world, as virus spores in the ground are blown into the atmosphere by such storms, similar tourban smog or acid rain.

Kuwait is being subjected to the hazardous effect of the dust fallout during the prevailing weathering and erosional factors. Moreover, Kuwait is witnessing urban regeneration, which facilitated the high-rise building to proliferate. Such an activity causes the inhabitants to be in direct contact with severe duststorms. Al-Hurban and Al-Ostad (2009) stated that there were excessive concentrations of toxic minerals in the atmosphere, where such amounts are considerably above the tolerant magnitudes for human beings health. Such findings were compared with the international and local standards and it was found that it is highly likely that people experiencing prolonged exposure of dust fallout in Kuwait would suffer from its harmful and hazardous effects.

6. Controlling Duststorms

The aim of dust emission control is to restrict or prevent movement of soil particles, i.e., preventing the soil particles from being airborne. In general, emission or entrainment of particulate matter of the earth's surface which results in sand and dust storms can be controlled by applying different kinds of dust suppressants or wind breakers, which will work on preventing wind erosion.

Dust suppressants are used in different sites: unpaved roads, agricultural fields, road shoulders, vacant land, construction sites, animal enclosures, landfills, military sites, and mining operations (Piechota, *et. al.*, 2002) (Figure 18). Such suppressants may include physical covers , e.g., vegetation, aggregate, mulches or paving; and chemical compounds as shown in Table 6 (Bolander, 1999).

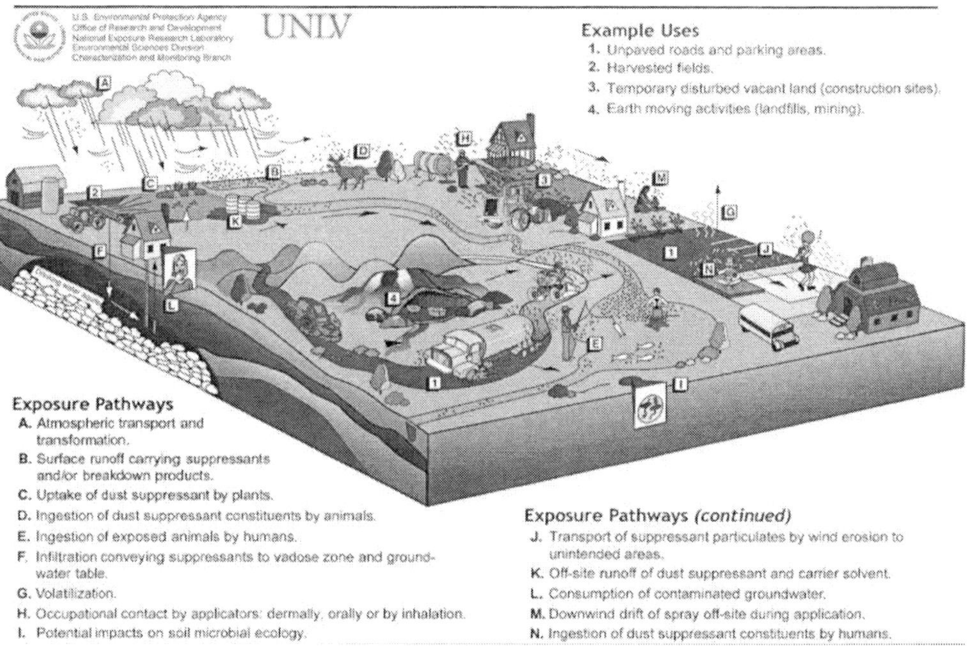

Figure 18. Conceptual model of the various uses of dust suppressants (Modified after Piechota, 2002).

As for chemical suppressants, water, either fresh, sea water or even reclaimed, can be applied alone as a dust suppressant, particularly on construction sites and unpaved roads, which are shortly disturbed lands. It is worth mentioning that the amount of water applied should taken into full consideration, as intense and heavy applications may turn the land into mud, by which the structure of the soil is destroyed and its ability of working as a sub-grade is damaged.

Table 6. Most commonly-used dust suppressants (Modified from Bolander, 1999)

Suppressant Type	Products
Water	Fresh and seawater
Salts and brines	Calcium chloride, magnesium chloride
Petroleum-based organics	Asphalt emulsion, cutback solvents, dust oils, modified asphalt emulsions
Non-petroleum based organics	Vegetable oil, molasses, animal fats, ligninsulfonate, tall oil emulsions
Synthetic polymers	Polyvinyl acetate, vinyl acrylic
Electrochemical products	Enzymes, ionic products (e.g. ammonium chloride), sulfonated oils
Clay additives	Bentonite, montmorillonite
Mulch and fiber mixtures	Paper mulch with gypsum binder, wood fiber mulch mixed with brome seed

Figure 19. Topical application of a dust suppressant using a large hose from a truck ended with a nozzle.

Salts and brines, of which most commonly used are calcium chloride and magnesium chloride, are applied to stabilize the soil by absorbing the moisture from the atmosphere, and hence it is essential when applying such a method to have adequate humidity (20-80%) (Bolander, 1999).

Figure 20. Topical application of a dust suppressant using a hand-directed spray hose.

Figure 21. Topical application of a dust suppressant using a spray bar mounted on a truck.

It is highly significant to choose the proper type of suppressant for the proper use. For instance, magnesium chloride and petroleum-based chemical suppressants are not recommended to be used for agricultural use due to their potential effect on the crops grown on the fields.

The mechanism in which dust suppressants work is to demolish the dust generation by changing the physical properties of the soil surface. This is because when the suppressant is applied, the soil particles will become no more susceptible for wind action, as they will be coated and aggregated together and hence become too heavy to be airborne particles.

Figure 22. Topical application of a dust suppressant using a spray gun mounted on a truck.

Dust suppressants could be applied either topically or mixed into the top layer of the soil (Piechota, *et. al.*, 2002). Topically, they can be applied using a spray bar on a truck or a large hose from a truck ended with a nozzle (Figure 19 & 20) for vacant lands and small plots. In cases of larger areas, the application of a dust suppressant is done by truck-mounted spray bars (Figure 21) and in some cases, water guns on trucks are used (Figure 22).

REFERENCES

Al-Hurban, A. and Al-Ostad, A., 2009. Textural characteristics of dust fallout and its potential effect on public health in Kuwait City and suburbs. *Journal of Environmental Earth Sciences,* Springer-Verlag 2009.

Al-Kulaib, A. A., 1984. The climate of Kuwait. *Directorate General of Civil Aviation Meteorological Department, Climatological Division.* State of Kuwait, 178p.

Al-Kulaib, A. A., 1985. *The climate of Kuwait. Climatological Section,* Meteorological Department, Directorate General of Civil Aviation, Kuwait (in Arabic).

Al-Ostad, A., 1994. A study of the three types of wind-blown dust in Kuwait. *Journal of Meteorology,* vol. 19, no. 189, May/June 1994 pp. 150-155.

Al-Ostad, A., 1998. *Aeolian sand and dust transport in southern Kuwait and relationships with climate (Ph. D. thesis).* Department of Geography, University of Birmingham, Birmingham, B15 2TT, England.

Anwar, M., Zaghloul, N., Gharib, I., and Al-Hashash, M., 1986b. Estimation of eolian sand transport rate at three strategic locations in Kuwait. *KISR Annual Research Report of Natural Environment,* pp. 117-120.

Anwar, M., Gharib, I., Al-Hashasha, M. and Zaghloul, N., 1986b. Surface modeling of dust fallout on the PROME sea area. *KISR, Annual Research Report of Natural Environment,* pp. 121-125.

Bagnold, R. A., 1941. *The physics of blown sand and desert dunes.* Edward Aenold, London, 2nd ed.

Bolander, 1999. *Dust palliative selection application guide,* United States Department of Agriculture.

Foda, M.; Khalaf, F. I. & Al-Kadi, A., 1985. Estimation of dust fallout rates in the northern Arabian Gulf. *Sedimentology,* 32, pp. 595-603.

Goudie, A. S., 1978. Duststorms and their geomorphological implications. *Journal of Arid Environments 1*, pp. 291-310.

Idso, S. B., 1976. *Duststorms. Scientific American* 235 (4): 108-11, 113-14.

Jackson, M. L.; Gillette, D. A.; Danielson, E. G.; Blifford, I. H.; Bryson, R. A. and Syers, J. K., 1973. *Global dustfall during the Quaternary as related to environments: Soil Science,* v. 116, pp. 135-145.

Jauregul, E., 1973. *The urban climate of Mexico City: Erdkunde,* vol. 27. Pp. 298-307.

Khalaf, F. I., Kadib, a., Gharib, I., Al-Hashash, M., Al-Saleh, S., Al-Kadi, A., Desouki, M., Al-Omran, L., Al-Ansari, L., Al-Houti, F. and Al-Mudhian, L., 1980. Dust fallout (Toze) in Kuwait "Mineralology, granulometry and distribution", ESS-12 final report. *Kuwait Institute for Scientific Research (KISR),* vol. 2. 221p.

In: Dust: Sources, Environmental Concerns and Control ISBN: 978-1-61942-547-7
Editors: Laurent B. Wouters and Michel Pauwels

Chapter 6

PHOTOCHEMICAL AND CLIMATE IMPLICATIONS OF AIRBORNE DUST

Jonathan E. Thompson
Department of Chemistry and Biochemistry,
Texas Tech University, Lubbock, US

ABSTRACT

Airborne mineral dusts suspended in earth's atmosphere can have dramatic effects on climate, regional photochemistry and human life. In this chapter, both the direct and indirect climate forcing effects of dust will be discussed. Physical models for treating light scattering and absorption by dusts will be considered. Since dust absorbs light in the ultraviolet and blue portion of the electromagnetic spectrum, its presence can reduce actinic flux and alter atmospheric photochemistry. This phenomenon will be considered from the perspective of common atmospheric pollutant gases. Finally, the chemistry occurring at / on dust surfaces will be considered and their significance discussed.

INTRODUCTION

Aeolian dusts are commonly dispersed into earth's atmosphere via the mechanisms of saltation and sandblasting in arid environments. These processes represent a major source of natural aerosol to earth's atmosphere and impact biogeochemical cycles significantly. Recent estimates suggest mass fluxes of 500 – 4500 Tg / yr of global emissions with North African, Asian, and Middle Eastern dusts accounting for ≈ 90% of the total emissions [1,2]. Figure 1 illustrates global estimates of PM_2 dust source strengths by continent. Once dispersed in the atmosphere, dust is believed to affect earth's energy balance by scattering and absorbing solar radiation [3,4,5,6]. Dust may alter atmospheric chemistry by providing reactants, catalysts, and / or surfaces for heterogeneous reactions to occur [7,8,9,10,11]. Dust can also be transported significant distances from source regions. In some cases, Asian dust can be transported to the west coast of North America. Also, Saharan dust regularly transits the Atlantic ocean [12,13,14,15,16,17,18]. An example of such long-range transport can be observed in Figure 2 which illustrates LIDAR observations of a large plume of Saharan dust at roughly 2 – 5 km altitude above the Caribbean Sea in August 2010. The left panels show the measurements of aerosol backscatter coefficient, aerosol depolarization ratio and extinction-to-backscatter ratio (lidar ratio) at 532 nm.

The aerosol backscatter coefficient is used to spatially map aerosol plumes, and in this case, illustrates a spatially extensive, optically thick plume. The depolarization ratio and the LIDAR ratios are intensive variables that vary with aerosol type but not concentration. Interestingly, the NASA Langley LIDAR instrument used combines four intrinsic aerosol optical properties to infer aerosol type (data shown in the upper right panel) [for details see [19]. For this plume, the depolarization ratio is elevated, indicating non-spherical particles characteristic of mineral dusts. Eight-day back-trajectories calculated using the NOAA HYSPLIT model, shown in the lower right panel confirm that the aerosol transited the Atlantic ocean from the Sahara desert.

While the dust in Figure 2 was measured aloft, dust deposition to the world's oceans is also believed to indirectly create a carbon dioxide sink by providing photosynthetic phytoplankton with micronutrients (e.g. iron) [20,21,22]. As can be seen from these examples – atmospheric mineral dust can affect biogeochemical cycles on a global scale. Understanding the entire dust cycle through all steps of production, transport, chemistry, and deposition/removal clearly is required to begin understanding the very complex feedback mechanisms present in the coupled earth - climate - biome system.

Providing definitive answers to all scientific questions within even one of these focus areas is far beyond the ability of this author, and well-beyond the scope of this chapter, or even this text (I regret it is truly unfortunate to disappoint the reader at such an early stage of this chapter). Nonetheless, this text will strive to present certain key issues and recent advances in the understanding of how dispersed dusts in the atmosphere may affect earth's climate and atmospheric chemistry.

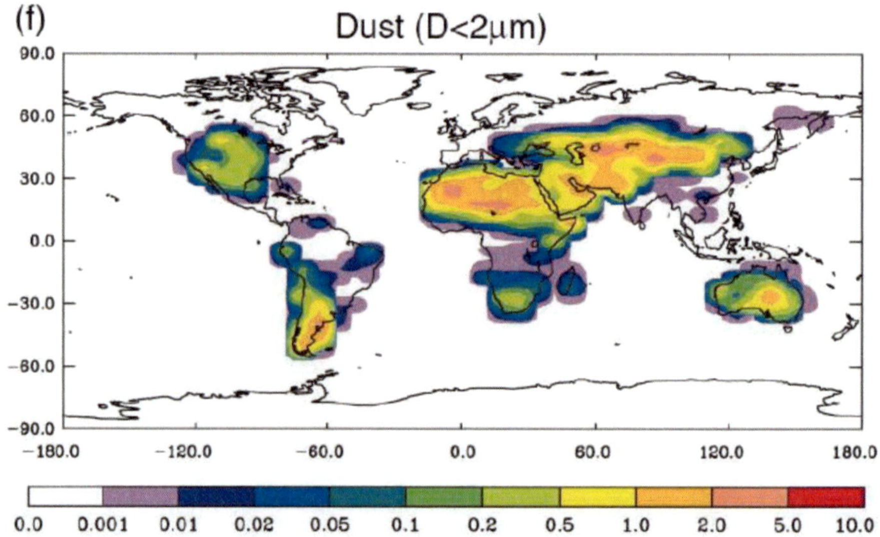

Figure 1. Map of annual average source strength of PM_2 dust in kg km^{-2} hr^{-1}. Figure courtesy of I.P.C.C. 2001 [113].

This work will not consider dust emission or deposition models [23,24,25,26,27,28], historical records of dust [29], or biogeochemical cycling (e.g. ocean fertilization) in detail.

The reader is directed to the previous references or an excellent recent review paper by Shao et al. [1] for a review of many of these topics. Also, references 20-22 discuss "*ocean fertilization*" and the results of phytoplankton based CO_2 sequestration. Instead, the first half of this chapter will discuss the basis for the direct and indirect radiative forcing effects of dust. The second half of the chapter will consider the effect of dust aerosols on atmospheric chemistry and consider chemical processing of dust in the atmosphere.

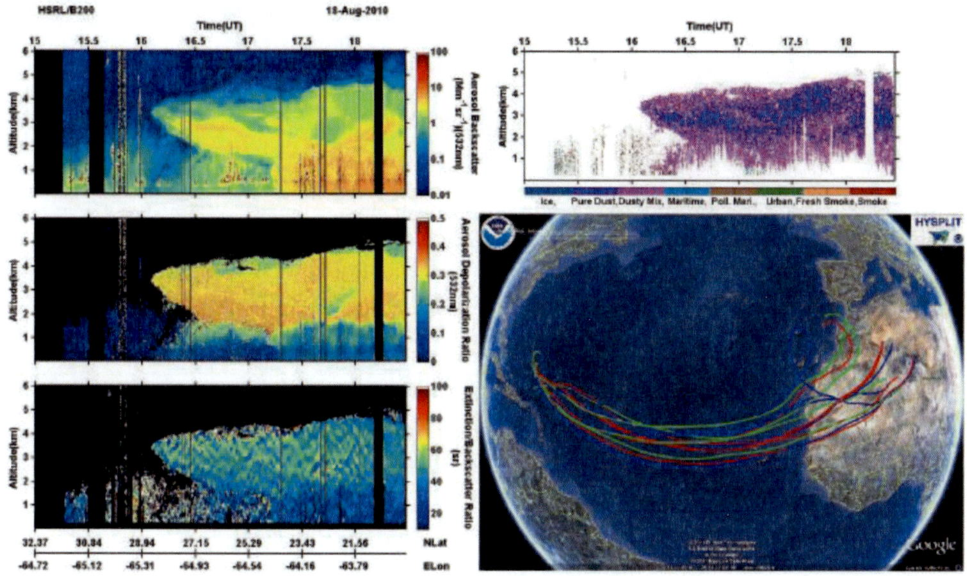

Figure 2. Aerosol backscatter coefficient, depolarization ratio, and extinction-to-backscatter ratio (Lidar ratio) at 532 nm as measured by the NASA Langley Airborne High Spectral Resolution Lidar (HSRL) [114] for August 18, 2010 over the Caribbean Sea. The authors acknowledge the NOAA Air Resources Laboratory (ARL) for the provision of the HYSPLIT transport and dispersion model and READY website (http://www.arl.noaa.gov/ready.php) used for this publication [115,116]. Figure graciously provided by S. Burton, R. Ferrare, and C.A. Hostetler of NASA.

PART 1. DUST DIRECT AND INDIRECT CLIMATE FORCING

1.1. A Brief Description of Direct and Indirect Forcing

Particles (D_p = 10 nm – 100 μm) suspended in earth's atmosphere can affect climate and weather patterns through two mechanisms described as *direct* and *indirect* climate forcing. Figure 3 illustrates these mechanisms. In direct forcing, dust scattering and absorption of solar radiation leads to changes in photon transport in the atmosphere and actinic flux. In indirect forcing, dispersed dust may affect the size distribution, number density, optical properties, or lifetimes of cloud drops. It has been suggested that mineral dust may suppress precipitation in certain non-polluted areas such as the Tibetan plateau [30]. Other authors have found a link between decreased tropical cyclone activity and increases in atmospheric dust transported over the North Atlantic from North Africa [31].

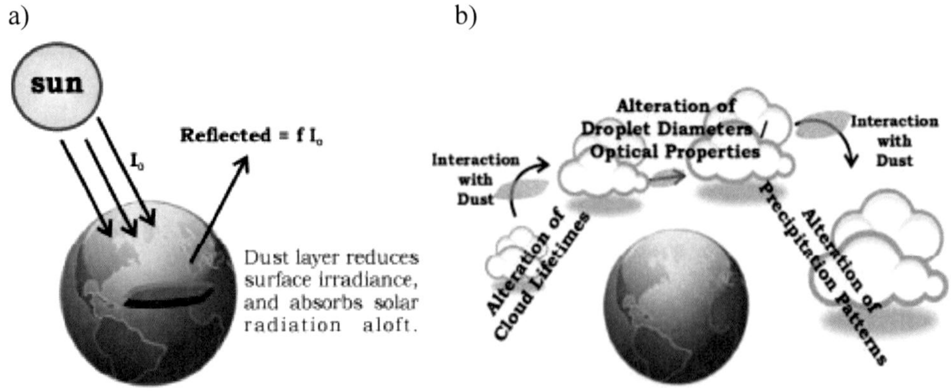

[a] Direct Forcing. Dust interacts with sunlight to affect radiative balance.

[b] Indirect Forcing and Climate Effects. Dust alters cloud optical properties or lifetimes. –Dust may affect weather patterns suppressing or enhancing precipitation.

Figure 3. Illustration of the aerosol direct and indirect effects on the climate system. The direct effect involves dust clouds scattering or absorbing solar radiation in the atmosphere. The indirect effect involves dust (or aerosols in general) influencing cloud properties. This could involve changes in cloud lifetime, precipitation patterns, or cloud droplet sizes that alter cloud optical properties.

These empirical examples of weather modification suggest mineral dusts may exert an influence on cloud properties. Additional evidence is found in results that suggest mineral dust promotes heterogeneous formation of ice crystals in the atmosphere at temperatures >-38° C [32,33,34,35,36,37,38]. Heterogeneous formation of ice on dust provides an alternate mechanism for ice to form at higher temperatures than what is required via homogeneous nucleation (i.e. -38 ° C).

Additionally, temperature increases of clouds influenced by dusts have been noted [39,40]. This also suggests a link to precipitation patterns as cloud top temperature can be used to infer cloud height.

At present, the magnitude of mineral dusts affect on radiative transport in the atmosphere is not well-known. The most recent Intergovernmental Panel on Climate Change (I.P.C.C.) report estimates mineral dust climate forcing from anthropogenic land surface modification is on the order of -0.1 ⊥ 0.2 W m^{-2} suggesting a very slight net cooling effect (for instance greenhouse gas warming ≈ +2.5 W m^{-2}). Determination of the global averaged indirect effect of mineral dust is challenging given the complex mechanism of action. However, estimation of the direct effect is a bit more straightforward. In 1995, Chylek and Wong [41] presented an oft cited equation for direct radiative forcing by aerosols:

$$\Delta F_R \frac{-S_0}{4} T_{atm}^2 (1-N)(1-a)^2 (2\beta\tau_{scat} - 4a\tau_{abs}) \tag{1}$$

In this equation, S_0 is the solar constant, T_{atm} atmospheric transmittance above aerosol layer, N is fractional cloud cover, a is albedo (reflectivity) of earth's surface, β is aerosol upscatter fraction and τ_{scat} and τ_{abs} the aerosol optical depths due to scatter and absorption. In this equation, β and the optical depth terms are the only dust related quantities. The optical depths depend on physical properties of the aerosol such as refractive index, size distribution, and mass concentration. The upscatter fraction (β) is the fraction of light scattered up (into space and away from earth). This depends on aerosol size distribution, refractive index, and solar zenith angle. The Intergovernmental Panel on Climate Change [113] report suggests a value for backscatter fraction of $\beta = 0.20 \pm 0.07$ for mineral dusts. The "backscatter fraction" is not the same as the "upscatter fraction" however, since the former does not properly consider solar zenith angle. However, this estimate is at least useful to give the reader an idea of the magnitude of this term. Figure 4 is also very useful to give the reader an idea of how the intensity of light scattered into a particular angle scales with refractive index and particle size. From figure 4, we see light scattering is often dominated by forward scatter (very near 0°), however, differences would exist if intensities were integrated over a range of angles that constitute "upscatter" (e.g. away from the earth). From this discussion it is then clear that the directionality and extent of light scattering and absorption determines the net direct effect of dust aerosols. However, it has now become necessary to understand what factors affect this. Section 1.2 below will outline light scattering models for understanding scattering and absorption of light by dust aerosols vis-a-vis climate forcing.

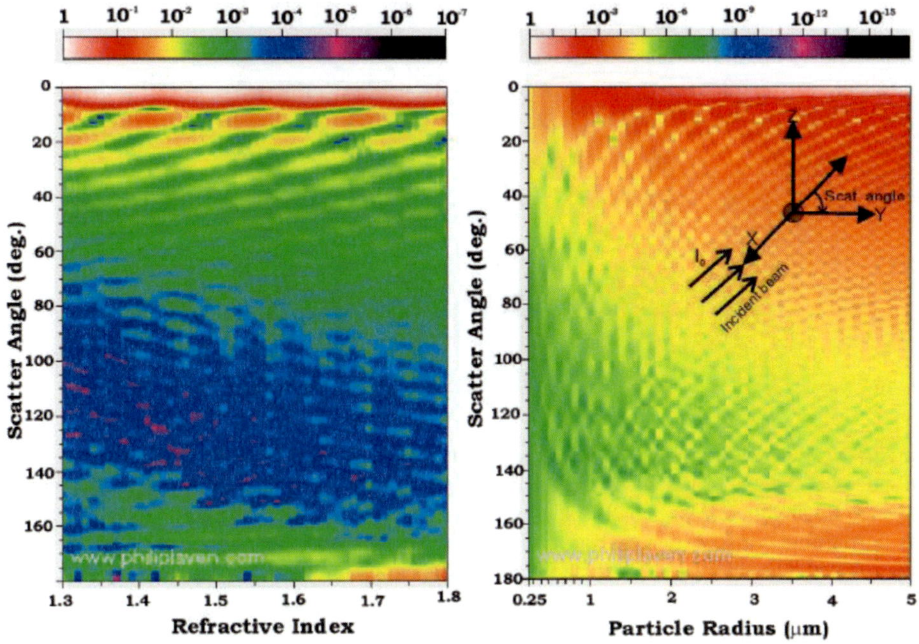

Figure 4. Illustration of how scattering intensity varies with scatter angle (see inset on right for definition) for differing refractive indices and particle radius. This calculation assumes spherical geometry, but dust is often not spherical and more advanced models must be used (see text below). For this calculation, unpolarized radiation at $\lambda = 530$ nm was assumed with a particle imaginary index of 0.0015i. The image on right assumes 1.52 for real portion of R.I. Images created with MiePlot software written by P. Laven.

1.2. Describing Light Scattering and Absorption by Particles

Figure 5 illustrates some of the effects by which a beam of electromagnetic radiation is redirected (or scattered) by an aerosol particle. Reflection, refraction, and diffraction of light comprise elastic scattering – in which the wavelength of light does not change. Absorption of light by the particle can also occur. In such a case, photon energy can be converted to heat.

a)

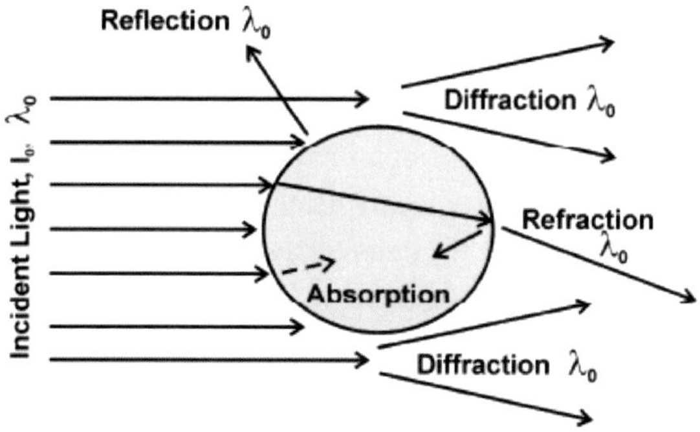

b)

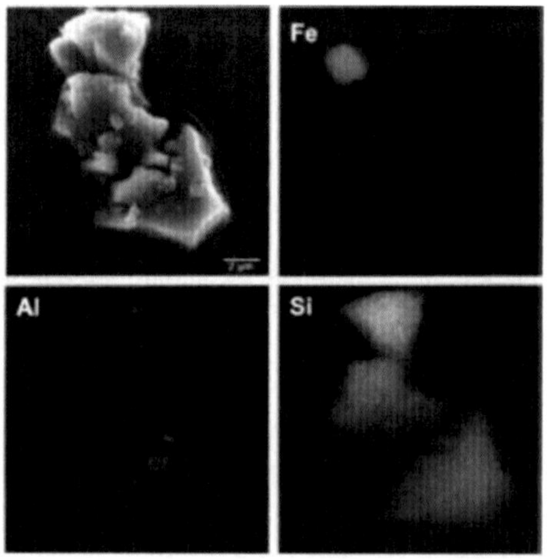

Figure 5. (A) Cartoon illustration of the interaction of light with an aerosol particle. The particle need not be spherical. As illustrated in the electron micrograph in the upper left corner of (B) mineral dust particles are frequently non-spherical. Elemental mapping of the particle for Fe, Al, and Si content reveals the particle is also not compositionally homogeneous. Non-spherical, inhomogeneous particles (such as mineral dust) require sophisticated methods for modeling optical properties. (B) has been reproduced from [111] with permission of publisher.

The sum of the effects of elastic scattering and absorption is called light extinction. The removal of light from a beam of collimated monochromatic radiation can be described by terms known as the extinction, scattering, and absorption cross sections for the particle (σ_{ext}, σ_{scat} and σ_{abs}) in question:

$$I_{ext} = I_0 \times \sigma_{ext}$$

where

$$\sigma_{ext} = \sigma_{scat} + \sigma_{abs} \qquad (2, 3)$$

In these equations I_0 and I_{ext} represent irradiances of the incident beam and that which is removed through extinction, respectively. The scattering / absorption / extinction cross-sections (σ) for a particle is defined as the product of a particle's physical cross-sectional area and a term represented by Q

$$\sigma_{ext} = Q_{ext} \times \pi r^2$$
$$\sigma_{scat} = Q_{scat} \times \pi r^2$$
$$\sigma_{abs} = Q_{abs} \times \pi r^2 \qquad (4, 5, 6)$$

In these equations r represents the particle radius and the unitless Q terms are the single particle extinction, scattering, or absorption efficiency parameters. At first glance it might seem the value for Q could not exceed 1, since this would imply a given particle could remove more light from a beam than what is incident upon it. However typical values of Q range from $< 1 - 4$.

Very small particles do not effectively interact with the oscillating field of electromagnetic radiation (Q <<1), while larger particles can and also diffract light yielding Q > 1. Single scatter albedo (SSA) is taken as the ratio of the scattering and extinction cross sections or the corresponding Q terms:

$$SSA = \frac{\sigma_{scat}}{\sigma_{ext}} = \frac{Q_{scat}}{Q_{ext}} \qquad (7)$$

The next natural question is how can one determine a value for Q? The exact value of Q for a particle is a complex function of particle size, shape, wavelength of light, and complex refractive index of the particle. Here, the refractive index (m) of a material is defined as the ratio of the speed of light in vacuum (c) to the speed of light in that material (v):

$$m = \frac{c}{v} \qquad (8)$$

Since the speed of light is lower in materials when compared to vacuum, the value of $m > 1$. The value of refractive index loosely scales with the density of the material under study and is typically 1.3 – 2 for most atmospheric aerosols. For particles that absorb light, refractive index is expressed as a complex number (m = n + ik) consisting of a real (n) and imaginary part (k) which describe the optical properties of a given particle at a certain wavelength. The real part helps describe the scattering behavior of the particle and the imaginary part (k) describes the extent of light absorption. The particle size parameter (α) accounts for size of the particle and is the ratio of a particle's circumference to the wavelength of probe light:

$$\alpha = \frac{2\pi r}{\lambda} \qquad (9)$$

For spherical scatterers, either the Rayleigh or Mie theory can be used to describe scattering depending on particle size parameter. However, as illustrated in figure 5 for Arizona test dust, mineral dusts are not spherical. For more complicated particle shapes more advanced physical theories such as discrete dipole method, the finite difference time domain (FDTD) method, or the so-called T-matrix method must be used to find Q terms and understand intensity of angular scattering. A detailed description of these methods is far beyond the scope of this work. Fortunately, in recent years there has been intense interest in modeling the optical effects of non-spherical particles by these advanced methods, and multiple textbooks have been developed describing the state of the art [42,43,44,45,46,47,48,49,50,51]. The optical properties of mineral dust aerosols have also recently been reviewed [52,53] and the *Journal of Quantitative Spectroscopy and Radiative Transfer* has recently devoted special issues to reporting on the proceedings of the annual Conference on Electromagnetic Scattering by Non-Spherical Particles (for instance JQSRT v.100(1-3) 1-496, JQSRT, v.106, 1-621, JQSRT v.109(8) 1335-1548, JQSRT, 2009, v.110(14-16) 1207-1780). These volumes are particularly useful to track developments in the field as they evolve. While we cannot consider all aspects of these models here, the key point is that light scattering and absorption is affected by particle shape and complex refractive index. This is demonstrated in figure 6 below adapted from Gasteiger et al. [54].

Figure 6 illustrates the results of a discrete dipole approximation (DDA) model for the particles illustrated in Figure 6I. The plots in II and III are extinction efficiency (Q_{ext}) terms vs. particle size parameters for the different test particles. As observed, particle shape plays an important role in determining maximum extinction efficiency, and the size parameter at which that occurs for both non-absorbing (II) and absorbing (III) particles. Particle heterogeneity such as that illustrated in figure 5B makes modeling particle optical properties from physical principles even more challenging, yet particle size, shape, orientation, and refractive index are the key variables required for model building. Clearly, all of the intricacies of particle shape are too complex to be considered for each individual atmospheric dust particle via techniques such as those employed in figure 6. To simplify, collections of dust particles in the atmosphere are often treated as being spheroidal with varying aspect ratios. This often yields light scattering models with acceptable accuracy. In fact, the unlabeled data traces present in figure 6 are representative of prolate spheroids with aspect ratios of 1.4 – 2.4. As can be observed, in certain cases numerical results for the spheroid model can approach those obtained through the more exact DDA method.

Dust refractive index will also vary with source / composition to an extent, however empirical generalizations can be made. Figure 7 illustrates a plot of the real (n) and imaginary (k) portion of refractive indices as a function of wavelength for mineral dust reported in the literature from a variety of global locations. This figure results from the compilation of measurements reported in the literature and a version of the figure (with details described) was originally presented in the review of Redmond et al. [5]. However, the figure has been updated with very recent results presented by Wagner et al. [55] for re-suspended Saharan dust, re-suspended Pullman and Amarillo soil types reported by Ma et al. [56], and values summarized in Ansmann et al. for the SAMUM-1 and SAMUM-2 experiment [57]. As observed, the real component of refractive index seems to be relatively constant $\approx$ 1.5, but does vary between 1.43 – 1.56 depending on source. Interestingly, this value (e.g. 1.5) lies very close to the refractive index expected for many atmospheric organic aerosols [58]. The imaginary portion of refractive index seems to be significantly more variable.

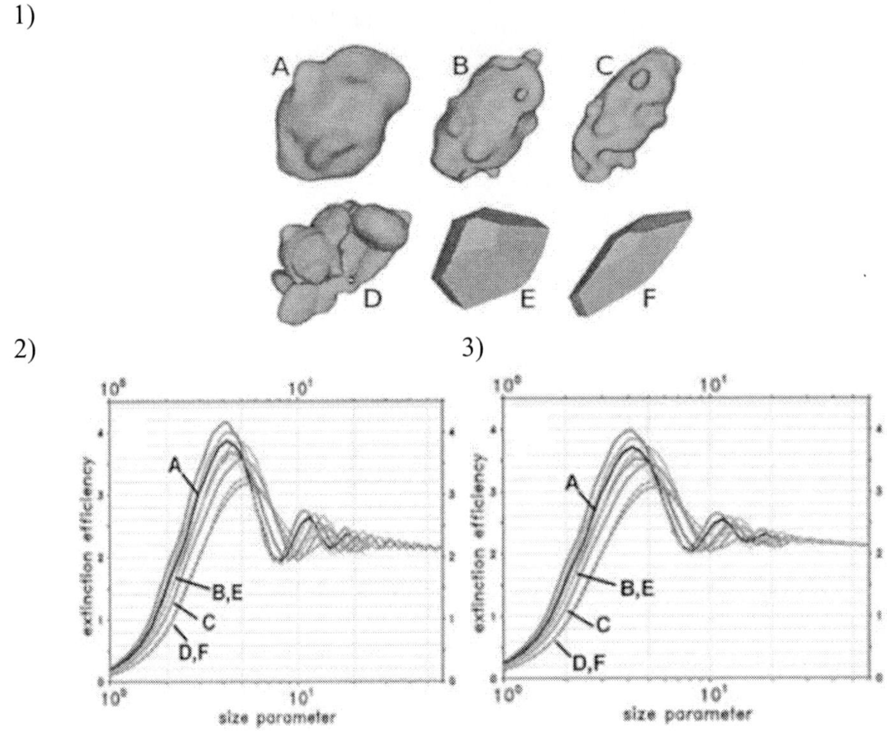

Figure 6. (1.) Irregularly shaped model particles used in discrete dipole modeling analysis. (2.) Plots of Q_{ext} vs particle size parameter for non-absorbing particles of 1.52 + 0.0i. (3.) Plots of Q_{ext} vs particle size parameter for light absorbing particles of 1.52 + 0.0172i. For II and III, the letters refer to observed particles from I. Figures reproduced from the work of Gasteiger et al. [54] with permission of publisher. The unlabeled data traces are representative of prolate spheroids with aspect ratios of 1.4 – 2.4.

This is likely a consequence of differences in dust chemical composition at each location. Light absorption by dust is often attributed to hematite content.

What is consistent within the dataset we have examined is k increases significantly with decreasing wavelength. This parallels the trend in k value for hematite which increases from ≈ 0.05 – 1 with decreasing wavelength over the visible spectral range. The studies considered in this work suggest $k \approx 0.003 - 0.01$ is typical for mid-visible wavelengths (550 nm). Figure 7 also illustrates a plot of single scatter albedo (SSA) as a function of wavelength for mineral dusts. Again, SSA is similar to the reflectivity of particles and is an important parameter describing their direct climate effects. For the mid-visible region, albedo values of ≈ 0.95 – 0.96 for 550 nm seems to be typical. However, albedo drops dramatically at wavelengths < 450 nm, which is consistent with increased light absorption at shorter wavelengths. Knowledge of dust refractive index and single scatter albedo is useful for modeling the direct climate impact of dispersed dusts.

However, given the prospects for surface chemical modification of dust (see section 2 below), formation of coatings, and uptake of water vapor, it is necessary to consider these effects. Li-Jones et al. [60] studied hygroscopic growth of aerosols (high and low relative humidity R.H.) at the West Indies and report that transported Saharan dust seemed to have a low hygroscopic growth factor of 1.0 – 1.1 when R.H. was increased to typical ambient levels from "dry" conditions. The authors conclude that water absorption is not important for

Saharan dust. Gasso′ et al. [61] have also commented on hygroscopic growth of dusts and suggest a dust hygroscopic growth factor of $\gamma = 0.23$ that fits the model:

$$\sigma_{sp}(RH) = k\left(1 - \frac{RH}{100}\right)^{-\gamma} \tag{10}$$

This suggests hygroscopic growth up to a factor of approx. 2 near saturation. The I.P.C.C. recommendation for relative humidity effects on dust is an enhancement of a factor of 1.3 ± 0.2 for 80% R.H. relative to the dry case [113]. Interestingly, in laboratory studies under well controlled conditions other authors have found that humidification of clay mineral particles (proxies of dust) can in certain cases lead to a slight decrease in optical extinction relative to the dry case [62]. For instance, Attwood and Greenslade [62] found that while montorillonite extinction increased by a factor of 1.4 at 90% R.H, Kaolinite extinction actually decreased with an f(RH) = 0.8, and illite clay was more or less unchanged. This also suggests that R.H. may not have a very large affect on the scattering / extinction properties of dust in the atmosphere. However, a study by D.A. Lack et al. [63] employed a very sensitive photo-acoustic spectrometer to study the effect of humidity changes on absorption of light by transported Saharan dust. These authors report increase in light absorption on the order of 1.5 fold when humidity increases to 73% compared to 25%. This result is significant for modeling light absorption by dispersed mineral dusts.

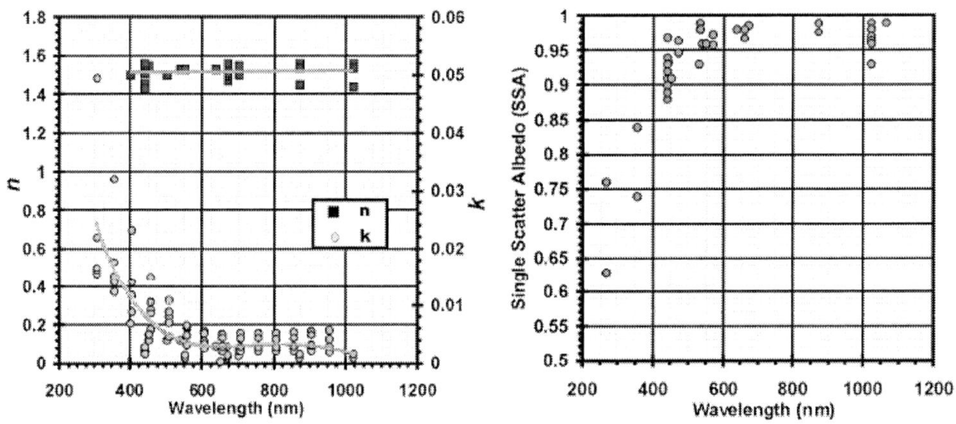

Figure 7. Left - Plot of real (n) and imaginary (k) components of refractive index for mineral dusts reported in the literature. Right – plot of single scatter albedo vs. wavelength for mineral dusts. Lines are meant solely to guide the reader's eye.

Part 2. Impact of Dust on Atmospheric Chemistry

2.1. Alteration of Actinic Flux by Dust: Light Absorption

In part 1 of this chapter, it was noted that light absorption by dust is stronger at shorter visible wavelengths. This more energetic electromagnetic radiation is the driving force behind many important photochemical reactions in the atmosphere. Many of these reactions are photolytic in nature and yield reactive radical products that subsequently participate in multiphase chemistry. Some of the more important reactions and wavelength ranges are summarized in table 1 below.

In these equations, light is treated as a reactant in the chemical process. If an absorbing aerosol (e.g. dust) reduces actinic flux, the kinetics of reactions 1-8 will be altered. This holds the potential for significant consequences to atmospheric composition since these reactions are in large part responsible for air quality near earth's surface. For instance, O (^{1}D) produced in reaction 1 subsequently can react with H_2O (g) to form the hydroxyl radical (OH) – a very powerful atmospheric oxidant. Also, the atomic oxygen (O) produced in reaction 2 or 3 can quickly react with atmospheric O_2 (g) to generate tropospheric ozone (O_3). It is believed this is the primary source of tropospheric ozone. Considering these reactions, we can then quickly understand how light absorption and scattering by dust may significantly impact gas-phase photochemistry.

Table 1. List of some important photolysis reactions occurring in Earth's atmosphere

$O_3 + h\nu(< 315\ nm) \rightarrow O\ (^1D) + O_2$
(Rxn. 1)
$O_3 + h\nu(300 - 700\ nm) \rightarrow O\ (^3P) + O_2$
(Rxn. 2)
$NO_2 + h\nu(290 - 430\ nm) \rightarrow NO + O\ (^3P)$
(Rxn. 3)
$HNO_3 + h\nu(< 315\ nm) \rightarrow NO_2 + OH$
(Rxn. 4)
$HONO + h\nu(< 400\ nm) \rightarrow NO + OH$
(Rxn. 5)
$HNO_3 + h\nu(< 315\ nm) \rightarrow NO_2 + OH$
(Rxn. 6)
$NO_3 + h\nu(< 650\ nm) \rightarrow NO_2 + O$ – or - $NO + O_2$
(Rxn. 7)
$H_2O_2 + h\nu(< 350\ nm) \rightarrow OH + OH$
(Rxn. 8)

The photolysis rate constant (J) for a particular molecule can be found by:

$$J = \int \sigma(\lambda,T)\phi(\lambda,T)F(\lambda)d\lambda \qquad (11)$$

where $\sigma(\lambda,T)$ (cm^2) is the absorption cross section of the molecule at a particular T and λ, $\phi(\lambda,T)$ is the reaction quantum yield, and $F(\lambda)$(photons cm^{-2} nm^{-1} s^{-1}) is the actinic flux at a wavelength. This equation then formalizes the previous discussion that alteration of actinic flux by absorbing dust aerosols will alter photochemistry. Further complications are encountered since dust effects actinic flux differently with altitude (above / below or in dust layer) and solar zenith angle. Liao et al. [64] used a discrete ordinate radiative transfer model considering both clear and cloudy sky conditions and found an absorbing mineral dust aerosol is expected to reduce the photolysis rate for reaction 1 ($J(O_3 \rightarrow O(^1D))$ by < 5% for 0° solar zenith angle. While reductions in photolysis rate constants were generally always expected, the effect was particularly evident at high solar zenith angles. For instance, when solar zenith angle increased to 60° a reduction in $J(O_3 \rightarrow O(^1D))$ of 5 – 10% was modeled.

The Photochemical Activity and Ultraviolet Radiation experiment (PAUR) which took place in June 1996 and May 1999 off the North African coast has provided significant data to study the effects of Saharan dust on actinic flux. di Sarra et al. [65] studied the effect of Saharan dust on ultraviolet irradiance on Lampedusa – a Mediterranean island located about 100 km off the North African coast. The authors found dust aerosols with an optical depth of 0.5 at λ = 415 nm reduced near UV irradiance (λ = 350 nm) by 20-25%. The calculations of Meloni *et al.* (also for Lampedusa in May 1999) suggest actinic flux in the NO_2 photodissociation band can be reduced up to 24% in the lower and middle troposphere due to Saharan dust [66]. These results suggest Saharan dust can significantly reduce actinic flux in photochemically relevant wavelength bands in the lower troposphere. Other authors have assessed the effect of mineral dust on photochemistry by measuring photolysis rate constants directly. For instance, D.S. Balis et al. [67] report a very slight (≈ 5%) reduction in $J(NO_2)$ for reaction 3 with a tripling of aerosol optical depth for two sites at Crete. The aerosol is believed to be mixed – but often heavily influenced by Saharan dust aloft. Jeong and Sokolik [68] performed extensive modeling of the effect of mineral dusts on photolysis constants for several different aerosol loading scenarios and light absorption characteristics. Model results indicate alteration of J values are strongly linked to the light absorbing properties of the dust aerosol, the dust size distribution and mixing state, and the spectral band used for photolysis reactions. This underscores the complexity of the problem. However, fractional change in J-values for gases absorbing in the blue and near UV such as $J(O_3 \rightarrow O(^1D))$ were on the order of -10- 31% (e.g. a reduction).

The authors predict a reduction in photolysis rate constant for gases such as NO_2 of -7 – 30% depending on exact circumstances of the model. Interestingly, the fractional changes in $J(NO_3)$ could be positive, and were only on the order of ±5%. Nitrate radical (NO_3) can be photolyzed by light throughout the visible spectrum, and the result speaks to the importance of considering spectral wavelength band when predicting change in J-terms. Aerosols that do not absorb light at a particular wavelength may actually increase actinic flux in certain atmospheric domains [69].

Ying et al. [70] modeled a maximum reduction in OH concentration of 60% in dust source regions near Mexico City, but predict only a 5-20% decrease in OH within the city owing to a lower concentration of dust. A comparison of measurement data and a model presented by these authors suggest dust may decrease the daily maximum $J(NO_2)$ at the surface by a few percent (≈ 5%) for Mexico City. Other authors have estimated the photochemical effects of dust decrease OH radical concentrations by 9 – 11.1% on average globally, however, over North Africa and downwind of source regions OH is expected to decrease by a much higher fraction - nearly 67% [71,72]. These results suggest mineral dusts can have significant impacts on photochemistry – particularly in source regions.

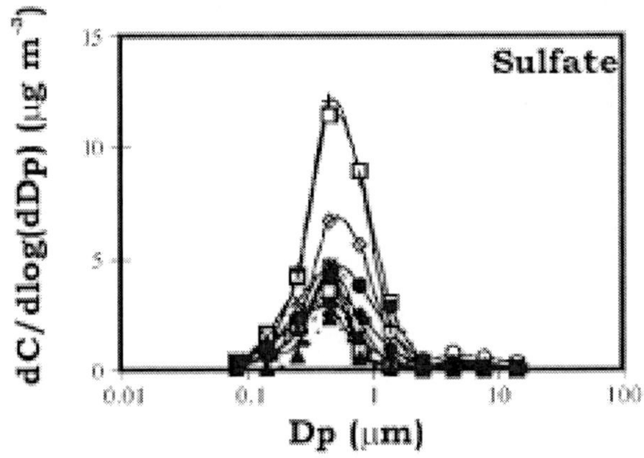

a)

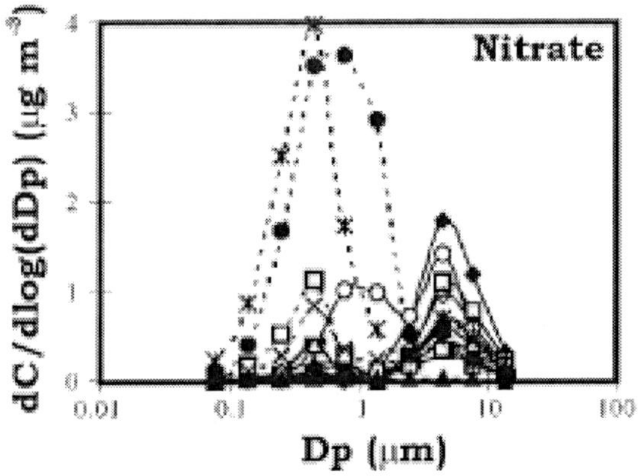

b)

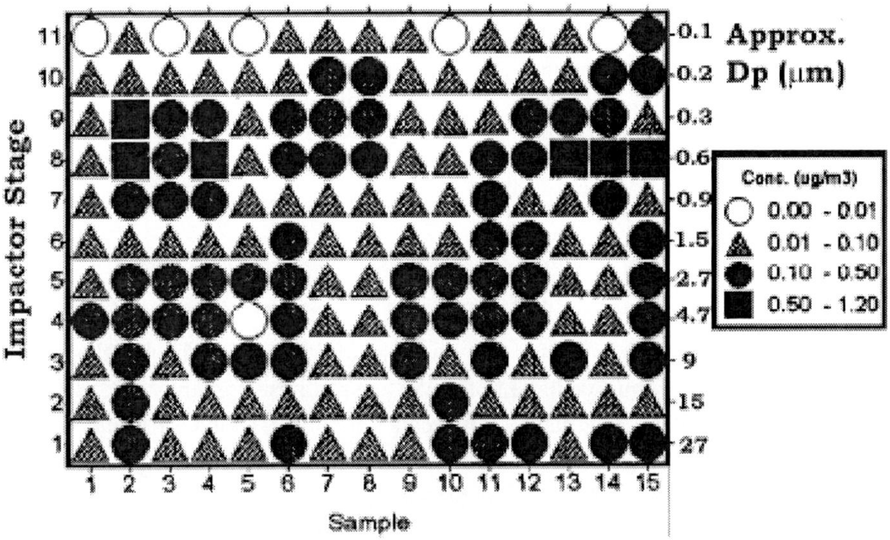

c)

Figure 8. (Continued).

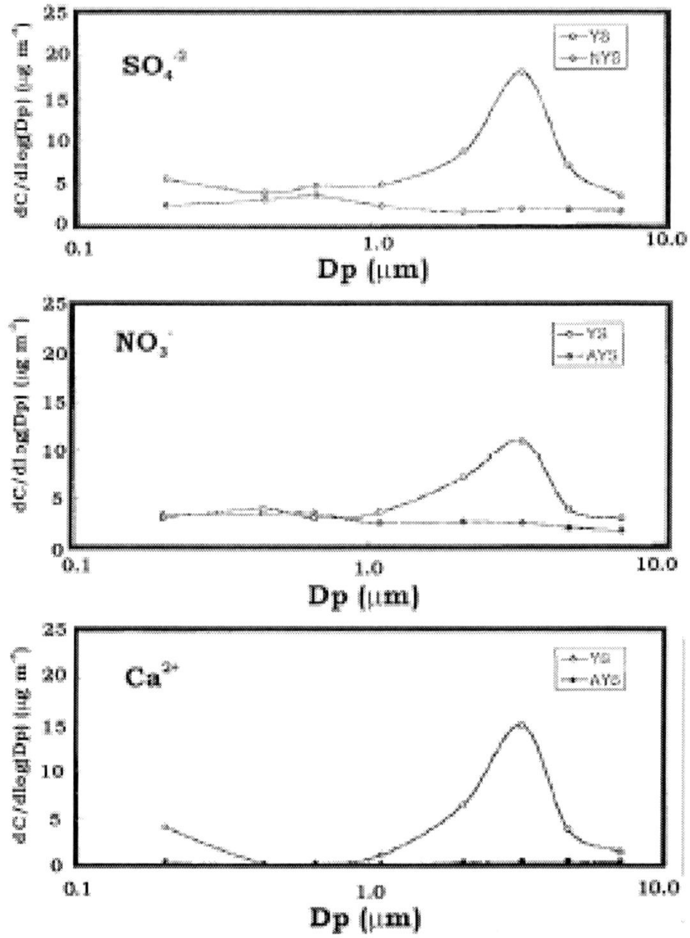

d)

Figure 8. Mass size distributions for (A) sulfate and (B) nitrate observed on different days at Newark, NJ USA. As observed the water soluble anions are often present in both coarse and fine modes. (C) Nitrate concentrations as a function of particle diameter (Dp) for a forest site in Indiana. Nitrate concentration profiles often exhibited bimodal behavior with a minima between modes at approx. 1.5 µm. (D) Mass distributions of sulfate, nitrate and calcium during a yellow sand storm event at Seoul, Korea. Sulfate and nitrate correlated well with calcium which is a major component of mineral dust. Reproduced with permission of publishers from Zhao and Gao [80], Pryor et al. [81], and Kim and Park [95].

2.2. Chemistry on Dust Surfaces

In the previous section we considered how absorption of light by dust may alter atmospheric photochemistry, but we have not yet discussed how dust may directly participate in heterogeneous atmospheric chemistry. This can occur by several different mechanisms [73].

Firstly, dust surfaces may act as a sink for gas-phase components through absorption / uptake or catalyzed decomposition. Also, the dust chemical components themselves may participate in chemistry. In this section we examine examples of each.

2.2.1. Properties of and Alteration of Mineral Dust Surfaces

The story begins with the realization that nitrates (NO_3^-) and sometimes sulfates (SO_4^{2-}) present in the atmospheric aerosol often exhibit a bimodal mass distribution [74,75,76,77,78,79].

Figure 8 illustrate this for sites in Newark, NJ, Indiana, USA, and Seoul, Korea [80,81]. The presence of these anions in super-micron particles could be considered a bit of a scientific mystery, since preponderance of thought suggested these anions form from oxidation of gas-phase precursors (SO_2 and NO_x) and therefore should be present in smaller, accumulation mode particles. Wu and Okada performed an electron microscopy study to find that nitrate often existed on individual sea salt and dust aerosols sampled in Japan and suggested heterogeneous chemical reactions on the surface of dust and sea-salt aerosols may be responsible [82]. This result is also confirmed by Kojima et al. [83] who suggest sulfate coatings on Asian dust resulted either from heterogeneous uptake and oxidation of SO_2 (g) or coagulation with cloud or fog droplets.

The general framework for chemical alteration of dust is believed to involve a heterogeneous replacement reaction in which sulfate or nitrate produced from oxidation of gas-phase SO_x or NO_x species replaces carbonate within a thin aqueous phase present at the particle surface [84,85,86,87,88,89,90,91,92].

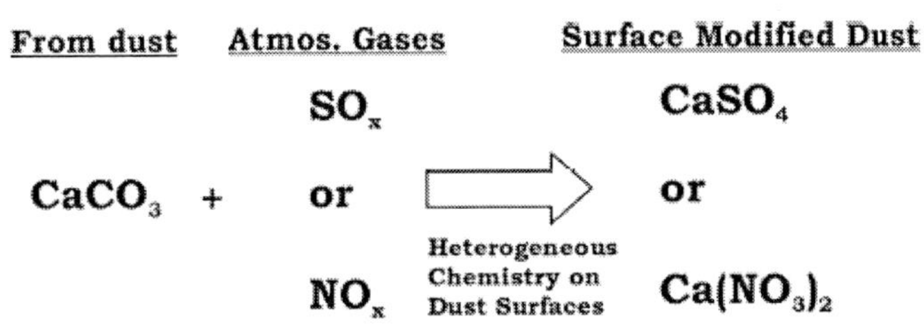

The presence of very thin layers of water on hematite has been confirmed via FTIR spectroscopy (see figure 9 below). This phase is believed to be very important for the replacement reactions because the products (e.g. nitrate salts) are quite hygroscopic and this can promote additional aqueous volume within which the reactions can proceed further [93]. This aqueous phase is clearly visible on the $CaCO_3$ particle that has been treated with nitric acid illustrated in figure 9 below. While much of this chemistry has been investigated in the laboratory, there is also evidence for alteration of ambient dust particle surfaces. Hwang and Ro [94] performed electron probe microanalysis on particles collected downwind of an Asian dust storm. This allowed the authors to analyze how the surface composition of dust particles may change when processed in the atmosphere. Figure 10 illustrates that a large number of dust particles do become surface modified, and the resultant coating is rich in nitrate relative to the particles core. The authors suggest the replacement reaction (nitrate or to a lesser extent sulfate for carbonate) proceeds readily on dust surfaces. This is in agreement with the work of Kim and Park [95] who found both sulfate and nitrate correlated well with calcium (see figure 8) during a yellow sand event in Seoul. A recent effort by Dall'Osto et al. considered atmospheric processing of Saharan dust [96]. These authors studied dust composition by aerosol time of flight mass spectrometry (ATOFMS) and suggest that native (fresh) Saharan dusts are neither rich in nitrate or sulfate. However, Saharan dust transported through the atmosphere to the Cape Verde islands ($\approx$ 450 km off West African coast) had been chemically processed to include internally mixed nitrate (but not sulfate). The same work also considers Saharan dusts transported via a more northerly route to Mace Head, Ireland. This dust was observed to be highly internally mixed, containing nitrate, sulfate, and methanesulfonate. While it appears dusts routinely are chemically modified in the atmosphere, the exact conditions required or the extent / frequency to which dust surfaces are modified may be a subject of debate. Other authors have reported that dust-laden air masses that mixed with anthropogenic pollutants over Beijing and Tianjin had characteristics consistent with being externally mixed [97]. It should also be noted that Sullivan et al. have recently found evidence supporting an analogous reaction for chloride involving HCl (g) [98].

$$2\ HCl\ (g) + CaCO_3\ (s) \rightarrow CaCl_2\ (aq) + CO_2\ (g) + H_2O\ (l)$$

These authors suggest 65% of sampled particles added chloride through this mechanism, and this chloride may represent 4 – 9% of particle mass.

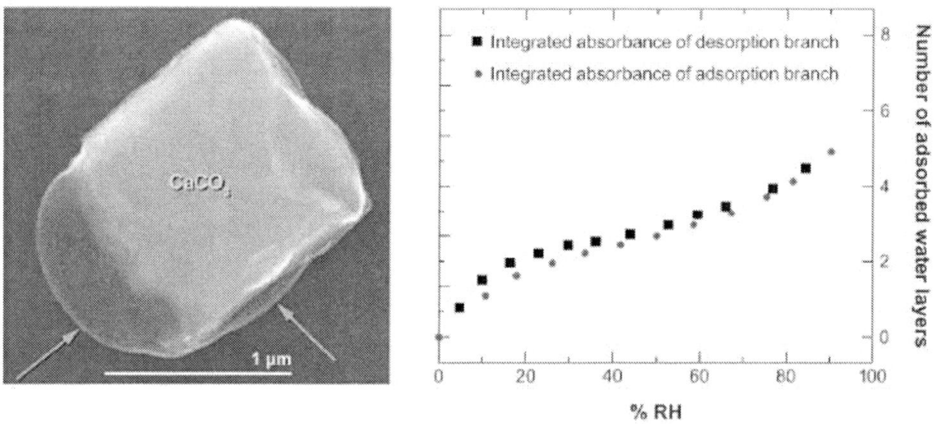

Figure 9. Left - Scanning electron micrograph of a $CaCO_3$ particle after treatment with HNO_3 (g) at 40% relative humidity (R.H.). The arrow points to regions in which hygroscopic $Ca(NO_3)_2$ has deliquesced forming an aqueous layer. Right – Adsorption of water vapor on hematite as a function of R.H. As observed, at minimum a monolayer of water is believed to exist at all atmospherically relevant R.H. values. Figures adapted from Cwiertny et al. [93] with permission of publisher.

Alteration of dust surfaces may seem solely like an academic pursuit – however, there are real environmental consequences of this chemistry. Dust may need to be considered to improve air quality modeling [99]. Also, modification of dust surfaces with nitrates may increase hydrophillicity substantially over the unmodified case [100], allowing dusts to acquire an aqueous surface phase when dispersed in the atmosphere. This may alter their ability to serve as aqueous condensation nuclei, affect dust atmospheric lifetime, and affect light absorption as discussed in the previous section. Additional research aimed at understanding the complex chemical kinetics of these processes is warranted.

2.2.2. Mineral Dust Surface Area: Additional Effects on Atmospheric Gases

Dispersed mineral dusts may also affect atmospheric chemistry by promoting or catalyzing atmospheric reactions. For instance, N_2O_5 can be hydrolyzed on dust surfaces to form HNO_3 and NO_2 [101].

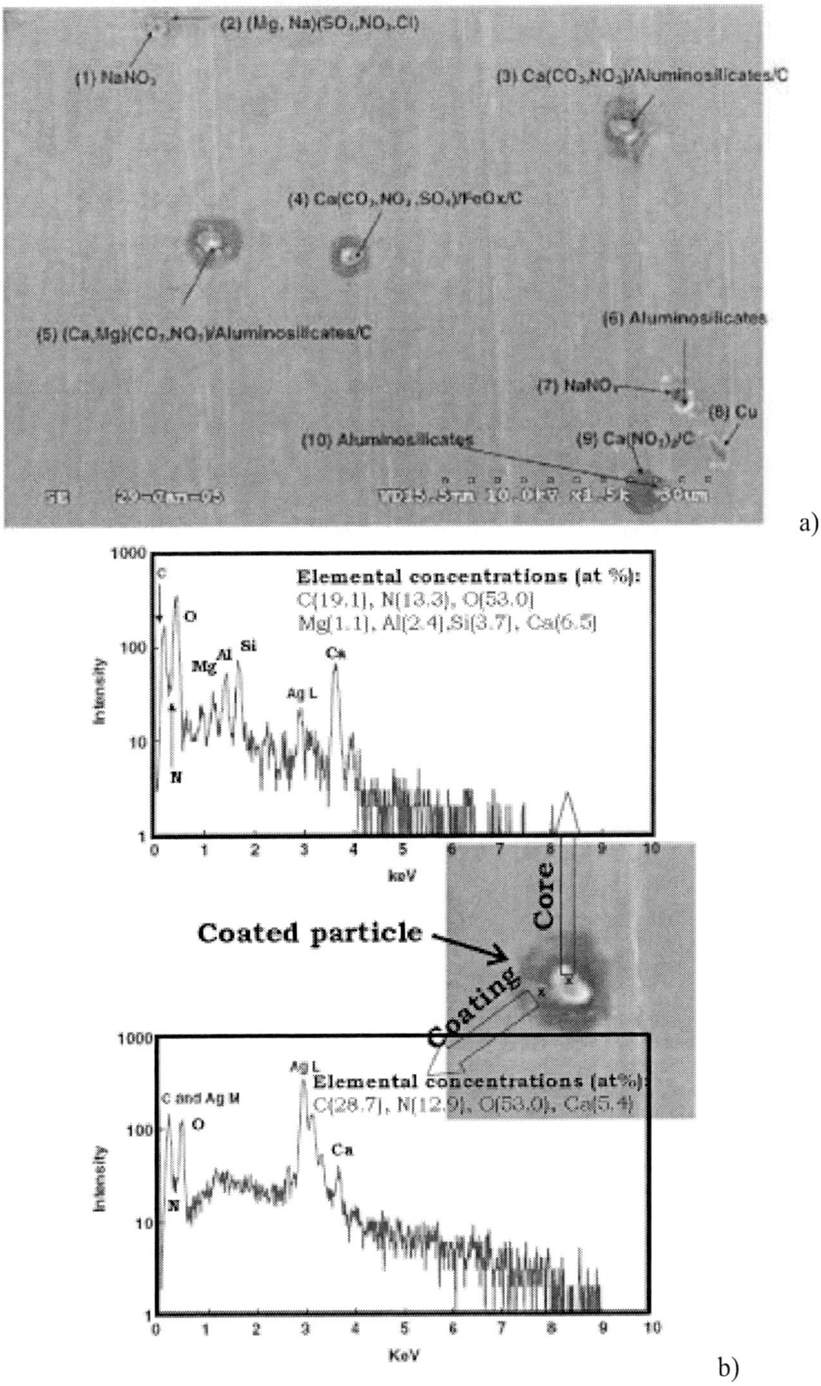

Figure 10. a) – secondary electron image of coarse particles collected in stage 3 (4 μm cut diam.) of a cascade impactor at Chuncheon, Korea on 11 Nov 2002 during an Asian dust storm event. Dust is believed to originate from the Gobi and Inner Mongolian deserts. The dark grey regions around bright cores are chemically modified regions of the particle. b)– X-ray spectrum and elemental concentrations for the particle core and chemically modified coating. While the core is rich in aluminosilicate material and/or $CaCO_3$, the coating is rich in calcium nitrate. Figures adapted from Hwang and Ro [94] with permission of publisher.

The work of Grassian et al. [102,103,104] also suggests proxies of mineral dusts have a catalytic effect on decomposition of ozone through a chemical mechanism such as:

O_3 (g) + surface → surface-O + O_2 (g)
O_3 (g) + surface – O → surface-O-O + O_2 (g)
surface-O-O → surface + O_2 (g)

In this sequence, surface-O indicates a surface oxide, and surface-O-O an unstable surface peroxide. Hence, the third reaction is considered to be fast and surface sites regenerated. Experimentally, the rate of ozone decomposition on dust appears to be strongly linked to relative humidity. This is illustrated in figure 11 below. The review of Cwiertny et al. [93] presents a more detailed chemical mechanism / kinetic model for these types of reactions that will not be considered here.

2.2.3. Mineral Dust: Soluble Iron and Fenton Chemistry

Another interesting aspect of mineral dust atmospheric chemistry is that dust frequently contains iron. Cycling of soluble iron through oxidation states (e.g. Fe^{2+} (aq) or Fe^{3+} (aq)) can play key roles in atmospheric REDOX chemistry. This REDOX cycling is known to mediate the formation of highly reactive species such as superoxide (O_2^-), hydrogen peroxide (H_2O_2), the hydroperoxy radical (HO_2), nitrite (NO_2^-), or the hydroxyl radical (OH) through Fenton chemistry [105]. For instance the decomposition of hydrogen peroxide to yield hydroxyl radical is catalyzed through iron:

$$\mathrm{Fe(II)} + H_2O_2 \longrightarrow \mathrm{Fe(III)} + OH + OH^-$$
$$\mathrm{Fe(III)} + O_2^- \longrightarrow \mathrm{Fe(II)} + O_2$$
$$\text{Net: } O_2^- + H_2O_2 \longrightarrow OH + OH^- + O_2$$

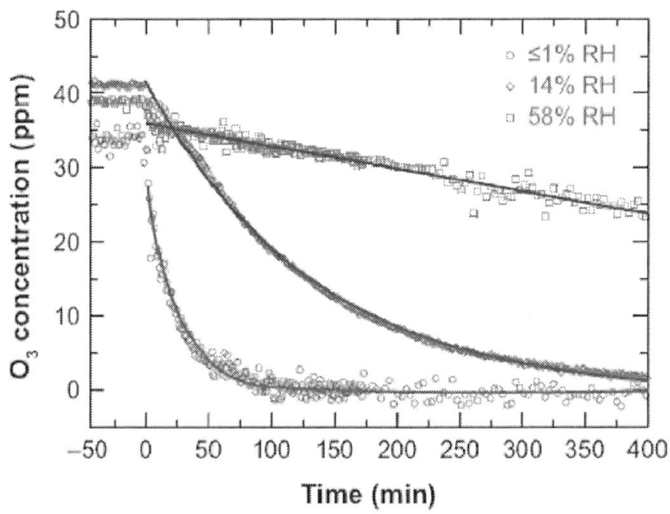

Figure 11. Kinetics of ozone decomposition on hematite as a function of relative humidity (R.H.). High R.H. values slow the reaction. Figure reproduced from the review of Cwiertny et al. [93] with permission of publisher.

Because of this, soluble iron may promote formation of highly reactive compounds that alter aerosol phase chemistry. This would be expected to have significant effects on the rates of oxidation of organic compounds within the aerosol phase [106,107,108,109] and consequently might affect yields of secondary organic aerosols (SOA).

In order for these chemical processes to occur, the iron in dust must be made soluble. The solubility of iron in aqueous phases from dust has been researched and solubility is known to increase significantly at low pH. This is evidenced by figure 12A and 12B as reported by Fu et al. [110]. Decreasing the pH to approx. 1 increases total dissolved iron by a factor > 3. Atmospheric waters are known to be acidic – often with pH of 3-5 so significant iron dissolution would be expected. Cwiertny et al. found that treatment of several test dust samples with acidic media caused 4-16% of total iron content of the sample to become dissolved within 24 hours [111].

a) b)

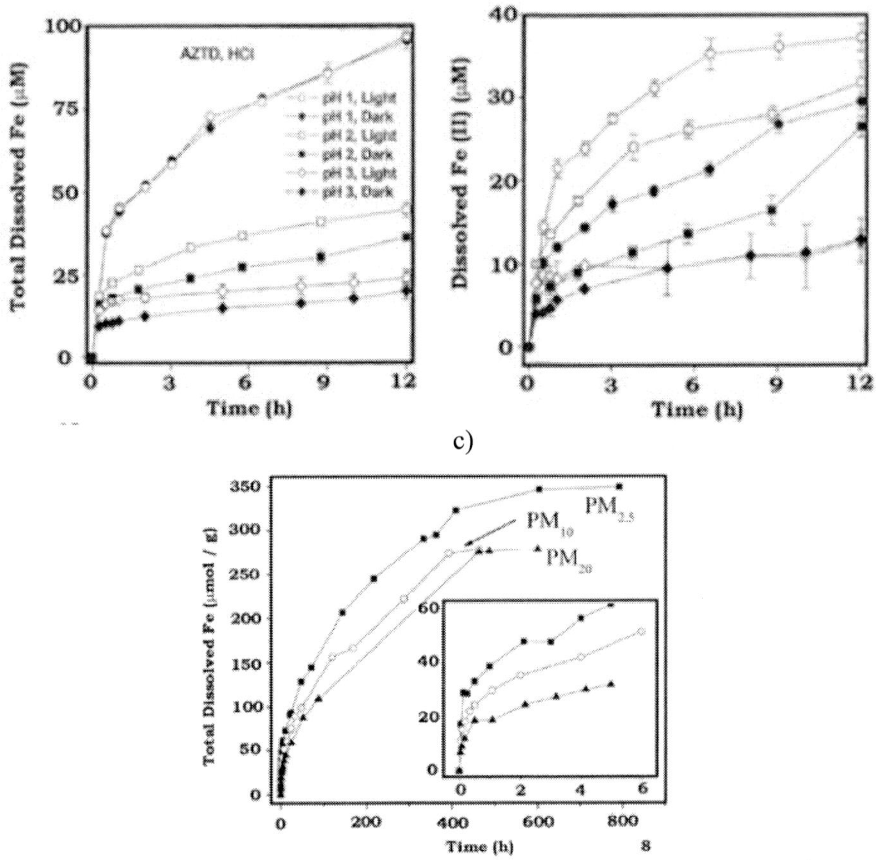

Figure 12. (A) Total dissolved iron and (B) Fe^{2+} as a function of time, pH, and irradiation for Arizona test dust. As observed, dissolution of iron is enhanced at low pH. Figures A and B from Fu et al. [110] reproduced with permission of publisher. (C) Kinetics of iron dissolution for Tibesti (Libya) dust at pH = 1 from Shi et al. [112] with permission of author. The inset shows the rapid initial solubilization of iron occurring within first 10 min. This may suggest significant kinetic limitations to iron dissolution in aqueous aerosol phases (and subsequent REDOX chemistry) do not exist for mineral dusts in earth's atmosphere.

These authors also proposed a REDOX reaction between Fe^{2+} and NO_3^- in which nitrate is reduced to nitrite via:

$$2\,Fe^{2+}\,(aq) + NO_3^- + 2H^+ \rightarrow 2\,Fe^{3+} + NO_2^- + H_2O$$

This is of interest for a variety of reasons, including the possibility of nitrite formation within the aqueous phase. This presumably could be a chemical route to nitration of aromatic organics in the aerosol phase. Another key aspect of iron dissolution is the kinetics of the process. If dissolution is very slow compared to the average residence time of dust, the REDOX chemistry described above may not occur. The rate of iron dissolution has been studied by Shi et al. [112] who observed an initial fast dissolution step, followed by a slower drift to steady state over several hundred hours (see Figure 12C below). As very small quantities of dissolved iron are required to exert the catalytic effects described above, this result suggests there may not be a kinetic limitation to the dissolution of iron within an aqueous phase on atmospheric dusts.

CONCLUSION

Aeolian dust represents a very large source of atmospheric particulate matter. Mineral dust can absorb and scatter solar radiation in the atmosphere, directly affecting climate and atmospheric photochemistry. Dust surfaces can promote ice formation in clouds aloft, directly participate in formation of atmospheric nitrate, and degrade ozone. Iron in dust also likely plays important roles in oxidation-reduction chemistry at dust surfaces.

It may now be useful to consider future directions of research in this field. In the opinion of this author, a significant body of observational data has now been reported for the optical properties of dry mineral dusts, and the key values used in radiative transport models can be constrained to acceptable ranges from the data.

In contrast, the hygroscopicity of dust has only been considered by a few investigators, but additional research in this field is certainly warranted – particularly since dust surfaces can be modified significantly during atmospheric transport.

Given the very large source strength of mineral dust, developing an accurate model for the effect of relative humidity on absorption, scattering and extinction of light is crucial to understanding dusts direct optical effects on weather / climate.

In terms of chemistry of the dusty atmosphere, significant progress has been made. For some key reactions, chemical mechanisms and kinetics have been described to the extent they can be incorporated into air quality models that include the effect of dust surfaces. However, at present there is not much understanding of how iron catalyzed processes may affect atmospheric chemistry on dust surfaces.

It is this author's opinion that considerable effort should be undertaken to explore these reactions given irons ubiquitous presence in mineral dust, and the very high mass flux of dust into the atmosphere. These processes could play very significant roles in altering the kinetics of aqueous phase atmospheric processing of both inorganic and organic molecules.

AUTHOR BIOGRAPHY

Jon Thompson (1976 -) is an Associate Professor in the Department of Chemistry and Biochemistry at Texas Tech University. He received his Ph.D. in Analytical Chemistry with J.D. Winefordner at the University of Florida in 2001. His current research interests include studying the chemical and optical properties of atmospheric aerosols. Additional interests include mentoring the next generation of *"students"* (here his children Jacob and Autumn) that he has been blessed with.

REFERENCES

[1] Shao, Y., Wyrwoll, K. H., Chappell, A., Huang, J., Lin, Z., McTainsh, G. H., Mikami, M., Tanaka, T. Y., Wang, X., Yoon, S. *Aeolian Research* 2011, 2, 181–204.

[2] Huneeus, N., Schulz, M., Balkanski, Y., Griesfeller, J., Prospero, J., Kinne, S., Bauer, S., Boucher, O., Chin, M., Dentener, F., Diehl, T., Easter, R., Fillmore, D., Ghan, S., Ginoux, P., Grini, A., Horowitz, L., Koch, D., Krol, M. C., Landing, W., Liu, X., Mahowald, N., Miller, R., Morcrette, J. J., Myhre, G., Penner, J., Perlwitz, J., Stier, P., Takemura, T., Zender, C. S. *Atmospheric Chemistry and Physics* 2011, 11, 7781-7816.

[3] Sokolik, I. N. and O. B. Toon *Nature* 1996, 381, 681-683.

[4] Zhao, C., Liu, X., Leung, L. R., Johnson, B., McFarlane, S. A., Gustafson Jr., W. I., Fast, J. D., Easter, R. *Atmos. Chem. Phys. Discuss*. 2010, 10, 9753-9799.

[5] Redmond, H. E., Dial, K. D.; Thompson, J. E. *Aeolian Research* 2010, 2(1), 5-26.

[6] Arimoto, R. *Earth-Science Reviews* 2001, 54, 29-42.

[7] Usher, C. R., Michel, A. E., Grassian, V. H. *Chem. Rev*. 2003, 103, 4883-4939.

[8] Chen, H., Navea, J. G., Young, M. A., Grassian, V. H. *J. Phys. Chem. A* 2011, *115* (4), 490–499.

[9] Rubasinghege, G., Elzey, S., Baltrusaitis, J., Jayaweera, P. M., Grasian, V. H. *J. Phys. Chem. Lett.* 2010, *1* (11), 1729–1737.

[10] Sullivan, R. C., Guazzotti, S. A., Sodeman, D. A., Prather, K. A. *Atmos. Chem. Phys.* 2007, 7, 1213–1236.

[11] Zhang, Y. and G. R. Carmichael *J. Appl. Meteor.*, 1999, 38, 353–366.

[12] Toon, O. B. *Nature* 2003, 424, 623-624.

[13] Uno, I., Eguchi, K., Yumimoto, K., Takemura, T., Shimizu, A., Uematsu, M., Liu, Z., Wang, Z., Hara, Y., Sugimoto, N. *Nature Geoscience* 2009, 20, 557-560.

[14] Fairlie, T. D., Jacob, D. J., Park, R. *J. Atmos. Environ.* 2007, 41(6), 1251-1266.

[15] Engelstaedter, S. and R. *Washington Journal of Geophysical Research* 2007, 112, D03103.

[16] Liu, J., Mauzerall, D. L., Horowitz, L. W., Ginoux, P., Fiore, A. M. *Atmos. Environ.* 2009, 43(28), 4327 – 4338.

[17] Lee, Y. C., Yang, X., Wenig, M. *Atmos. Environ.* 2010, 44(30), 3728-3738.

[18] Bory, A., Wolff, E., Mulvaney, R., Jagoutz, E., Wegner, A. Ruth, U., Elderfield, H. *Earth and Planetary Science Letters* 2010, 291(1-4), 138-148.

[19] Burton, S. P., Ferrare, R. A., Hostetler, C. A., Hair, J. W., Obland, M. D., Rogers, R. R., Cook, A. L., Harper, D. B., Froyd, K. D. *Atmospheric Measurement Techniques Discussions* 2011, 4(5), 5631-5688.

[20] Cassar, N., Bender, M. L., Barnett, B. A., Fan, S., Moxim, W. J., Levy II, H., Tilbrook, B. *Science* 2007, 317 (5841), 1067-1070.

[21] Mahowald, N. M., Engelstaedter, S., Luo C.,Sealy, A., Artaxo, P., Benitez-Nelson, C., Bonnet, S., Chen, Y., Chuang, P. Y., Cohen, D. D., Dulac, F., Herut, B., Johansen, A. M., Kubilay, N., Losno, R., Maenhaut, W., Paytan, A., Prospero, J. M., Shank, L. M.; Siefet, R. L. *Annual Rev. Mar. Sci.* 2009, 1, 245–278.

[22] Boyd, P. W., Jickells, T., Law, C. S., Blain, S., Boyle, E. A., Buesseler, K. O., Coale, K. H., Cullen, J. J., De Baar, H. J. W., Follows, M., Harvey, M., Lancelot, C., Levasseur, M., Owens, N. P. J., Pollard, R., Rivkin, R. B., Samiento, J., Schoemann, V., Smetacek, V., Takeda, S., Tsuda, A., Turner, S., Watson, A. J. *Science* 2007, 315(5812), 612-617.

[23] Kohfeld, K. and S. P. *Harrison Earth-Science Reviews* 2001, 54 (1-3), 81-114.

[24] Xuan, J., Liu, G., Du, K. *Atmospheric Environment* 2000, 34(26), 4565-4570.

[25] Xuan, J. and I. N. Sokolik *Atmospheric Environment* 2002, 36(31),4863-4876.

[26] Wells, K. C., Witek, M., Flatau, P., Kreidenweis, S. M., Westphal, D. L. *Atmospheric Environment* 41(31), 2007, 6585-6597.

[27] Laurent, B., Marticorena, B., Bergametti, G., Mei, F. *Global and Planetary Change* 2006, 52(1-4), 121-141.

[28] Rajot, J. L., Gomes, L., Alfaro, S. C., Gaudichet, A. J. *Aerosol Science* 2000, 31(1), 428-429.

[29] Xu, J. et al. *Quaternary Research* 2010, 73(1), 33-38.

[30] Han, Y., Fang, X., Zhao, T., Bai, H., Kang, S., Song, L. *Atmos. Environ.* 2009, 43(3), 568-574.

[31] Evan, A. T., Dunion, J., Foley, J. A., Heidinger, A. K., Velden, C. S. *Geophys. Res. Lett.* 2006, 33, L19813, doi:10.1029/2006GL026408.

[32] DeMott, P. J., Sassen, K., Poellot, M. R., Baumgardner, D., Rogers, D. C., Brooks, S. D., Prenni, A. J., Kreidenweis, S. M. *Geophys. Res. Lett.* 2003, 30(14), 1732.

[33] Cziczo, D. J., Murphy, D. M., Hudson, P. K., Thomson, D. S. J. *Geophys. Res.* 2004, 109, D04201.

[34] Archuleta, C. M., DeMott, P. J., Kreidenweis, S. M. *Atmos. Chem. Phys.* 2005, 5, 2617–2634.

[35] Richardson, M. S., DeMott, P. J., Kreidenweis, S. M., et al. *J. Geophys. Res.* 2007, 112(D2), D02209.

[36] Mohler, O., Field, P. R., Connolly, P., Benz, S., Saathoff, H., Schnaiter, M., Wagner, R., Cotton, R., Kramer, M., Mangold, A., Heymsfield, A. J. *Atmos. Chem. Phys.* 2006, 6, 3007–3021.

[37] Kanji, Z. A. and J. P. D. Abbatt *J. Geophys. Res.* 2006, 111, D16204.

[38] Kulkarni, G. and S. Dobbie *Atmos. Chem. Phys.* 2010, 10, 95–105.

[39] Kl. L. and T. Holzer-Popp *Atmos. Chem. Phys. Discuss.* 2010, 10, 6167–6197.

[40] Sassen, K. *Geophys. Res. Lett.* 2002, 29(10), 1465.

[41] Chylek, P. and J. Wong *Geophys. Res. Lett.* 1995, 22, 929-931.

[42] Mishchenko, M. I., Hovenier, J. W., Travis, L. D. *Light Scattering by Nonspherical Particles: Theory, Measurements, and Applications*; Academic Press: San Diego, CA, 2000.

[43] Taflove, A. and S.C. Hagness Computational electrodynamics: the finite-difference time-domain method; Artech House, Boston, MA, 2000.

[44] Borghese, F., Denti, P., Saija, R. Scattering From Model Nonspherical Particles. Theory and Applications to Environmental Physics; *Springer,* Berlin, Germany, 2007.

[45] Jin, J. *The Finite Element Method in Electromagnetics*; Wiley, New York, NY, 2002.

[46] Babenko, V. A., Astafyeva, L. G., Kuzmin, V. N. Electromagnetic Scattering in Disperse Media: Inhomogeneous and Anisotropic Particles; Praxis, Chichester, UK, 2003.

[47] Mishchenko, M. I., Travis, L. D., Lacis, A. A. Scattering, Absorption, and Emission of Light by Small Particles; Cambridge University Press, Cambridge, UK, 2002.

[48] Li, L. W., Kang, X. K., Leong, M. S. Spheroidal *Wave Functions in Electromagnetic Theory*; Wiley, New York, NY, 2002.

[49] Doicu, A., Eremin, Y. A., Wriedt, T. Acoustic and Electromagnetic Scattering Analysis using Discrete Sources. Academic Press, San Diego, CA, 2000.

[50] Doicu, A., Wriedt, T., Eremin, Y. A. Light Scattering by Systems of Particles. Null-field Method with Discrete Sources: Theory and Programs; *Springer*, Berlin, Germany, 2006.

[51] Wriedt, T. Generalized Multipole Techniques for Electromagnetic and Light Scattering; *Elsevier*, Amsterdam, Netherlands, 1999.

[52] Nousiainen, T. Journal *of Quantitative Spectroscopy and Radiative Transfer* 2009, 110, 1261-1279.

[53] Redmond, H. E., Dial, K. D., J. E. Thompson *Aeolian Research* 2010, 2(1), 5-26.

[54] Gasteiger, J., Wiegner, M., Grob, S., Freudenthaler, V., Toledano, C., Tesche, M., Kandler, K. *Tellus* 2011, 63B, 725-741.

[55] Wagner, R., Ajtai, T., Kandler, K., Lieke, K., Linke, C., Müller,T., Schnaiter, M., Vragel, M. *Atmos. Chem. Phys. Discuss.* 2011, 11, 21363-21427.

[56] Ma, L., Zobeck, T. M., Hsieh, D. H., Holder, D., Morgan, C. L. S., Thompson, J. E. *Aeolian Research* 2011, 3(2), 235-242.

[57] Ansmann, A., Petzold, A., Kandler, K., Tegen, I., Wendisch, M., Uller, D. M., Weinzierl, B., Uller, T. M., Heintzenberg, *J. Tellus* 2011, 63B, 403–442.

[58] Redmond, H. and J. E. Thompson *Phys. Chem. Chem. Phys.* 2011, 13, 6872-6882.

[59] Sokolik, I. N. and O. B. Toon *J. Geophys. Res*. 1999, 104 (D8), 9423–9444.
[60] Li-Jones, X., Maring, H. B., Prospero, J. M. *J. Geophys. Res.* 1998, 103(D23), 31,113-31,121.
[61] Gasso′ S., Hegg, D. A., Covert, D. S., Collins, D., Noone, J., Ostrom, E., Schmid, B., Russell, P. B., Livingston, J. M., Durkees, P. A., Jonsson, H. *Tellus* 2000, 52B, 546–567.
[62] Attwood, A. R. and M. E. Greenslade *Aerosol Science and Technology* 2011, 45(11), 1350-1359.
[63] Lack, D. A., et al. *Geophys. Res. Lett.* 2009, 36, L24805, doi: 10.1029/2009GL041002.
[64] Liao, H., Yung, Y. L., Seinfeld, J. H. *J. Geophys. Res*. 1999, 104(D19), 23697 – 23707.
[65] Di Sarra, A., Cacciani, M., Chamard, P., Cornwall, C., DeLuisi, J. J., Di Iorio, T., Disterhoft, P., Fiocco, G., Fuá, D. Monteleone, F. *J. Geophys. Res.* 2002, 107(D18), 8135.
[66] Meloni, D., Di Sarra, A., Fiocco, G., Junkermann, W. *J. Geophys. Res*. 2003, 108(D10), 4323, doi:10.1029/2002JD003293.
[67] Balis, D. S., Zerefos, C. S., Kourtidis, K., Bais, A. F., Hofzumahaus, A., Kraus, A., Schmitt, R., Blumthaler, M., Gobbi, G. P. *J. Geophys. Res*. 2002, 107(D18), 8138, doi:10.1029/2000JD000136.
[68] Jeong, G. R. and I. N. Sokolik *J. Geophys. Res.* 2007, 112, D21308, doi:10.1029/2007JD0008442.
[69] Casasanta, G., Di Sarra, A., Meloni, D., Monteleone, F., Pace, G., Piacentino, S., Sferlazzo, D. *Atmos. Environ.* 2011, 45(24), 3937-3943.
[70] Ying, Z. M., Tie, X. X., Madronich, S., Li, G. H., Massie, S. *Atmos. Environ.* 2011, 45(15), 2549-2558.
[71] Bian, H. and C. S. Zender *J. Geophys. Res.* 2003, 108(D21), 4672, doi:10.1029/2002JD003143.
[72] Martin, R. V. et al. *J. Geophys. Res*. 2002, 107(D18), 4351, doi:10.1029/2001JD001480.
[73] Usher, C. R., Michel, A., Grassian, V. H. *Chemical Reviews* 2003, 103, 4883 – 4940.
[74] Lundgren, D. A. *J. Air Pollut. Control. Assoc.* 1970, 20, 603-608.
[75] Kadowaki, S. *Atmos. Environ*. 1977, 11, 671-675.
[76] Hara, H., Kato, T., Matsushita, H. *Bull. Chem. Soc.* Japan 1989, 62, 2643-2649.
[77] Appel, B. R., Kothny, E. L., Hoffer, E. M., Hidy, G. M., Wesolowski, J. *J. Environ. Sci. and Technol.* 1978, 12, 418-428.
[78] Hara, H., Honda, K., Nagara, K., Goto, A. *J. Chem. Soc*. Japan. 1983, 8, 1221-1225.
[79] Xie, R., Jackson, K. A., Martin Seip, H., McLeod, C. W., Wibetoe, G., Schofield, M. J., Anderson, D., Hanssen, J. E. *J. Environ. Monit*. 2009, 11, 336-343.
[80] Zhao, Y. and Y. Gao *Atmos. Environ*. 2008, 42, 4063-4078.
[81] Pryor, S. C., Barthelmie, R. J., Jensen, B. *Geophys. Res. Lett.* 1999, 26(6), 691-694.
[82] Wu, P. M. and K. Okada *Atmos. Environ*. 1994, 28(12), 2053-2060.
[83] Kojima, T., Buseck, P. R., Iwasaka, Y., Matsuki, A., Trochkine, D. *Atmos. Res*. 2006, 82(3-4), 698-708.
[84] Krueger, B. J., Grassian, V. H., Iedema, M. J., Cowin, J. P., Laskin, A. *Anal. Chem.* 2003, 75, 5170-5179.
[85] Krueger, B. J., Grassian, V. H., Cowin, J. P., Laskin, A. *Atmos. Environ.* 2004, 38, 6253-6261.

[86] Zhuang, H., Chan, C. K., Fing, M., Wexler, A. S. *Atmos. Environ.* 1999, 33, 4223-4233.
[87] Zhang, D. and Y. Iwasaka *Atmos. Environ.* 1999, 33, 3213-3223.
[88] Song, C. H. and G. R. *Carmichael Atmos. Environ.* 1999, 33, 2203-2218.
[89] Ro, C. U., Hwang, H. J., Kim, H., Chun, Y. S., Van Grieken, R. *Environ. Sci. and Technol.* 2005, 39, 1409-1419.
[90] Preszler-Prince, A., Kleiber, P. D., Grassian, V. H., Young, M. A. *Physical Chemistry Chemical Physics* 2007, 9, 3432-3439.
[91] Preszler-Prince, A., Kleiber, P. D., Grassian, V. H.; Young, M. A. *Physical Chemistry Chemical Physics* 2007, 9, 622-634.
[92] Liu, Y., Gibson, E. R., Cain, J. P., Wang, H., Grassian, V. H., Laskin, A. *J. Physical Chemistry* A 2008, 112, 1561-1571.
[93] Cwiertny, D. M., Young, M. A., Grassian, V. H. *Anu. Rev. Phys. Chem.* 2008, 59, 27-51.
[94] Hwang, H. and C. U. Ro *Atmos. Environ.* 2006, 40(21), 3869-3880.
[95] Kim, B. G. and S. U. Park *Atmos. Environ.* 2001, 35(18), 3191-3201.
[96] Dall'Osto, M., Harrison, R. M., Highwood, E. J., O'Dowd, C., Ceburnis, D., Querol, X., Achterberg, E. P. *Atmos. Environ.* 2010, 44(26), 3135-3146.
[97] Song, C. H., Maxwell-Meier, K., Weber, R. J.; Kapustin, V., Clarke, A. *Atmos. Environ.* 2005, 39(2), 359-369.
[98] Sullivan, R. C, Guazzotti, S. A., Sodeman, D. A., Tang, Y., Carmichael, G. R., Prather, K. A. *Atmos. Environ.* 2007, 41(34),7166-7179.
[99] Hodzic, A., Bessagnet, B., Vautard, R. *Atmos. Environ.* 2006, 40(22), 4158-4171.
[100] Shi, Z., Zhang, D., Hayashi, M., Ogata, H., Ji, H., Fujiie, W. *Atmos. Environ.* 2008, 42(4), 822-827.
[101] Mogili, P. K., Kleiber, P. D., Yong, M. A., Grassian, V. H. *Atmos. Environ.* 2006, 40(38), 7401-7408.
[102] Michel A. E., Usher C. R., Grassian V. H. *Geophys. Res. Lett.* 2002, 29, 1665; doi: 10.1029/2002GL014896.
[103] Michel A. E., Usher C. R., Grassian V. H. *Atmos. Environ.* 2003, 37, 3201–3211.
[104] Usher, C. R., Michel, A. E., Stec, D., Grassian, V. H. *Atmos. Environ.* 2003, 37(38), 5337-5347.
[105] Wardman, P. and L. P. *Candeias Radiation Research* 1996, 145(5), 523-531.
[106] Gelencser, A. Hoffer, G. Kiss, E., Tombacz, R. Kurdi, and L. Bencze *J. Atmos. Chem.* 2003, 45, 25–33.
[107] Ruppert, G., Bauer, R., Heisler, G. *J. Photochemistry and Photobiology A: Chemistry* 1993, 73(1), 75-78.
[108] Araña J, Tello Rendón, E., Doña Rodríguez J. M., Herrera Melián, J. A., González Díaz, O., Pérez Peña, J. *Chemosphere* 2001, 44(5), 1017-1023.
[109] Chang, J. and J. E. Thompson *Atmos. Environ.* 2010, 44(4), 541-551.
[110] Fu, H., Cwiertny, D. M.; Carmichael, G. R., Scherer, M. M., Grassian, V. H. *J. Geophys. Res.* 2010, 115, D11304, doi:10.1029/2009JD012702.
[111] Cwiertny, D. M., Baultrusaitis, J., Hunter, G. J., Laskin, A., Scherer, M. M., Grassian, V. H. *Journal of Geophysical Research* 2008, 113, D05202, doi:10.1029/2007JD009332.
[112] Shi, Z., Bonneville, S., Krom, M. D.; Carslaw, K. S., Jickells, T. D., Baker, A. R., Benning, L. G. *Atmos. Chem. Phys.* 2011, 11, 995–1007.

[113] Intergovernmental Panel on Climate Change (I.P.C.C), Climate Change 2001. *The Scientific Basis.* Chapter 5. Cambridge University Press, Cambridge, UK, 2001.

[114] Hair, J. W., Hostetler, C. A., Cook, A. L., Harper, D. B., Ferrare, R. A., Mack, T. L., Welch, W., Izquierdo, L. R., Hovis, F. E. *Applied Optics* 2008, 47(36), 6734-6752, doi: 10.1364/AO.47.006734.

[115] Draxler, R. R. and G. D. Rolph, 2011. HYSPLIT (HYbrid Single-Particle Lagrangian Integrated Trajectory) Model access via NOAA ARL READY Website (http://ready.arl.noaa.gov/HYSPLIT.php). NOAA Air Resources Laboratory, Silver Spring, MD.

[116] Rolph, G. D., 2011. Real-time Environmental Applications and Display sYstem (READY) Website (http://ready.arl.noaa.gov). NOAA Air Resources Laboratory, Silver Spring, MD.

In: Dust: Sources, Environmental Concerns and Control ISBN: 978-1-61942-547-7
Editors: Laurent B. Wouters and Michel Pauwels

Chapter 7

DUST EXPLOSIONS: PROTECTION OF SILOS BY VENTING

Alberto Tascón[1] and Pedro J. Aguado[2]
[1]Departamento de Agricultura y Alimentación,
Universidad de La Rioja, Spain
[2]Departamento de Ingeniería y Ciencias Agrarias, Universidad de León, Spain

ABSTRACT

It is well established that dust explosions represent a serious hazard in many process industries and silo facilities. Many catastrophic cases have been reported, and statistics show that the number of incidents per year is still significant. Venting is the most common protective measure used in silos in order to prevent the appearance of unacceptably high internal pressures. However, the installation of venting devices is not always simple, and technical complications frequently arise. Furthermore, the cost of protection by venting must be taken into account. Vent areas in silos should be large enough to prevent damaging overpressures, but not so large that the use of venting becomes impracticable. Thus, vent area sizing is a critical issue. To calculate vent area size, standards EN 14491 (2006) and NFPA 68 (2007) are commonly used in Europe and North America, respectively, but in certain situations, they are contradictory.

The aim of this study was to analyze dust explosion venting in silos by comparing the above venting standards and also conducting CFD simulations. The pressures and associated vent areas in these numerical simulations were compared to those contemplated in the standards. In addition, some calculations were carried out for venting devices with inertia. The simulated explosion pressures showed the expected trends for the associated vent areas and agreed reasonably well with the values contemplated in NFPA 68 (2007).

For low overpressure values, the differences were significant compared to those vent areas predicted by standard EN 14491 (2006). Numerical simulations could serve as a powerful tool for helping engineers to design explosion protection and calculate the structure of vented silos. However, the results of numerical simulations should be extrapolated with caution since they are absolutely dependent on the flow characteristics and the dust concentration of the initial dust cloud.

1. Introduction

Many materials react to oxygen. One example of this process is a piece of wood that burns. The rate of such a chemical reaction determines its consequences. When the material is split into small particles, i.e. dust, the speed of the process is significantly increased and becomes violent. Dust explosions occur when combustible dust is dispersed in the air as a cloud and reacts with oxygen in the presence of an ignition source, generating a violent combustion reaction in the form of flame propagation [1]. The materials that can cause dust explosions include agricultural and food products (grain, malt, flour, maize starch, sugar, wood, etc.), synthetic organic materials (plastics, pigments, pesticides, etc.), metals (aluminum, magnesium, etc.) and coal. A comprehensive list of combustible dusts and their characteristics have been reported elsewhere [2]. A silo is a large vertical container used for storing granular or powdered material, usually built from reinforced concrete, stiffened metal plate or polyester. Many of the materials normally stored in such containers produce combustible dusts. When an explosion occurs within a constant volume, for example a silo, there is a rapid and significant increase in pressure due to the production of hot gases, typically reaching 7-10 bar (700-1000 kPa). The first dust explosion reported in the literature occurred in 1785 in the wheat flour warehouse of a bakery in Italy [3]. Since then, a significant number of major dust explosions and their catastrophic consequences have been reported [1, 4]. Some authors have studied the frequency of dust explosions [4]. However, not all dust explosion events, especially the minor ones, are reported: some studies indicate that only about 15% of the dust explosions which have occurred have been recorded. Thus, the statistics are indicative rather than quantitative. One of the most recent studies on dust explosion frequencies was carried out by the Chemical Safety and Hazard Investigation Board (CSB) in the USA [5]. The CSB report, which excluded incidents in coal mines and grain elevators, identified nearly 300 dust fires and explosions in US industrial facilities from 1980 to 2005, resulting in 119 fatalities and 718 injuries. The statistics collected in the study showed that the number of events per year appears to be holding steady, and may even be increasing. A high percentage of dust explosions involve silos, particularly those used in the agricultural and food industries. This can be deduced from the statistics from the USA presented by Schoeff [6], and those from Germany reported by Jeske & Beck [7]. The most recent major silo accidents occurred in Blaye (France) in 1997 and in Wichita (Kansas) in 1998. The former killed 11 people and destroyed the majority of the reinforced concrete cells, the over-cell gallery and one of the two handling towers of the port storage facility [4]. The latter occurred at the largest grain elevator in the world at that time, which consisted of 246 circular reinforced concrete silos, and damaged many of the discharging tunnels, silo cells and galleries of the elevator complex, killing 7 people and injuring 10 more [8].

2. An Overview of Dust Explosion Protection in Silos

The measures taken to protect against explosions should adhere to three principles, given here in order of priority: prevention of explosive dust cloud formation, avoidance of ignition, and mitigation of explosion effects [9].

However, bulk materials inevitably form dust clouds in the silo during filling or empting of the vessel, posing an explosion hazard. Thus, fighting potential ignition sources that could trigger such dust clouds is essential. However, due to the countless and complex ways in which dust explosions can occur, it is often impossible to completely eliminate the risk of a dust explosion. Moreover, it should be noted that in many of the cases analyzed in the literature, the primary explosion occurred outside the silo cells, in conveyors, bucket elevators, ducts or other auxiliary equipment, but later spread with increasing violence throughout the system; this severely complicates explosion prevention at an industrial plant. Therefore, in many situations, it is also necessary to consider mitigation measures to reduce the damage that might be produced.

Dust explosion mitigation strategies in silos include containment, venting and suppression, in addition to isolation in order to avoid spreading to other parts of the installation [10]. Containment [11, 12] is appropriate for small and robust vessels and for bins set into a building where venting is not possible, but not for large silos or elevators. Suppression [11, 13] is also a good option for small vessels, although it is important to ensure that the suppressant does not compromise the hygienic conditions required, especially in the food industry, or breach other environmental and toxicological regulations. Thus, venting [14, 15] is the most common protective measure used in silos, except when toxic dusts are involved, owing to its more favorable technical and economic features; indeed, it is usually the only option in large silos.

Venting consists of the release of the explosion through an opening of the appropriate size. Vent devices open at a pre-determined pressure to prevent the explosion pressure in the protected vessel from reaching unacceptably high levels. Correctly sized vent openings can ensure that sufficient quantities of gas (and particulates) are released to avoid destructive pressures. The maximum overpressure generated by an explosion in a vessel protected by venting is the reduced explosion pressure (P_{red}). Explosion venting for silos requires installing vent devices in the upper sidewall, always above the maximum height of the material contained in the silo, or in the roof.

Vent areas in silos must be large enough to prevent unacceptably high pressures, but not so large that the use of vents becomes impracticable. Vent area sizing is, therefore, a critical issue. The recommendations given in standards EN 14491 [14] and NFPA 68 [15], commonly used in Europe and North America, respectively, can be used to size the vent area. Abbasi & Abbasi [4], Barton [16] and Eckhoff [1] have compiled comprehensive lists of other methods for dust explosion vent design. However, standards EN 14491 (2006 edition) [14] and NFPA 68 (2007 edition) [15] are, in certain situations, somewhat contradictory [17].

For silos, an alternative calculation method can be found in Annex A of the German VDI 3673 Guidelines (2002 edition) [18]. Although the European standard EN 14491 (2006 edition) is based on this previous standard, it does not include the alternative method provided therein. These alternative formulae can be used to determine the vent area required in silos for pneumatic conveying and for free-fall filling. The vent areas calculated using these alternative equations are notably smaller than those calculated using the general equation in EN 14491 or VDI 3673 [17].

The above explosion venting standards propose empirical formulae that rely on experimental data. In spite of the extensive research and development conducted over the last four decades [1, 19], our basic understanding of the processes inside and outside vented vessels, including dust cloud generation and cloud burning, is still unsatisfactory [20].

Venting neither prevents nor extinguishes an explosion; it merely limits the explosion overpressure. Thus, undesirable external effects, including flame, blast waves and flying debris, are to be expected. The above standards on venting advise on these hazardous effects and provide methods for estimating them [14, 15]. Furthermore, the flow exiting from the vent opening produces recoil forces in the direction opposite to the flow. Such recoil forces should be considered in the structural design of the vented equipment and its supports. In addition, the above venting standards include formulae to estimate the magnitude of these forces and their duration [14, 15].

Venting devices include bursting discs and panels, explosion doors and other less usual systems [21]. Common materials are metals, alloys and plastic membranes. Bursting discs and panels are almost inertia-free devices and an efficiency of 1 is usually considered in calculations. In contrast, other venting devices, for example explosion doors, have an efficiency of less than 1 due to inertia or design features. One of the most recent developments in this field has been the flameless venting device, which is very useful for equipment set into a building [1].

The venting of silos is not always simple and complications frequently arise. Many silos, especially the large steel cylindrical silos used in the agricultural and food industries, have a low-strength roof and upper wall. This renders the installation of vents technically difficult because the required vent area increases when the pressure that the vessel is able to resist decreases. In addition, the cost of protection by venting should be taken into account as it may represent an important burden. Protection costs may involve a significant percentage of the total cost of silo facilities [17].

Standards on explosion venting provide a simple method to calculate vent areas for most practical applications. However, the need for a differentiated approach to assess vent sizes is becoming generally accepted, in view of the very different levels of turbulence, dust concentration, and homogeneity of dust distribution found in industry. It is foreseeable that conventional empirical equations will gradually be complemented by comprehensive computer models that are able to predict propagation of dust explosions in the variety of complex situations encountered in industrial facilities [20].

3. The DESC Project

The Dust Explosion Simulation Code (DESC) project was funded by the European Commission under the Fifth Framework Programme and involved 11 partners. The main purpose of the project was to develop a simulation tool to predict the course of dust explosions based on Computational Fluid Dynamics (CFD). The DESC programme has been commercially available from GexCon since 2005. Further details on the DESC project can be found elsewhere [22].

DESC is based on the previously existing CFD programme for gas explosions know as FLame ACceleration Simulator (FLACS). It is a finite volume code in which transport equations for the mass, momentum, enthalpy, fuel, mixture fraction, turbulent kinetic energy and the turbulent energy dissipation rate of the fluid flow are solved on a structured Cartesian grid. Solid objects are mapped onto this grid using a distributed porosity concept [23]. The modeling approach adopted involves the extraction of the necessary input combustion parameters from pressure-time histories measured in standardized tests in 20-liter explosion vessels [24, 25]. Scenarios taking into account the geometry of the problem, the characteristics of the dust cloud, the grid, the ignition point and ignition energy, monitoring points and output variables etc., are defined in the code pre-processor, and simulation results are presented in the post-processor. The combustion and turbulence models implemented in the program, and the approach adopted for DESC development, are those described by Skjold *et al.* [24, 25]. Experimental verification was carried out by comparing numerical simulations with several experiments described in the literature, and reasonably good agreement was obtained [22, 24, 25].

4. SIMULATION OF VENTED DUST EXPLOSIONS IN SILOS

This chapter presents some simulations of vented dust explosions in silos conducted using a commercial CFD-code, the DESC (Dust Explosion Simulation Code), in order to predict pressure build-up over time. Subsequently, explosion pressures and the vent sizes necessary to deal with them were compared with the values predicted in standards EN 14491 [14] and NFPA 68 [15]. The silos simulated were cylindrical flat-bottomed silos with mechanical feeding systems - a very common type of silo used to store grain and other granular and powdered products, especially in the agricultural and food industries. The silos were 6 m in diameter (D) and either 9, 12 or 15 m in height (L). To simplify input of the geometry data, the usual conical roof of this type of silos was transformed into a prolongation of the cylindrical silo wall with the same volume and a flat roof.

Eight inertia-free pressure relief panels were considered in the flat roof, each sized 1 m^2. The total vent area could be varied from 0 m^2 to 8 m^2 and vent panel activation pressure (P_{stat}) was set to 0.1 bar. The ignition was located at 0.50 m above the silo bottom and ignition energy was set to 10 kJ, which is the standard energy used in explosion severity tests [26, 27].

To observe variations in essential parameters during the explosion, several monitoring points were selected along the vertical axis of the cylinder, one every 2 m, and eight others were situated on a plane near the roof, one for every 45°.

The grid defined in DESC was three-dimensional, with cells 0.2 m in size. Information on the sensitivity of the results with respect to the grid resolution defined in DESC can be found elsewhere [28].

Maize starch (P_{max} = 8.63 bar and K_{st} = 149 bar.m/s) was used for all simulations, entering the corresponding explosion characteristics as supplied by GexCon on the DESC program's fuel file. The silo was considered completely filled by a dust cloud - in principle the most dangerous of situations according to standards EN 14491 [14] and NFPA 68 [15].

The most challenging task was to define the initial characteristics of the dust cloud generated in the silo: level of turbulence, air velocity and dust concentration. It is well known that such parameters play an important role in the ignition, propagation and violence of dust explosions [1]. However, there is a significant lack of published data in this area for large vessels under real industrial conditions. Hauert, Vogl and Radant [29, 30], carried out comprehensive experiments using maize starch and wheat flour inside a 12 m^3 steel silo at the FSA (Forschungsgesellschaft für Angewandte Systemsicherheit und Arbeitsmedizin, Research Centre for Applied System Safety and Industrial Medicine) as part of the European

CREDIT (Cooperative Research of Exploding Dusts on Industrial Terrains) project [31]. Hauert *et al.* used the Laser-Doppler-Anemometry (LDA) technique to measure the flow field in the silo. Dust concentration was measured by a probe developed by the FSA and based on the principle that the intensity of a light beam is attenuated when it penetrates a cloud containing solid particles. Measurements were taken under three different feeding conditions: a pneumatic conveying system with axial or with tangential release and a mechanical system (free-fall after a screw conveyor). The measuring equipments were placed at different heights and also in different positions in radial directions inside the 12 m^3 silo.

The approach adopted for the simulations presented in this chapter consisted in selecting the dust concentrations, mean air velocities and RMS turbulence velocities measured in the FSA 12 m^3 silo when fed mechanically with maize starch. The data were obtained from the CREDIT Final Technical Report [30], averaging the values for the entire volume of the FSA 12 m^3 silo. Thus, a simplified homogeneous initial dust cloud which completely filled the silo was considered in the simulations. More detailed information on these aspects is available elsewhere [28]. Table 1 shows the initial dust cloud conditions selected for the simulations (condition set 0).

Table 1. Initial dust cloud conditions selected for the simulations

Parameters	Condition set 0	Condition set 1	Condition set 2
Dust concentration	300 g/m^3	500 g/m^3	500 g/m^3
Air velocity	0.15 m/s	0.15 m/s	0.15 m/s
Turbulence intensity	112%	80%	104%

The results of the DESC simulations are provided as graphs and text files. Special interest was paid to the overpressures, which determine the structural analysis of a vented silo. Figure 1 shows a comparison of the overpressure time histories for four different vent areas in a silo 6 m in diameter and 12 m in height. Such overpressures were recorded at a monitoring point situated over the z-axis at 5.10 m from the silo bottom. Note that the pressure in bars denotes gauge pressure, i.e., pressure above atmospheric pressure. Pressure distribution in the volume of the silos during the explosions was uniform, i.e., the pressures recorded at different monitoring points were almost identical.

Negative pressure phases over the pressure-time curve appeared in some of the numerical simulations; the maximum values reached were about -0.02 bar. These underpressures seemed to follow certain trends: they increased with the height to diameter (L/D) ratio of the silo; for the same silo geometry they increased with vent area size; and they increased as vent activation pressure decreased.

Such underpressures could induce strong compression of the silo walls and a buckling failure of the structure. Therefore, it may not be sufficient to only take the maximum overpressure reached by vented explosions into account when designing a silo structure. However, neither venting standard EN 14491 [14] nor NFPA 68 [15] warn of these underpressures when venting is provided, and no method seems to be available in the literature for predicting these potential underpressures. This issue is different to that of the underpressures developed in vessels fitted with explosion doors, which are described in EN 14491 and NFPA 68 and controlled by using vacuum breakers.

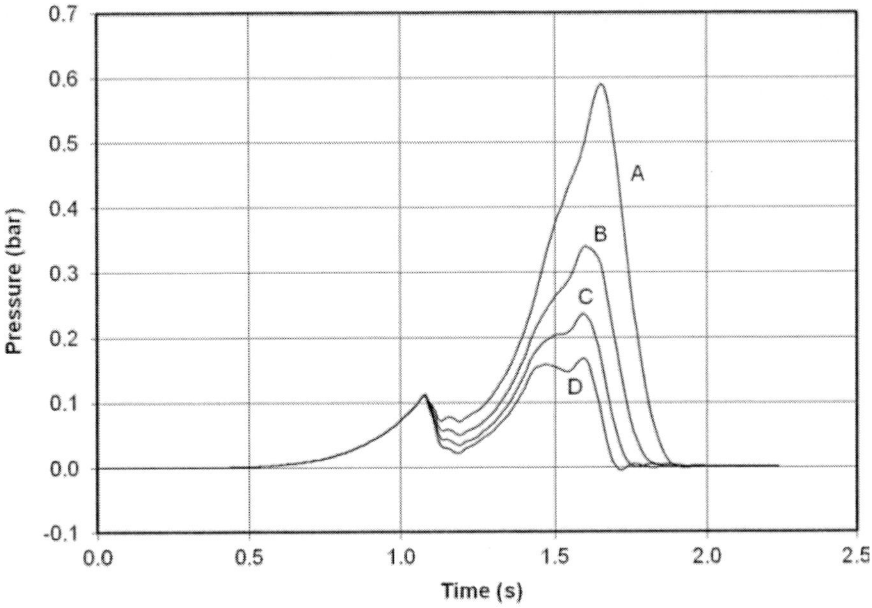

Figure 1. Overpressure time histories. Maize starch. D=6m. L=12 m. P_{stat}=0.1 bar. Condition set 0. Vent area = 4 m^2 (A), 5 m^2 (B), 6 m^2 (C) and 7 m^2 (D).

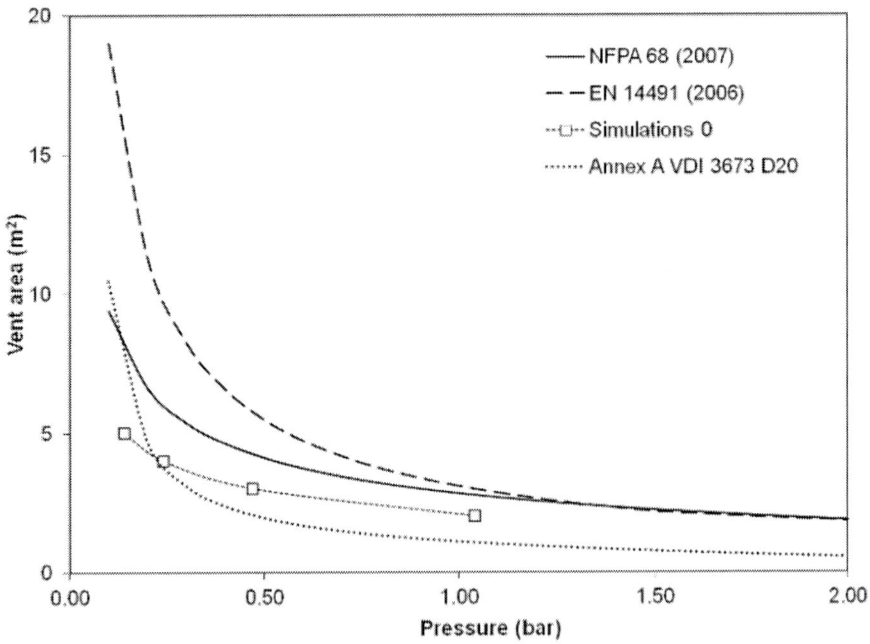

Figure 2. Correlation of the reduced explosion overpressure with the vent area. Comparison of results obtained by computer simulation with the values contemplated in EN 14491 [14], NFPA 68 [15] and Annex A-VDI 3673 [18]. Condition set 0. Maize starch. D = 6 m. L = 9 m. P_{stat} = 0.1 bar.

Figure 2 shows the correlations of the maximum reduced explosion overpressures (P_{red}) obtained by computer simulation using the DESC code with the corresponding vent areas for silos 9 m in height. Figure 3 shows the same for silos 12 m in height, and Figure 4 shows the same for silos 15 m in height. The results were compared to the vent areas required by standards EN 14491 [14] and NFPA 68 [15]. The overpressures plotted are the maximum values recorded at any monitoring point during the explosion simulation. In addition, the vent areas proposed by the alternative method presented in Annex A of VDI 3673 (2002) [18] were included; these areas were calculated for a feed pipe diameter of 0.20 m.

With respect to EN 14491 [14], the general equation for isolated enclosures with P_{red}<1.5 bar was used to calculate the vent areas. For NFPA 68 [15], the general equation for calculating the minimum necessary vent area with low or medium turbulence was followed; and the correction for L/D values of >2 was included when necessary.

The trends of the curves for the DESC simulations were similar to those obtained using standards EN 14491 [14] and NFPA 68 [15]. However, the overpressure values obtained by DESC were lower than those predicted by the standards. For low overpressure values, the differences were very significant compared to those vent areas predicted by standard EN 14491. The differences between the DESC simulations and standards EN 14491 [14] and NFPA 68 [15] may be explained by the initial dust cloud conditions considered. For the simulations presented in this chapter, dust concentration, air velocity and turbulence were set to those measured during the mechanical filling process of a small, real silo [30]. In contrast, in the explosion test series on which the present standards are partially based [32], the dust clouds were generated by the discharge of pressurized dust containers. This method attempted to reproduce the conditions used in laboratory explosion tests, in which the mean velocity and the RMS turbulence are much higher than into a silo [30].

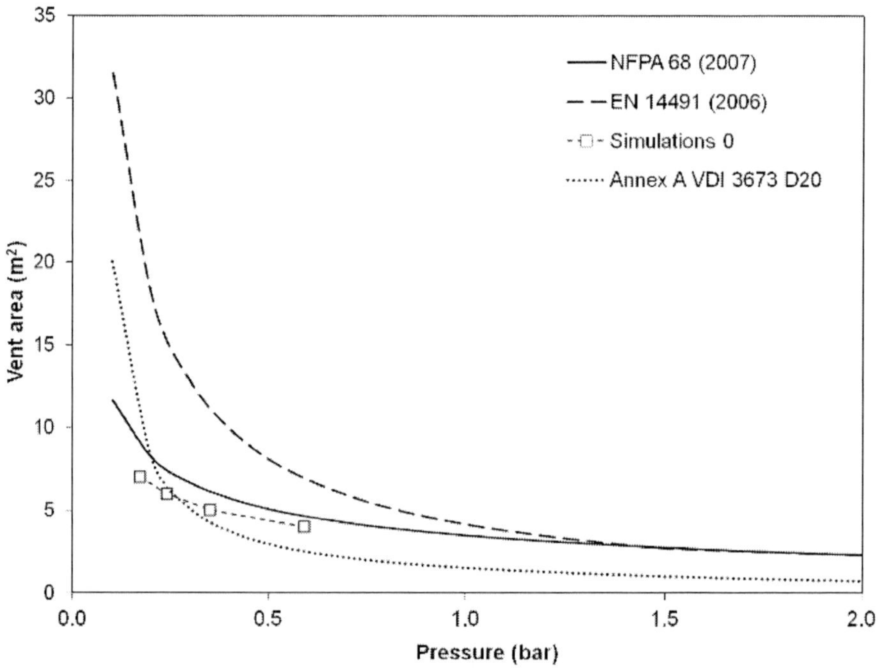

Figure 3. Correlation of the reduced explosion overpressure with the vent area. Comparison of results obtained by computer simulation with the values contemplated in EN 14491 [14], NFPA 68 [15] and Annex A-VDI 3673 [18]. Condition set 0. Maize starch. D = 6 m. L = 12 m. P_{stat} = 0.1 bar.

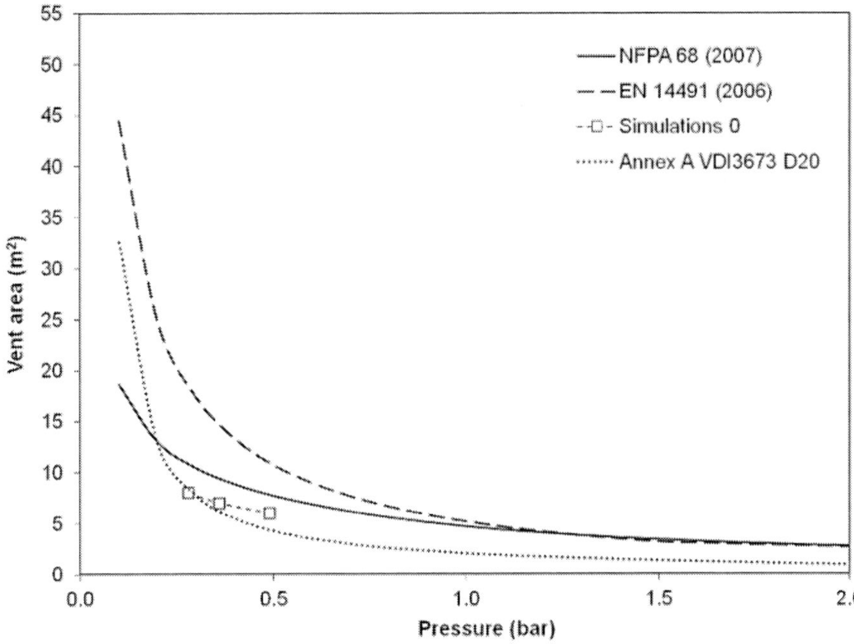

Figure 4. Correlation of the reduced explosion overpressure with the vent area. Comparison of results obtained by computer simulation with the values contemplated in EN 14491 [14], NFPA 68 [15] and Annex A-VDI 3673 [18]. Condition set 0. Maize starch. D = 6 m. L = 15 m. P_{stat} = 0.1 bar.

The significant differences between EN 14491 [14] and NFPA 68 [15] for low overpressures may be explained by the vent area correction for elongated vessels which is required by EN 14491 when P_{red} < 1.5 bar and L/D > 1. This correction imposes a notable increment in vent area if P_{red} is very low [17]. The equations in NFPA 68 also introduce a correction for elongated vessels when L/D > 2 but it imposes more moderate increments of vent area than EN 14491 (2006) when P_{red} is very low [17].

4.1. Influence of Dust Cloud Characteristics on Simulations

In later simulations, the influence of the initial dust cloud conditions on the reduced explosion overpressure was studied. For this purpose, simulations were carried out for three different sets of conditions: condition set 0, in which the values reported by the FSA researchers, averaged for the entire volume of the experimental silo, were used; condition set 1, in which a higher dust concentration and a lower turbulence were assumed; and condition set 2, in which the dust concentration was the same as in condition set 1 but turbulence was increased. Table 1 describes the initial dust cloud conditions selected for these three groups of simulations.

A higher concentration for the initial dust cloud (500 g/m^3) might correspond to silos made of corrugated steel plates, which can hold large dust deposits, thicker than those found on polished or smooth walls. Such layers of settled dust would be dispersed in the air by turbulence produced by the filling process and the explosion itself, increasing the average dust concentration inside the silo. Corrugated steel plates are widely used in silos for the storage of grain and similar materials because of their better mechanical behaviour. The reduction in turbulence was supported by the reasoning that in silos much larger than the 12 m^3 FSA silo, the ratio of the volume to the capacity of the mechanical feeding system implemented often increases, reducing the average turbulence inside the vessel [28]; a value of 0.80 (80%) was selected for condition set 1. In condition set 2, the dust concentration was the same as in condition set 1, but the relative turbulence intensity was increased by 30% to study its effect on the results.

A comparison of the reduced overpressures and the related vent areas obtained for the three sets of conditions is presented in Figure 5. According to the results, the dust concentration had a strong influence on the reduced explosion overpressure. For condition set 1 (500 kg/m^3), the overpressure increased markedly, despite a lower turbulence than in condition set 0 (300 kg/m^3). It is important to remember that the optimum concentration is approximately 750 kg/m^3 for the maize starch used in the simulations. In a silo 6 m in diameter and 9 m in height, an average increment of 55% was found in the reduced explosion pressure from condition set 0 to condition set 1 (minimum of 46%; maximum of 63%, standard deviation of 7.15).

By comparing condition sets 1 and 2, it was possible to analyze the influence of turbulence level on explosion violence, as the dust concentration remained constant. In a silo 6 m in diameter and 9 m in height, an average increment of 10% was found in P_{red} from condition set 1 to condition set 2 (minimum of 7%; maximum of 13%, standard deviation of 2.24).

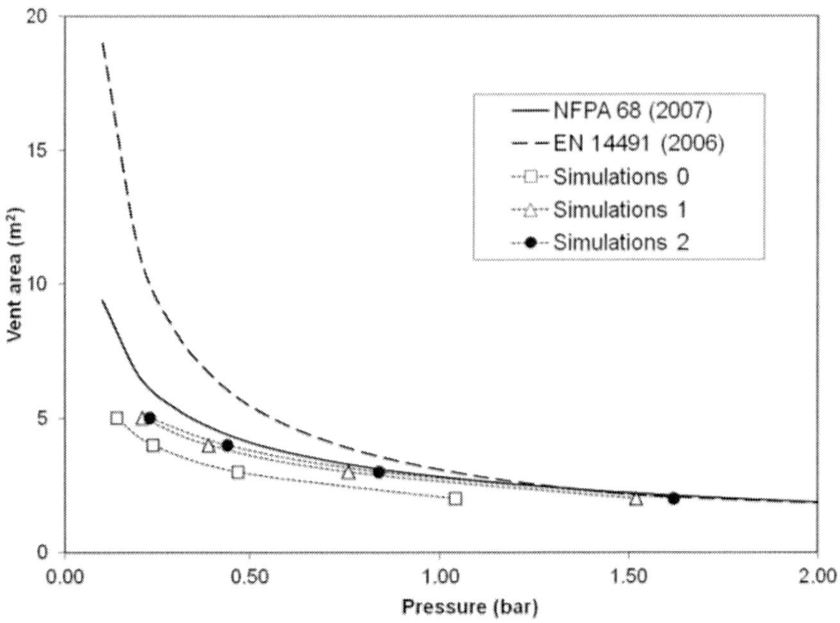

Figure 5. Correlation of the reduced explosion overpressure with the vent area. Comparison of results obtained by computer simulation with the values contemplated in EN 14491 [14] and NFPA 68 [15]. Condition sets 0, 1 and 2. Maize starch. D = 6 m. L = 9 m. P_{stat} = 0.1 bar.

For condition sets 1 and 2, the results matched the values predicted by NFPA 68 [15] very well. The overpressure values obtained by DESC were lower than those found in NFPA 68 for silos 9 m in height, and slightly greater for silos 12 m and 15 m in height. When the P_{red} value was small, the results obtained were very far from the values considered by EN 14491 [14], even for condition set 2, which was the most unfavorable one.

These results would seem to confirm the widespread belief that vent areas calculated according to EN 14491 [14] are very conservative compared to what might really be necessary for most large silos [33, 34]. However, some potential scenarios could involve conditions of temporarily high turbulence and dust concentration levels, for example a secondary dust explosion.

When isolation measures have been implemented, the likelihood of secondary explosions is remote, and it seems clear that EN 14491 considerably oversized vent areas for large isolated silos with low turbulence levels and non-homogeneous dust clouds.

4.2. Influence of Inertia

Simulations were also carried out for venting devices with inertia in a silo 6 m in diameter and 15 m in height. It was assumed that half of the silo top was designed to act as a vent. Thus, the total vent area of the silo was 14.14 m^2. The mass of this venting device was set to 20 kg/m^2, which is equal to the mass of a 2.5 mm thick steel plate. The activation pressure of the venting device was set to 0.03 bar.

Figure 6 shows the correlations of the maximum reduced explosion overpressures (P_{red}) obtained by computer simulation using the DESC code with the corresponding vent areas. The results were compared to the vent areas required by standard NFPA 68 [15]. Both condition sets 0 and 1 are shown in Figure 6.

It should be noted that standard EN 14491 [14] does not provide any method to calculate vent areas if venting devices with inertia are used. Instead, it proposes experimental determination of the efficiency of such devices in comparison with inertia-free vent panels, which have a venting efficiency equal to 1.

In contrast, standard NFPA 68 [15] does provide a methodology to calculate vent areas for vents with inertia. The vent area calculated following the general equation is adjusted for vent mass when the vent mass exceeds a threshold mass (M_T), which depends on the volume, the reduced explosion pressure, the number of vent panels and the deflagration index K_{st}. Furthermore, where vent mass is greater than 40 kg/m^2 or K_{st} is greater than 250 bar·m/s, standard NFPA 68 provides an alternative calculation method in Annex G for the correction factor due to increased vent panel mass, included for informative purposes only. For the simulations presented in this chapter, the mass of the vent device considered (20 kg/m^2) was much lower than the threshold mass M_T or than 40 kg/m^2. Thus, no correction of the vent area calculated from the general equation was needed.

Figure 6 shows that the results of the simulations matched the values predicted by NFPA 68 [15] very well indeed. For condition set 1, the explosion overpressure obtained by DESC was 0.151 bar. The vent area calculated for this P_{red} following the standard was 14.30 m^2, almost equal to the vent area considered in the simulations (14.14 m^2). For condition set 0, the difference was slightly higher: $P_{red} = 0.096$ bar and a vent area of 18 m^2.

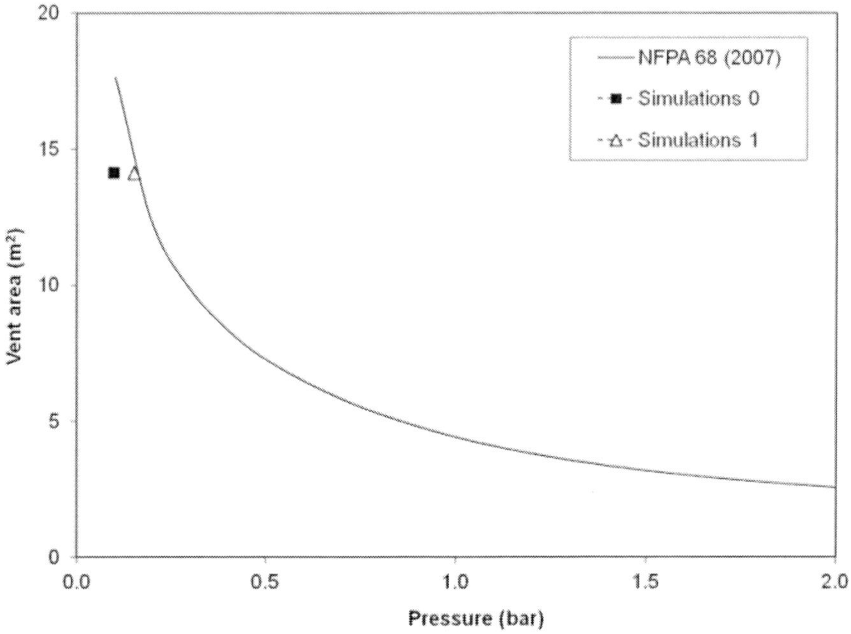

Figure 6. Correlation of the reduced explosion overpressure with the vent area. Comparison of results obtained by computer simulation with the values contemplated in NFPA 68 [15]. Condition sets 0 and 1. Maize starch. D = 6 m. L = 15 m. P_{stat} = 0.03 bar.

CONCLUSION

The DESC program has proven to be a useful tool for studying the propagation of dust explosions and the performance of protection systems. However, the results obtained by computer simulation should be extrapolated with caution since they are dependent on the initial dust cloud conditions assumed. Thus, the correct characterization of dust cloud properties in the actual equipment currently used in industry should be a priority for researchers in order to improve the reliability of CFD simulations. The uncertainties related to the input data are one of the weak points of numerical simulations.

Confidence in any comprehensive computer code has to be built on extensive validation against full-scale explosion experiments. Experimental verification is still necessary to completely validate the results presented in this chapter. Furthermore, there are uncertainties associated with the chosen approach for the original development of DESC; these are discussed by Skjold [22, 24, 25].

On the other hand, computer simulations offer a number of advantages: they generate a great deal of information for any point of interest, the initial conditions can easily be varied, and simulations are much cheaper than performing real explosion experiments.

The vent areas and associated overpressures obtained by DESC agreed quite well with those contemplated by American standard NFPA 68 [15]. When the reduced explosion overpressure was low ($P_{red} \leq 0.3$ bar), the vent areas calculated according to numerical simulations were much smaller than those considered by European standard EN 14491 [14].

Some vented dust explosions seemed to develop negative pressures, which could lead to buckling failures of the silo structure. Since standards on venting do not provide any method for estimating this, numerical simulations could be a very useful tool in this respect.

Improving the design of vented silos will require further research to provide a basic understanding of the venting and combustion processes and the characteristics of dust clouds. Numerical simulations could be a powerful tool for designing explosion mitigation methods and calculating the structure of vented silos.

ACKNOWLEDGMENTS

Acknowledgements are due to the Education Council of the Community of Castilla y León (Spain) and the European Social Fund, for funding this work via projects LE010B05 and LE013A07 and via a research grant. The authors also express their deepest gratitude to the late Dr. Frank Hauert (FSA) for providing valuable information on the CREDIT Project.

REFERENCES

[1] Eckhoff, R. K. (2003). *Dust Explosions in the Process Industries* (3rd edition). Boston, MA: Gulf Professional Publishing/Elsevier.

[2] Beck, H., Clienke, N., & Möholmann, C. (1997). *BIA-Report 13/97: Combustion and explosion characteristics of dusts*. Sankt Augustin, Germany: Hauptverband der Gewerblichen Berufsgenossenschaften.

[3] Cardillo, P. (2001). Some historical accidental explosions. *Journal of Loss Prevention in the Process Industries, 14,* 69-76.

[4] Abbasi, T., & Abbasi, S. A. (2007). Dust explosions - Cases, causes, consequences, and control. *Journal of Hazardous Materials, 140,* 7-44.

[5] Blair, A. S. (2007). Dust explosion incidents and regulations in the United States. *Journal of Loss Prevention in the Process Industries, 20,* 523-529.

[6] Schoeff, R. W. (1984). *Dust explosion statistics: U.S.* Manhattan, KS: Dept. of Grain Science & Industry, Kansas State University.

[7] Jeske, A., & Beck, H. (1989). Evaluation of dust explosions in the Federal Republic of Germany. *EuropEx Newsletter, 9,* 2-4.

[8] OSHA (2003). Regulatory Review of OSHA's Grain Handling Facilities Standard [29 CFR 1910.272]. 2003. Available from: *http://www.osha.gov/dea/lookback/grainhandlingfinalreport.html#6*

[9] Directive 1999/92/EC of the European Parliament and of the Council of 16 December 1999 on minimum requirements for improving the safety and health protection of workers potentially at risk from explosive atmospheres. *Official Journal of the European Union, L23, 28.01.2000,* 57-64.

[10] EN 1991-4 (2006). *Eurocode 1 - Actions on structures: Part 4 - Silos and tanks.* Brussels, Belgium: European Committee for Standardization.

[11] NFPA 69 (2008). *Standard on Explosion Prevention Systems.* Quincy, MA: National Fire Protection Association.

[12] EN 14460 (2006). *Explosion resistant equipment.* Brussels, Belgium: European Committee for Standardization.

[13] EN 14373 (2005). *Explosion suppression systems.* Brussels, Belgium: European Committee for Standardization.

[14] EN 14491 (2006). *Dust explosion venting protective systems.* Brussels, Belgium: European Committee for Standardization.

[15] NFPA 68 (2007). *Standard on explosion protection by deflagration venting.* Quincy, MA: National Fire Protection Association.

[16] Barton, J. (2002). *Dust Explosion Prevention and Protection: A Practical Guide.* Rugby, UK: Institution of Chemical Engineers.

[17] Tascón, A., Aguado, P. J., & Ramírez, A. (2009). Dust explosion venting in silos: A comparison of standards NFPA 68 and EN 14491. *Journal of Loss Prevention in the Process Industries. 22,* 204-209.

[18] VDI 3673 (2002). *Pressure venting of dust explosions.* Düsseldorf, Germany: Verein Deutscher Ingenieure.

[19] Bartknecht, W. (1989). *Dust explosions: Course, Prevention, Protection.* Berlin, Germany: Springer Verlag.

[20] Eckhoff, R. K (2005). Current status and expected future trends in dust explosion research. *Journal of Loss Prevention in the Process Industries,* 18, 225-237.

[21] García-Torrent, J. (2003). *Seguridad industrial en atmósferas explosivas.* Madrid, Spain: Laboratorio Oficial J.M. Madariaga.

[22] Skjold, T. (2007). Review of the DESC project. *Journal of Loss Prevention in the Process Industries, 20,* 291-302.

[23] Arntzen, B. J. (1998). *Modelling of turbulence and combustion for simulation of gas explosions in complex geometries, Dr. Ing. Thesis.* Trondheim, Norway: Norwegian University of Science and Technology.

[24] Skjold, T., Arntzen, B. J., Hansen, O. R., Storvik, I. E., & Eckhoff, R.K. (2006). Simulation of dust explosions in complex geometries with experimental input from standardized tests. *Journal of Loss Prevention in the Process Industries, 19,* 210-217.

[25] Skjold, T., Arntzen, B. J., Hansen, O. R., Taraldset, O. J., Storvik, I. E., & Eckhoff, R. F. (2005). Simulating dust explosions with thc first version of DESC. *Process Safety and Environmental Protection, 83,* 151-160.

[26] EN 14034-1 (2004). *Determination of explosion characteristics of dust clouds: Part 1, Determination of the maximum explosion pressure p_{max} of dust clouds.* Brussels, Belgium: European Committee for Standardization.

[27] ASTM E1226-10 (2010). *Standard Test Method for Explosibility of Dust Clouds.* West Conshohocken, PA: ASTM International.

[28] Tascón, A., Ruiz, Á., & Aguado, P. J. (2011). Dust explosions in vented silos: Simulations and comparisons with current standards. *Powder Technology, 208,* 714-724.

[29] Hauert, F., Vogl, A., & Radant, S. (1996). Dust cloud characterization and its influence on the pressure-time-history in silos. *Process Safety Progress, 15,* 178-184.

[30] Hauert, F., & Vogl, A. (1995). *CREDIT-Project: Measurement of Dust Cloud Characteristics in Industrial Plants, Final Technical Report.* Mannheim, Germany: FSA.

[31] Gibson, N. (1996). Problems in the control of dust explosions: an overview of the CEC CREDIT project. *Journal of Loss Prevention in the Process Industries, 9,* 291-302.

[32] Bartknecht, W. (1986). Pressure venting of dust explosions in large vessels. *Plant/Operations Progress, 5,* 196-204.

[33] Eckhoff, R. K. (1990). Sizing of dust explosion vents in the process industries: Advances made during the 1980s. *Journal of Loss Prevention in the Process Industries, 3,* 268-279.

[34] Eckhoff, R. K. (1993). Influence of initial and explosion-induced turbulence on dust explosions in large vented silo cells. *Safety Science, 16,* 511-525.

In: Dust: Sources, Environmental Concerns and Control ISBN: 978-1-61942-547-7
Editors: Laurent B. Wouters and Michel Pauwels

Chapter 8

CONTAMINATION OF DUST PARTICLES BY HEAVY METALS: THE ROLE OF SOURCES AND TRANSPORTATION PATHWAYS

***Aurela Shtiza*[1,*] *and Artan Tashko*[2]**
[1]Katholieke Universiteit Leuven,
Geo-Institute, Heverlee, Belgium
[2]Universiteti Politeknik, Fakulteti Gjeologji-Miniera,
Tirana, Albania

ABSTRACT

The dust composition due to the impact of industry and traffic in the some Albanian cities are subject of this investigation. 42 dust samples from the house lofts or green house gutters, where undisturbed dust deposition has occurred for a period of about 30 years, were investigated from the vicinity of three industrial sites (i.e. Burrel: a ferrochromium smelter; Elbasan: a metallurgical complex; Porto-Romano: a former chemical plant).

The characterization of the dust samples consisted in assessing the elemental and speciation analysis, by means of atomic absorption spectroscopy (AAS) and X-ray diffraction. The elemental concentration of the dust samples from these industrial sites, were compared with local background dust concentrations. The geochemical signature indicated that the contamination of dust by Cr, Fe, Ni, Co, Mn and to a lesser extent by Zn and Pb in the samples from the industrial sites (42 in total) is apparent when total concentrations are compared with the background values (three samples). The time these industrial sites have been active, played an important role in the geochemical signature of the dust accumulated, especially for chromium, iron, cobalt, nickel. An additional source dominated of aluminium, potassium, and sodium indicates that a natural source, not linked to the industrial processes, but most probably from pulverized clay-rich soils also contributes to the composition of the dust particles in the investigated sites. Additionally the composition of the dust indicates that the industrial activities are the main source of the contamination especially with the heavy metals. The composition of dust particles is ruled

* E-mail address: aurela_shtiza@yahoo.com; Tel: +3216327797; Fax: +3216322980

by the intensity of the industrial activity as confirmed by XRD and SEM-EDX, while the distribution is dominated from the prevailing winds. In the dust particles collected outside the ferrochromium smelter/stalk chimney, was seen that ferrochromium was the main species relating to the industrial activity, chromite and quartz were also detected most likely relating to the technological process use in the industrial plants as well as to its performance.

Maps for the spatial distribution show the spreading of the contaminants in relation to the pollution sources and the influence of the prevailing winds in transportation and deposition in distance from these investigated industrial sites. Although with a limited data set of samples, it is clear that the prevailing winds are the dominant factor in the transportation of the dust particles in distance from the industrial site.

Moreover, from the air contamination history, as recorded by the age of the dust samples it can be concluded that the air quality has been severely affected, mainly during the 1980's, due to the intensive industrial activity in these industrial sites. Knowing that the health related issues, due to the quality of air are strongly related to asthma, these studies might be an initial step in applying some on-site risk management measures minimise the dust release and to filter the gases and dusts released from these industrial sites. Obviously these conclusions may be useful in the long term health studies in these industrial sites as well as to determine and regulate the air quality in the Albanian cities.

Keywords: Dust samples, (attic) dust chemistry, heavy metals, environment, metallurgical complex, industry, Albania (Elbasan, Burrel)

INTRODUCTION

Air pollution is the presence of contaminant or pollutant substances in the air that do not disperse properly and interfere with human health or welfare, or produce other harmful environmental effects (OECD, 2001). Problems with the air pollution in urban areas have been known, from a long time, but the attitude in changing the environmental behavior is conditioned from the overall industrial and economical development as well as pressure from the concerned communities. Demand for raw materials and finished products drives the economy, once this demand has a flat pattern, than the environmental protection becomes part of the equation, because then, there are the financial means to invest for the environmental protection (Shafic, 1994). If development in the early phase of the industrial revolution was perceived as driven from the demand and supply, now the environmental protection is an essential part, which shall be considered from industry, in order to achieve the sustainable development goals starting from a local and then go into a global scale. Pollution in major cities tends to increase during the industrial build up and development phase, then they reach a maximum and then there is a reduction, as abatement technologies are developed (Fenger, 1999). In the western and developed countries, the air pollution is in the last stage with reduced levels of air contaminants (such as sulphur dioxide and soot), while in developing counties there are severe problems with air quality and there are in most of the cases no country standards on what levels have to be achieved. In most of the industrialized world urban air pollution is now monitored routinely with WHO, UNEP dedicated programs. However in the focus are also the projected developments for the year 2030 in outdoor air pollution, especially particulate matter and ozone, and on urban air quality (OECD, 2008).

The anthropogenic airborne particulate matter comes from a variety of sources, which include, but are not restricted to traffic, industry, heating and additional domestic and commercial uses (Nriagu and Pacyna, 1988; Fenger, 1999; Han and Naeher, 2006). The quality of air both indoors and outdoors is closely related to problems of the respiratory system and cardiovascular diseases. Moreover the contamination by airborne is more serious in the developing countries, especially in those currently facing rapid industrialization and changes in the land use like the countries of the former Eastern Europe (i.e. like Albania).

Dust deposited in the house roofs is defined as a unique archive for the chemical element contents of the solid particles in the air from the time when the roof was last refurbished until the sampling time (Sajn, 2005). Moreover it is a selective sample medium since it only accumulates chemical elements in particle phases. According to Nriagu et al., (1988), at any particular time and location, the amount of an element in the atmosphere depends on: 1. the intensity of the production processes or proximity to the source; 2. the amount of element released; 3. the degree of mixing determined by meteorological factors; 4. the aging history; 5. the rate and mobility mechanism of the suspended particulate matter. Moreover the rate at which airborne particles interact with the environment is greatly affected by the particle size as shown in Sheppard (1995). The large number of relevant impacting variables often leads to large variations in the geographical distribution of elements in the atmosphere as shown by Gosar and Sajn (2001); Sajn (2005). They used dust from house attics to study the spatial distribution of mercury and other element concentrations in Slovenia. They demonstrated that the attic dust chemistry provided valuable information with regard to the historical state of the atmosphere.

Although already heavily industrialized in many areas, the production policies and many of the items produced throughout the former Soviet Union and Eastern Europe created an industrial system in many ways incompatible with modern European standards of production, market competition and quality. While there are short term environmental benefits derived from the closure of most of the dust making factories, such as the Elbasan combine in Albania, the fact remains that the majority of industrial technology in Eastern Europe relies upon old, inefficient and uncompetitive machinery and processes, as well as resources which, while ‘cheap’, are of an extremely volatile nature when factored into the environmental equation Burrous, (1991); Hallstrom, (1999).

While in Europe the daily and yearly limits of fine (PM_{10} and $PM_{2.5}$) particle releases are regulated and addressed in the European Directive (EC/30/99), in Albania to our knowledge no such legislation exists. Therefore the identification of pollution sources by investigating dust particles has started only recently Tashko et al, (2005); Shtiza et al, (2005; 2007, 2009); Mazreku et al, (*in press)*; Mazreku and Tashko, (*in press*). These investigations address the chemical and mineralogical composition of dust particles settled in the house roofs and the main transportation pathways that influence their dispersion.

In light of the above information, the aims of the present study were: 1. Assess contamination and dust release in two cites in central Albania; 2. Make an inventory of the spreading of dust particles; 3. Identify the sources and transportation trends contributing to the contamination levels near or in distance from the source; 4. Address measures to be undertaken in order to minimize the release and exposure for humans and biota. This reconnaissance study aims to use dust from house lofts as a geochemical medium for the identification of air pollution sources in the vicinity of 2 industrial sites in Albania and estimate the impact of dust precipitations especially rich in heavy metals in the surrounding environment. Assuming that in time the dust samples represent "undisturbed" sedimentation as well as the chemical element contents in attic dust are proportional to their content in air, an estimation can be made with regard to the quality of air as well as the spatial distribution of the contaminants in relation to the single or various pollution sources as well as the historical variation of the dust emissions.

Overall Regional and Industrial Setting

The cities of Elbasan and Burrel are located 35 km (respectively SE and NE) from Tirana (Figure 1, A). They occur in horizontal valleys at similar altitudes 100-400 m and the prevailing local winds are mainly relief controlled. In Elbasan they have a southwest and northeast direction while in Burrel they have a north southwest direction (Pumo et al., 1990) as shown in Figure 1, B, C.

Elbasan was developed as an industrial city in the late 1960's initially with a cement factory (1968) and starting from 1976 with a metallurgical complex. The cement factory situated seven km northeast of Elbasan city started in 1968 by making use of Cretaceous-Eocene limestone and the Miocene clay deposits occurring in the surrounding. The metallurgical complex is situated north of the Quaternary depositions nearby the Shkumbini River. The metallurgical complex with a surface of 3 km^2 consists of different units (i.e. steel, iron smelter, Ni-Co, ferrochromium, refractory bricks). The iron smelter and steel plant started their activity in 1976, recording a maximum activity during the 1980s, making use of the iron and nickel laterites (which occur in the eastern part of the country) as raw materials. The ferrochromium smelter of Elbasan, which started in 1989, was composed of three furnaces. In Burrel only one ferrochromium smelter, consisting of three furnaces, was active continuously for around 20 years (1979 until 2000) since it was located in the vicinity of ultramafic formations and chromium ore exploitation locations.

The units were operational for 24h and in none of them has a filtering system ever been operational. Thus the quantity of the gases released from each ferrochromium smelter was estimated to be around 55000 m^3/h, with a content of about 1.8% Cr2O3 (ISPTK, 1995). Studies form ISPM, (2000); Banja, (2002), indicated that the iron smelter industry is the major contributor for the dust emissions in the Elbasan region. In 2001 its contribution was estimated at about 67% of the general dust emission in the area (Banja, 2002). The industrial activities were reduced drastically after 1990 due to financial constraints related to the political changes which took place in Albania in this period.

Sampling

According to the ISPM (2000), the dust content in the ferrochromium smelters were 2-5 times higher in volume/m^3 than the allowed standards for air quality in the working places. However the dusts in the earlier study were not characterized geochemically. Additional studies from Shallari et al, (1998), Dhimo et al, (1999); Sallaku et al, (1999); ISPM, (2001); Shtiza et al, (2004; 2005, 2009) demonstrated that the activity of the industrialized area has an impact on the total heavy metal concentration in the top-soils around industrialized areas in Albania especially due to the deposition of waste materials. Moreover dust precipitation is recognized as an important additive to the contamination by heavy metals (Steinnes et al,

1997; Kim et al, 1998; Shallari et al, 1998; Shtiza et al, 2005; Tashko et al., 2005) in top-soils, sediments and plants occurring near the industrial sites.

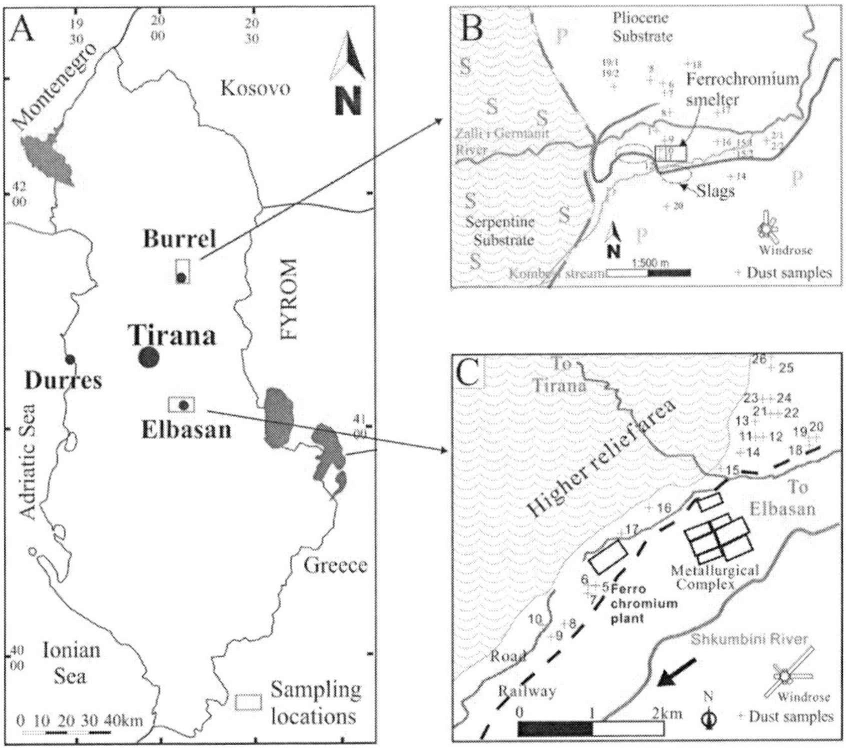

Figure 1. A. Index map of Albania; B. Sampling locations for the dust samples collected in the Burrel Ferrochromium smelter (samples 10, 11, 12) on the gutter connections of a green house (sample 2) and the rest of the samples in the house lofts in the vicinity of the smelter; C. Sampling locations for the dust samples collected in the house lofts in the vicinity of metallurgical complex in Elbasan except samples 18, 19, 20 which were collected on the gutter connections of a green house.

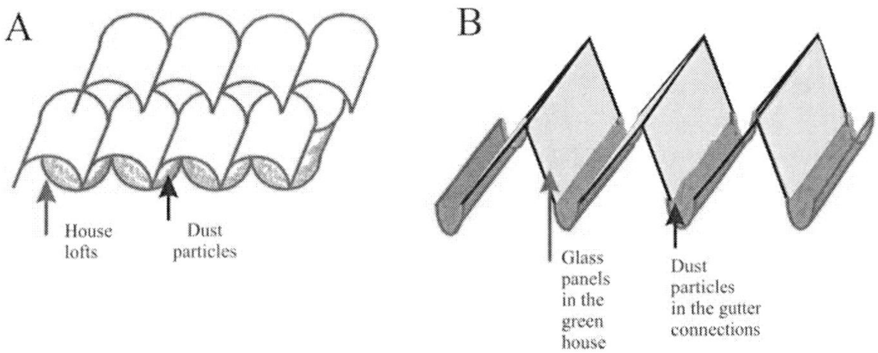

Figure 2. Illustration of the dust deposition settings. A. House lofts; B. Green house gutter.

Thus in this study, the industrial activities of the metallurgical complex (Elbasan) and a ferrochromium smelter (Burrel) are investigated with respect to the release of unfiltered gases rich in dust particles, which deposit (in the roofs, and green house gutters) of houses nearby the industrialized area and have been undisturbed over the last years.

For each of the samples, when possible, information about the "age" of the dust (last time when the lofts were refurbished) was collected from the inhabitants. The sampling was carried out with a brush from the loft surfaces and green houses, in the place where they overlay each-other. The weight of the samples varied between 2 and 5g. The samples were stored in plastic bags, sealed, labeled and transported to the laboratory for analysis.

Local Background Samples

In order to compare the results with dusts not directly related to the industrial activity, background dust samples were collected in distance from the industrial area. Three attic dust background samples were collected from some isolated buildings at distances up to 20 km north of the Elbasan metallurgical complex. The selection of these samples as background was based on a desk study, taking into account the distance from the industrialized area, the prevailing wind direction and the absence of nearby roads. The industrial impact in the selected background sampling spots is minimized due to the distance and elevation about 700 m above the industrialized area.

Dust Samples

22 attic dust samples were collected during the summer of 2003 and 2004, in houses situated along the main road in the NE-SW direction corresponding to the direction of the prevailing winds in Elbasan region. The sampling surface covers an area of 5 km by 0.5 km. 17 dust samples were collected in different fieldwork campaigns in the surrounding of the Burrel smelter. The selection of the sampling locations was based on the presence of houses as well as on the assumption that accumulated dust had not been disturbed over recent years.

METHODOLOGY

Sample Pre-treatment

The pre-treatment of the dust samples consisted in oven-drying for 3 days at 40°C and separating the loft spots (when present) and organic (leaf) particles, by means of tweezers before treating the samples further. The additional pre-treatment of the dust samples consisted in the desegregations by rolling, homogenisation and sieving. Finally the fraction below 2 mm was manually crushed in a porcelain mortar to ensure a representative homogeneous powder sample.

Total Concentrations and Mineralogy

The results obtained from the 4-Acid digestion (2.5ml HNO_{3conc}, 5ml HF_{conc}, 1.5ml $HClO_{4conc}$, 2.5ml HCl_{conc}) which resulted to have a low insoluble residue (<1%, apparently devoid of phases bearing heavy metals as confirmed by XRD and SEM-EDX analysis) was used to measure the total concentrations in dust samples (Shtiza et al., 2005). Total concentrations of Cr, Ni, Co, Zn, Cu, Pb, Fe, Ca, Mg and Al were measured with an FAAS (Flame Atomic Absorption Spectroscopy) Varian AA6 and Varian AA-1475. The detection limit was for most elements above 1 mg/kg. To evaluate the accuracy in sample handling, treatment and analysis, 30% of the collected samples were analyzed in duplicate. The analytical precision was better then 5% for all elements at the 95% confidence level. To check the accuracy of the digestion method, due to the absence of a dust standard, three replicates of the certified soil standard GBW07411 were used. The heavy metals concentrations in mg/kg were: Cr, 54.5±2.3 (certified 59.6±5.0); Co, 12.5±0.9 (certified 11.6±1.4); Cu, 70.2±4.3 (certified 65.4±4.7); Ni, 20.8±3.5 (certified 24.2±2.1); Zn, 3700± 100 (certified 3800±100). The mineralogical determinations carried out on the dust samples allowed correlation of the emitted dust particles with the presence of the industrial sites. In addition the prevailing winds in the study areas were of relevant importance since they were the main carrier for transportation of the dust particles.

RESULTS

Total Concentrations

The high average concentrations of Fe, Cr, Ni and Co recorded in Elbasan compare to the high values of Cr and less Fe and Ni from Burrel indicate that the industrial activities play an important role in the release of heavy metals in the atmosphere (Table 1). However, the values recorded are highly variable if the minimal and maximal concentrations are taken into account. Moreover the anthropogenic metal impact on dust samples may be determined by comparing the element concentrations in dust samples with the natural local background of the same element (Table 1).

Table 1. Mean, minimum and maximum concentrations of the major and heavy metals in the dust samples from local background signature as well as from Burrel and Elbasan industrial sites. In brackets is given the number of samples analyzed

	Local Background (3)			Burrel (17)			Elbasan (22)		
	Mean	Minimum	Maximum	Mean	Minimum	Maximum	Mean	Minimum	Maximum
					wt%				
Ca	6,67	5,43	8,96	3,34	1,10	9,17	6,77	3,86	13,80
Mg	1,51	1,14	1,71	4,00	0,93	8,51	2,76	1,55	2,92
Al	2,84	2,33	3,11	3,65	2,02	5,19	4,54	2,92	6,20
Fe	1,67	1,55	1,79	5,82	2,10	30,14	10,10	3,89	33,64
K	1,33	0,94	1,59	0,78	0,14	1,20	N.A	N.A	N.A
Na	1,64	1,23	1,84	0,49	0,07	0,78	N.A	N.A	N.A
					mg/kg				
Cr	160	114	232	2416	138	17809	1603	547	3845
Ni	171	123	248	418	40	873	1012	448	2308
Zn	190	180	200	612	99	4605	495	237	1465
Cu	132	103	155	110	41	374	144	62	376
Co	15	11	20	37	0	97	110	52	218
Pb	65	45	78	N.A	N.A	N.A	187	42	462
Mn	521	484	594	896	502	1561	N.A	N.A	N.A

As can be seen in Figure 3 only Al, Ca, Cu and less Mg contents may be considered as of natural origin in the dust samples analyzed from Burrel and Elbasan. The average Cr, Fe, Co and Ni contents of attic dust samples show values six times higher than in the natural background. Even the minimal contents of these elements are twice as high as the natural local background samples.

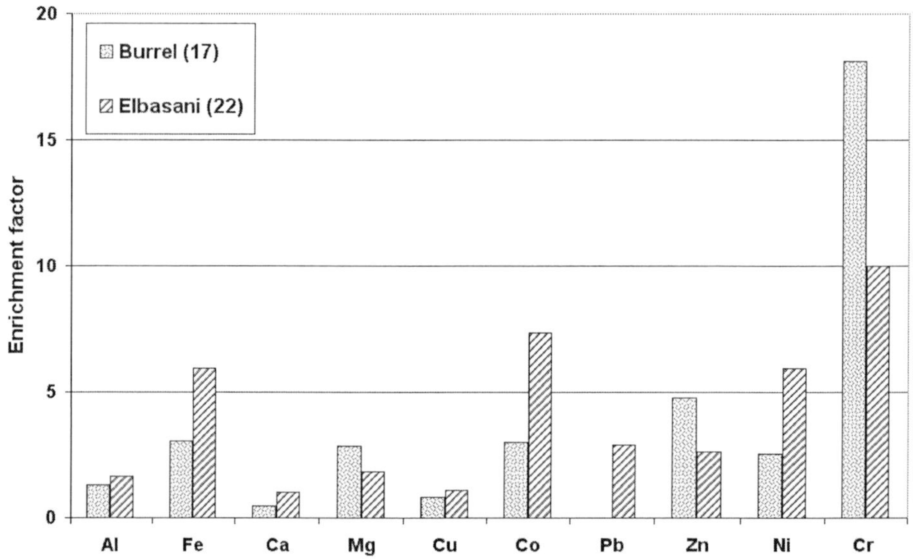

Figure 3. Enrichment factors calculated as the mean total concentration in the dusts normalized to the local background concentrations. In brackets is given the total number of samples.

Spatial Distribution

The spatial distribution patterns in Elbasan display the impact of Fe, Cr, Ni, Co and Pb on a local scale. The samples 18, 19 and 20 which are sampled in a green house gutter indicate the highest concentration of Fe and this might be also due to the material composition with which the greenhouse has been constructed.

As expected in Burrel the highest Cr and Ni concentrations were recorded in the vicinity of the smelter chimneys while the highest concentrations in the eastern part of the smelter might be related to the impact of wind in this direction. Elevated Zn concentration values were recorded within many of the dust particles and could be due to the smelting and ferrochromium production. The very high Zn content in the dust sample 2 (5386 mg/kg), however, is so anomalous that other pollution sources should be taken into account. A possible explanation is that Zn was used in the greenhouse construction material, but no supportive information is available on that.

XRD analysis of dust samples (data not given) showed the presence of quartz, calcite, iron and chromium oxides, nichromite syn., komacite, magnetite, carbon and forsterite syn. Some of these minerals rather reflect contribution of additives in the production process while others clearly can be linked to byproducts of smelting and other metallurgical activities.

The question whether, in some wind directions more elevated concentrations of heavy metals occur, cannot be answered with this limited database. Also the lower Cr-concentration to the north of smelter in Burrel could be of less relevance since some of the roofs of these houses were refurbished 3 to 5 years ago. Consequently, the dust collected there reflects only the accumulation during this limited period, which corresponds to a period of declining smelter activities in Burrel. In Elbasan the contamination is spreading around but whether the impact of the prevailing winds is of relevance cannot be assessed accurately due to the overlapping of the dominant winds with the sampling locations. Thus more dust samples covering the other directions will be of use in understanding the impact of wind in the transportation of dust particles.

DISCUSSION

Industry in general is one of the main contributors in the contamination of air by heavy metals (Nriagu and Pacyna, 1988; Fenger, 1999). Some of the common practices, which are widely used in developed countries to regulate the air pollution are: emission standards, air quality standards, emission taxes and cost benefit analysis. The soundest way of regulating the emission from a source, a sector or the entire country is to determine pollution levels in a particular ambient air and regulate their dispersion accordingly (Fenger, 2009). In Albania, a first step in setting air quality standards for the country shall be initiated in order to prepare additional legislations for interested companies, which would like to invest in the country. Various authors have investigated the overall history of the contamination in Albania, in the last decades. However these investigations consist in mainly inventorying the contamination and recommending the necessary industrial modifications needed to ameliorate the present situation. As it can be seen the carried out investigations, the contaminated samples are normalized with local background samples, since the national standards are absent. Scientist in other eastern European countries such as Armenia, are struggling with the same absence of national standards for lead concentration in house dusts (Petrosyan et al, 2004). The absence of adequate national standards for the air quality, make the task of the researchers and scientists, rather difficult. The setting of national standards will help in the future development of the country in a sustainable manner and will be attractive to industries, which aim to work in a sustainable manner.

The results from Elbasani site (Shtiza ct al, 2005, 2007, 2009) are supported from new studies recently finalized in Elbasan from Mazreku and Tashko, (in press); Mazreku et al, (in press); Tashko et al, (2005), could relate the age of the dust, dust composition with the production history of the separate units of the metallurgical complex in Elbasan. Additionally, Tashko et al (2007) investigated in total 42 dust samples from the triangle Tirana (24 samples) - Elbasan (14 samples) - Durres (4 samples). Within this triangle and the surrounding one third of the Albanian population is living. The results indicated high concentrations in heavy metals indicating that in the early decades (1960-1980) industry was the main cause of concern for the air quality, while nowadays the tendencies in migration, development, infrastructure changes as well as the changes in traffic attitude are the main factors driving the air contamination and the precipitation and accumulation of the dust particles.

Table 2. The results of the contamination of dust in the surrounding of Elbasan city by (Mazreku et al, in press)

	Elbasan (18)		
	wt/%		
	Mean	Minimum	Maximum
Ca	9.73	3.23	13.54
Mg	N.A.	N.A.	N.A.
Al	7.65	3.55	10.76
Fe	3.81	3.02	6.15
K	0.78	0.58	1.01
Na	N.A.	N.A.	N.A.
	mg/kg		
Cr	869	214	3510
Ni	522	333	738
Zn	286	179	622
Cu	67	15	160
Co	N.A.	N.A.	N.A.
Pb	183	87	748
Mn	693	576	850

The experience from developed countries, indicate that quality of air in smelters, both indoors and outdoors is closely related to problems of the respiratory system and cardiovascular diseases (Han and Naeher, 2006; Johnsen et al, 2007). Therefore the presence of filters in the industrial sites, will minimize their particle release. In Albania, from general country of city statistics it is very difficult to extract the impact of a single or a complex contamination site on the overall population. Therefore the results of these studies conducted in each of the sites in Albania, can be used as a starting point in initiating toxicological investigations in these sites, in order to assess the human exposure and take the necessary action to ameliorate the existing situation. The authorities shall propose mitigation measures in the frame of the legal framework. To improve the action of the concerned industries, various incentives, such as economical (tax rebates); administrative reduced inspection rates); reputational (such as achieving recognized industry standards, environmental management systems) shall be considered by the local authorities to improve the actual environmental situation in the concerned sites.

CONCLUSION

There is a long way from emissions to air quality. However, based on the results of the investigated sites, is possible to depict the following conclusions and recommendations:

- By means of geochemistry and mineralogy was possible to associate the activity of a single- or a complex-site as the main source contributing to the elevated chemical composition, especially for the heavy metals, in the dust particles. Although this study can be considered as screening exercises, the strong link between the history of contamination and the dust composition could be established. Therefore dust deposition, originating from industrial activities shall be taken into account when investigating contamination in general.

- The availability of contemporary and improved inventories of heavy metal emissions remains an important component in the understanding of the global transported the fate of these heavy metals once they are released into the environment. The availability of the information on point sources (i.e. single industrial units) or mega sites (i.e. military, harbours, metallurgical complexes) and the tracking of emissions over time remains a challenge for all people dealing with data emissions on local (Gosar and Sajn 2001; Tashko et al., 2005; Sajn, 2005; Shtiza et al, 2005, 2009; Tashko et al, 2007) or global inventories (Wilson et al., 2006; Pacyna et al., 2006).
- The modelling activities, which can give insights into these issues, rely on such inventories which cover different spatial distribution grids. The work reported in this research contributes to several of the above mentioned needs with respect to local anthropogenic, heavy metal emission datasets released in the environment for certain periods of time. However more samples should be collected in these sites in order to have a better overview of the transportation pathways and distribution patterns.
- The local results of this investigation can be extended further than the regional borders in order to obtain larger inventories with same or different contaminants.
- However the data also indicate that there is scope for a more detailed study of pollution caused by dust particles, with special reference to human toxicological risks. Heavy metal screening in blood in very sensitive populations (especially in woman and children and elderly people) in the vicinity of the investigated sites shall be initiated. The authorities shall consider these screening initiatives in sites where mega sites industrial activities were continuous for decades and where the density of the population is very high. Elbasan, Tirana and Durres shall be in the top-priorities in the short and long term monitoring initiatives of the responsible authorities.

RECOMENDATIONS

Regulatory Issues

All relevant regulatory bodies must consider whether the existing regulations are appropriate to protect humans and the environment from the inhalation or deposition of dust particles. They must publish their review and details of how they will address any regulatory gaps. We recommend that researchers and regulators looking to develop methods to measure and monitor airborne particles coordinate with those who are working on the measurement of pollutant particles from sources such as vehicle emissions.

Need to Monitor

Additionally the data generated from separate investigations shall be stored in a generic database, which can be accessible to interested parties and Albanian institutions. This database can be used as screening database to access easily the available information and identify the need for additional sampling locations as well as taking necessary actions to act and see what the effect is in time of the measures, which are undertaken by industry.

Industrial Sites

The installation of the filter systems in the industrial plants will minimize the release of the dust particles, therefore will consist in minimizing the exposure. Additionally a series of life cycle assessment and life cycle inventory can be developed for the dust particles released by the separate industrial activities. To have public credibility these studies need to be carried out or reviewed by an independent body (e. g. Institute of Public Health, ISHP). Also the initiative of concerned industrial installations which undertake monitoring initiatives shall be simulated and supported by the regulatory bodies and Albanian authorities.

At the Work Place

We recommend that the Institute of Public Health and in general the Health authorities monitor the dust release from industrial sites. This should be made with the purpose to protect the health of the workers and the community living in the vicinity of industrial sites.

Acknowledgments

A. Shtiza benefited from the partial financial support of the K. U. Leuven, East European Initiatives, Project 3E000659. Additional acknowledgements are addressed to all the residents in the surrounding of the Burrel and Elbasan sites who provided access and permission, to collect samples in the roofs of their houses.

References

Banja Sh., 2002. Studim i bërë nga Prefektura e qarkut, *Zyra Rajonale e Mjedisit, Elbasan.* (In Albanian).

Burrows B., 1991. Into the 21st century: a handbook for a sustainable future. *UK: Adamantine Press.*

Dhimo Ll., Haxhimanka A., Xhule M., Cara F., Pluger W.L., Lausmann M., 1999. Raport i studimit "Vlerësimi Gjeoambiental " në Rajonin e Elbasant. Projekt i përbashkët Shqiptaro-Gjerman. *Sherbimi Gjeologjik Shqiptar*. Unpublished report. (in Albanian).

Gosar M., Sajn R., 2001. Mercury in soil and attic dust as a reflection of Idrija mining and mineralization (Slovenia). *Geologija*. 44: 137-159.

Fenger J., 1999. Urban air quality. *Atmospheric Environment*. 33: 4877-4900.

Fenger J., 2009 Air pollution in the last 50 years – From local to global. *Atmospheric Environment*. 43: 13-22.

Hallstrom, L.K. 1999. Industry versus ecology: environment in the new Europe. *Futures* 31: 25-38.

Han X., and Naeher L.P., 2006. A review of traffic related air pollution exposure assessment studies in the developing world. *Environment International*. 32: 106-120.

ISPM., 2000. Monitorimi i nivelit të ndotjeve në vendet e punës të uzinave të Ferrokromit në Elbasan dhe Burrel për periudhën Tetor 1998-1999. (Monitoring contamination levels at the workplace in the ferrochromium plant in Elbasan and Burrel in the period October 1998-1999). *Institute for Studies in Processing Technology.* (in Albanian).

ISPM., 2001. Përcaktimi i ndotjeve në ajër, ujë dhe tokë për hekurin, kromin, bakrin. (Environmental impact of the industrial zone, in land and air in the Elbasani city). Pp. *Institute for Studies in Processing Technology* report. 56-61. (in Albanian)

ISPTK., 1995. Njohja e gjendjes dhe permiresimet e mundeshme ne sistemin e pluhurkapjes nga furrat. (Knowing the actual situation and possible mitigation measures for the smelters). *Institute for Studies and Processing of Technological Chremistry.* Pp. 2-15 (in Albanian).

Johnsen H.L., Soyseth V., Hetland S., Benth J.B., Kongerud J., 2007. Smelter workers environment – a hazard for respiratory disease? *Respiratory Medicine* 3: 126.

Kim K.W., Myung J.H., Ahn J.S., Chon H.T., 1998. Heavy metal contamination in dusts and stream sediments in the Taejon area, Korea. *Journal of Geochemical Exploration.* 64: 409-419.

Mazreku A., Tashko A., Civici N., In press. Ndikimi mjedisor i zones industriale ne tokat dhe ajrin e qytetit the Elbasanit. (Environmental impact of the industrial zone, in land and air in the Elbasani city). *Buletini i Shkencave Gjeologjike.* (In Albanian with English abstract). Pp. 1-12.

Mazreku A., Tashko A., In press. Impakti mjedisor i metalurgjise ne tokat e zones se Elbasanit. (Environmental impact of metallurgy on the Elbasani area soil). *Buletini i Shkencave Tenike Universiteti .Politeknik i Tiranes.* (In Albanian with English abstract).

Nriagu J.O., Pacyna J.M., 1988. Quantitative assessment of worldwide contamination of air, water and soils by trace elements. *Natura* 333: 134-139.

Nriagu J.O., Pacyna J.M., Milford J.B., Davidson C.I., 1988. Distribution and characteristic features of chromium in the atmosphere. In: J.O. Nriagu and E. Nieboer Eds., *Chromium in the natural and human environment.* John Wiley & Sons. New York. Pp. 125-172.

OECD, 2001: *Glossary of Statistical terms from OECD.*

OECD 2008. Air Pollution in Ed. *OECD Environmental Outlook to 2030.* Pp. 177-195.

Pacyna E.G., Pacyna J.M., Steenhuisen F., Wilson S., 2006. Global anthropogenic mercury emission inventory for 2000. *Atmospheric Environment.* 40: 4048-4063.

Petrosyan V., Orlova A., Dunlap C.E., Babayan E., Farfel M., Von Braun M., 2004. Lead in residential soils and dust in a mining and smelting district in northern Armenia: a pilot study. *Environmental Research* 94: 297-308.

Pumo E., Krutaj F., Lamani F., Gruda Gj., Kabo M., Demiri M., Meçaj N., Pano N., Qiriazi P., Jaho S., Sala S., Aliaj Sh., Spaho Sh., Melo V., 1990. Physical Geography of Albania. *Albanian Academy of Sciences.* Volume 1. Pp. 175-177. (In Albanian).

Sajn R., 2005. Using attic dust and soil for the separation of anthropogenic and geogenic elemental distributions in an old metallurgic area (Celje, Slovenia). *Geochemistry: Exploration, Environment, Analysis.* 5: 59-67.

Sallaku F., Shallari S., Wegener H.R., Henningsen P.F., 1999. Heavy metals in industrial spatial of Elbasan. *Bulgaria Agricultural Science.* 3: 85-92.

Shafic S., 1994. Economic development and environmental quality: an economic analysis. *Oxford Economic papers.* 46: 757-773.

Shallari S., Schwartz C., Hasko A., Morel J.L., 1998. Heavy metals in soils and plants of serpentine and industrial sites of Albania. *The Science of Total Environment.* 209: 133-142.

Sheppard S.C., 1995. A model to predict concentration enrichment of contaminants on soil adhering to plants and skin. *Environmental Geochemistry and Health* 17: 13-20.

Shtiza A., Swennen R., Tashko A., 2004. Importance of the geological substrate in environmental studies: soil contamination around chromium smelters (Albania). *Proceedings of 5th International Symposium on Eastern Mediterranean Geology.* Thessaloniki, Greece pp. 1026-1029.

Shtiza A., Swennen R., Tashko A., 2005. Chromium and nickel distribution in soil, active river and overbank sediments and dust around the Burrel chromium smelter (Albania). *Journal of Geochemical Exploration* 87: 92-108.

Shtiza A., Tashko A., Swennen R., Van den Brande A. 2009. Impact of metallurgy on the geochemical signature of dusts, soils and sediments in the vicinity of Elbasan complex (Albania). *Central European Journal of Geosciences.* 1/1: 63-83.

Shtiza A., Tashko A., Elezi M., Swennen R. 2007. Poster presentation: Mapping the spatial distribution and inventory of heavy metal emissions by means of dust samples: case studies from two industrial sites in Albania. *Proceedings of DustConf. 23-24 April 2007*, Maastricht, The Netherlands.

Steinnes E., Allen R.O., Petersen H.M., Rambek J.P., Varsog, P., 1997. Evidence of large scale heavy-metal contamination of natural surface soils in Norway from long-range atmospheric transport. *The Science of Total Environment.* 205: 255-266.

Tashko A., Elezi M., Shtiza A., 2005. Pluhurat e cative si regjistrues kohor të ndotjes së ajrit në zonën industriale të Elbasanit (Attic dust as time record of air pollution in the industrial zone of Elbasan). *Buletini i Shkencave Gjeologjike.* 1: 59-68 (In Albanian with English abstract).

Tashko A., Mazreku A., Caci I., 2007. Historia e ndotjes se ajrit ne zonen Tirane-Durres-Elbasan, bazuar ne studimin e pluhurave te cative (History of the air contamination in the area Tirana-Durres-Elbasan, based on the investigation of the dust samples). *Unpublished report.* Pp. 1-33. (In Albanian).

Wilson S.J., Steenhuisen F., Pacyna J.M., Pacyna E.G., 2006. Mapping the spatial distribution of global anthropogenic mercury atmospheric emission inventories. *Atmospheric Environment.* 40: 4621-4632.

In: Dust: Sources, Environmental Concerns and Control ISBN: 978-1-61942-547-7
Editors: Laurent B. Wouters and Michel Pauwels

Chapter 9

ASSESSING THE QUALITY OF URBAN ENVIRONMENT BY THE ELEMENTAL CONCENTRATIONS OF FOLIAGE DUST

E. Simon[1], A. Vidic[2], M. Braun[3], I. Fábián[3] and B. Tóthmérész[1]

[1]Department of Ecology, University of Debrecen, Debrecen, Hungary
[2]Department of Conservation Biology, Vegetation Ecology and Landscape Ecology, University of Vienna, Vienna, Austria
[3]Department of Inorganic and Analytical Chemistry, University of Debrecen, Hungary

ABSTRACT

The sources and pathways of dust play an important role for the global climate and biogeochemical cycles. In that context, the assessment of the elemental concentration of foliage dust elevated to an interesting topic of many scientific studies. Foliage dust may indicate heavy metal contamination and the source of contamination from atmospheric deposition in urban areas. Foliage dust contains heavy metals that may have harmful effects on human health; thus, elemental contents of foliage dust are useful to assess air pollution.

The concentrations of heavy metals in dust samples may vary across cities depending on the density of industrial activities. Our aim is to summarize the sources of dust and heavy metal concentration in dust samples. In a case study we demonstrated that dust is a useful indicator to assess the level of pollution in urban environment. We studied the elemental concentrations in foliage dust along an urbanization gradient in Vienna, Austria.

Samples were collected from urban, suburban and rural areas. We analyzed 10 metals in samples from all three areas: Al, As, Ba, Co, Cu, Fe, Pb, S, Sr and Zn. We found that the elemental concentrations of foliage dust were significantly higher in the urban area than in the rural area for aluminium, barium, iron, lead, phosphor and selenium. Urbanization changed significantly the elemental concentrations of foliage dust and the applied method proved itself as a useful way to monitor the environmental load.

Introduction

Urban and industrial centers may pollute the water, air and soil causing a decline in environmental quality [Charlesworth et al. 2003]. The levels of air pollutants are increasing at an alarming rate in many countries [Singh et al. 2005]. Air pollution has negative effects on forests [Caroll et al. 2003], water quality [Fenn et al. 2003] and human health [WHO 1987]. Various chemicals may emit into air from natural and/or anthropogenic sources, respectively [Theakston 2000]. The nitrogen dioxide [Pujadas et al. 2000], ozone [Cisneros et al. 2010], sulphur dioxide [Cape et al. 2003] and particulate matter [Wang et al. 2008] are the classical air pollutants. The effects of potential airborne pollutants as harmful ecological effects have increased which may accumulate in ecosystems [Cape et al. 2003]. These pollutants may be present in the atmosphere in form of gases or in the particulate phase, solids or adsorbed on the surface of particles [Cape et al. 2003]. The levels of street dust contamination began to attract special attention in the last decades, particularly the impact of high lead (Pb) concentrations in blood of children who live in urban areas [Charlesworth et al. 2003]. A large amount of road dust is caused by heavy traffic and the fast movement of vehicles. Kumar et al. [2001] reported that road dust is composed of 41% vehicular emission and 6% of resuspended particulate matter (PM) from traffic junction. However, level of contamination of road dust is affected by factors as volume, speed and type of vehicles, road type, neighboring environments and meteorological conditions [Duong and Lee 2011]. Street dust is characterized by short residence time and it is a complex mixture of different materials [Robinson et al. 2000] because it may include eroded urban soils which may contain historical metal pollution [De Miguel et al. 1997]. So, street dust may be a useful indicator of heavy metal contamination from atmospheric deposition [Christoforidis and Stamatis 2008].

Plants are suitable bioindicators to assess the level of dust contamination because their leaves capture the dust. The leaf surface, the size of stomata and stomatal density are the most important factors controlling the amount of captured dust and heavy metal [Abbruzzese et al. 2009]. So, trees are able to store dust on their leaf surfaces and trichomes [Lohr and Peason-Mims 1996]. Due to the fact that leaves are polluted by particulate matter (PM) near the roadside and especially in urban areas, they are good accumulators of the atmosphere dusts [Lu et al. 2008]. However, plants are able to decrease the level of air pollution, which was demonstrated in earlier studies [Lohr and Peason-Mims 1996]. It was shown that the atmospheric dust level was lower (75%) in wooded areas than in non-vegetated and populated areas. In urban areas, heavy metal containing dust can be accumulated in human body via directly inhalation, ingestion and dermal contact absorption [Wei and Yang 2010]. As nearly half of the worldwide human population live in urban agglomerations [Shi et al. 2008], the determination of origin, distribution and amount of dust, analyses of heavy metals in dust and possible ways to decrease the amount of dust are important questions in studying air pollution.

Heavy Metals in Dust

Many studies about dust have focused on its elemental concentration [Al-Khashman 2004]. The rapid growth of urbanization and industrialization resulted in a strong increase of automobile and factory emission [Kim et al. 1998]. Nevertheless, the urban environment is becoming the main habitat for humans. So, studies of dust are important to determinate the origin, distribution and level of heavy metals in dust [Duzgoren-Aydin et al. 2006]. Heavy metals in dust may originate from various types of sources. From these, the vehicle emission is one of the major sources of heavy metals in urban environments [Duong and Lee 2011].

Many studies are referring to the concentration of As, Cd, Cr, Cu, Fe, Ni, Pb and Zn which are widespread elements in urban dust. The total concentrations of these metals were determined from different areas [Apeagyei et al. 2011, Duong and Lee 2011, Wei and Yang 2010]. These studies indicated that the analyzed metals were associated with higher road traffic in urban and rural areas, respectively [Apeagyei et al. 2011]. Generally, the traffic emission is the main source of metals in dust in urban area [Wei and Yang 2010]. On the other hand, the anthropogenic heavy metal particles derived from road construction and automobiles [Meza-Figueroa et al. 2007].

In case of Pb, the main source was leaded gasoline [Thomas 1995]. An earlier study reported that another main Pb source were leaded paints from the 1970s [Jacobs et al. 2002]. Cu, Zn, and Cd originated from car components such as tire abrasion and lumbricants [Meza-Figueroa et al. 2007]. Tire treads and tire dust contain large amount of Zn while the brake housing dust and crushed brake pads contain high concentration of Fe and Cu [Adachi and Tainoshob 2004, Schauer et al. 2006]. The corrosion of cars and chrome plating of motor vehicle parts may result high concentration of Ni and Cr in dust in urban environment [Christoforidis and Stamatis 2009]. Finally, diesel fuel may contain markedly amounts of As. It emits through fossil fuel combustion, use of pigments and pesticides in urban environment [Tsai et al. 2003].

CASE STUDY

The aim of our study was to assess the level of environmental pollution in urban, suburban and rural areas among gradients of urbanization in Vienna, Austria [Simon et al. 2011]. Trees may remove air pollutants by absorbing gaseous pollutants through their leaf stomata. Therefore the foliar dust may be helpful to detect toxic elements in the air [Alfani et al. 2000, Morani et al. 2011]; thus, we determined the elemental concentration of foliage dust samples. Our hypothesis was that the concentration of heavy metals in foliage dust is lower in the rural area than in the urban area because of the higher level of traffic and industrial activity in the latter.

Vienna is a most populated city in Central Europe. It is characterized by vehicular traffic and industrial activity, which causes a high pollution pressure on the biota [Krommer et al. 2007]. The increase of sulphur dioxide (SO2), nitrogen oxides (NOx), ozone (O_3) and trace elements were reported in this region [Zechmeister and Riss 2009, Zechmeister et al. 2009].

Sampling sites were located in and around the city of Vienna (Austria). Vienna is an important administrative and cultural centre of Austria. About 2 million people are living in this region and the population density is about 4000 inhabitants per km^2. Climatically Vienna is located in a transitional zone with an average annual temperature of 10-20°C and rainfall of 1000-2000 mm. The traffic emission is the most important pollution source caused by numerous roads and motorways with heavy traffic.

Two urbanization gradients were chosen to study the effects of urbanization on elemental concentration in various habitat and soil types. One gradient represented habitats with higher rates of disturbance (floodplain-forests) and the other one represented habitats with steady conditions (oak-hornbeam-forests). Each gradient extended over a distance of approximately 10 km. The dynamic gradient proceeded along the Danube River in remnant floodplain forests. All sampling areas along this gradient were characterized by their formerly high habitat dynamics and display specific soil parameters and vegetation types. High soil water content (>25%), alkaline pH-value (~8) and remainings of sandy sediments or rubble are typical characteristics of alluvial forest soil (Ellenberg and Leuschner, 2010). Typical tree species of floodplain forests like *Fraxinus excelsior* or *Populus alba* were present at these sampling areas. Along both gradients 3 areas were selected: urban, suburban and rural area. The urban area was located in the *Vienna Prater,* near to the city centre. The suburban area was in the district of *Leopoldstadt*, further away from the *Vienna Prater*. The rural area was along the *Donau-Oder-Kanal* within the national park *Donauauen*. The stable urbanization gradient started from the urban area of *Schönbrunn*, over the suburban area of the *Pötzleinsdorfer Schlosspark* to the rural area of *Michaelerwald* within the Biosphere Reserve *Wienerwald*. Old oak-hornbeam forests were present at all three sampling areas. The brown forest soil showed lesser soil water content (less than 18%) and higher ph-values, which became more acid (~6).

While *Acer pseudoplatanus* occurred in all three investigated areas along the dynamic gradient, *Populus alba* occurred only in the rural area. Nevertheless, leaves from both species were sampled for further analyses. In case of the stable gradient, leaves from *Carpinus betulus* and *Quercus robur* were chosen.

Both species occurred in all sampling areas. Sampling was performed during August 2009. Four sampling sites were selected in each of the areas and five trees were chosen randomly. After collection, the leaf samples were pooled according to sampling sites within the investigated areas. Samples were collected in plastic bags and stored at +4°C in the dark before preparation for analysis [Margitai and Braun 2005].

The foliar dust was washed down from leaves by deionised water. The leaves were put into a 500 ml plastic box and 250 ml of deionised water was added. The samples were shaken for 10 minutes. The dust containing suspension was filtered through a 150 μm sieve. Again, the leaves were washed with 50 ml deionised water and the resulting suspension was filtered and added to the samples. This 300 ml of dust containing suspension was transferred into a microwave oven where the volume of water was reduced to 20-30 ml. Then the suspension was transferred into 50 ml glass beakers and the rest of water was evaporated at 105°C. The beakers were reweighted to determine the dry weight of dust. Samples were prepared for analysis in the same vessels. They were digested using 5 ml 65% (m/m) nitric acid and 2 ml 30% (m/m) hydrogen-peroxide at 80°C for 4 hours. Digested samples were diluted to 10 ml using 1% (m/m) nitric acid. The elemental analysis was performed by inductively coupled plasma optical emission spectrometry (ICP-OES) using IRIS Intrepid II XSP instrument. We used six-point calibration procedure with multi-element calibration solution (Merk ICP multi-element standard solution IV).

In the next step, two canonical discriminant functions were performed for the dynamic gradient based on the concentrations of the measured elements of foliage dust. In the case of foliage dust, the canonical discriminant functions were marginally significant ($P = 0.06$). The first discriminant function showed a significant negative correlation with S ($r = -0.142$), but positively correlated with Ba (r = 0.213) and Sr (r = 0.199). The concentrations of S decreased in the urban area, while the concentrations of Ba and Sr increased (Figure 1). Negative correlation was found between the second discriminant function and the concentration of Pb (r = -0.548), Fe (r = -0.400), Al (r = -0.270), Cu (r = -0.235), Co (r = -0.150) and As (r = -0.082). This means that the concentrations of these elements increased in the urban area. No significant differences were found in the concentrations of Al, As, Co, Cu, S, Sr and Zn along the urbanization gradient ($P > 0.05$). The concentrations of Al, Fe and Pb were significantly higher in the urban areas than the in rural areas ($P < 0.05$). Significant differences were not found in the concentrations of these elements when the urban and suburban areas were compared. Significant differences were not found between the foliage dust concentrations of *A. pseudoplatanus* and *P. alba* in the cases of all elements ($P > 0.05$) [Simon et al. 2011].

Based on the concentrations of the measured elements of foliage dust, again two canonical discriminant functions were conducted for the stable gradient. In the cases of foliage dust of *Carpinus betulus* and *Quercus robur*, the canonical discriminant functions were significant ($P < 0.01$) (Figure 2). In the case of *Carpinus betulus* Sr (r = 0.716), Zn (r = 0.407), Cu (r = 0.310), As (r = 0.275) and Co (r = 0.216) was correlated positively with the first discriminant function. The second canonical function showed negative correlation with S (r = -0.263), but negatively correlated with Pb (r = 0.335), Fe (r = 0.284), Al (r = 0.272) and Ba (r = 0.127) (Figue 2A). This means that the concentrations of S increased in the urban area, while the concentrations of other elements decrease in the urban area. The Sr (r = -0.482) and As (r = -0.066) negatively correlated with the first discriminant function while the S (r = 100) and Pb (r = 0.094) positively correlated in the case of dust of *Quercus robur*. The second canonical discriminant function positively correlated with Co (r = 0.321), Al (r = 0.278), Cu (r = 0.217), Fe (r = 0.185) and Ba (r = 0.068), but the correlation with Zn was negative (r = -0.121) (Figure 2B).

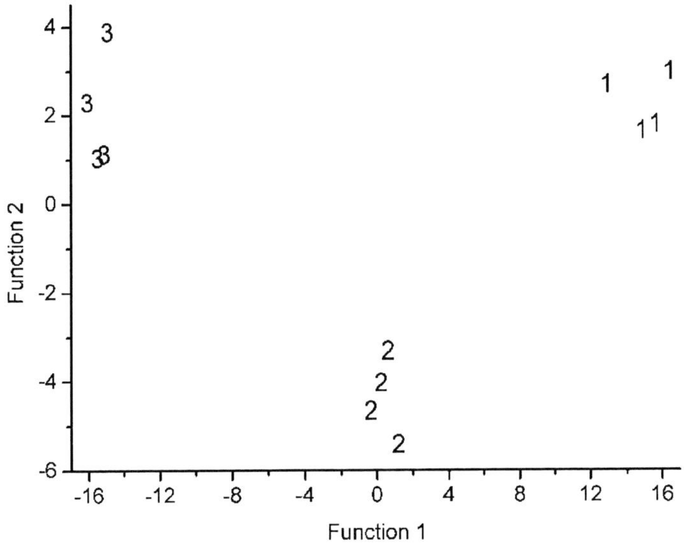

Figure 1. Canonical discriminant analysis of different areas based on elemental concentrations (mg kg^{-1}) of foliage dust of *Acer pseudoplatanus* along the dynamic gradient. Notations: 1 = urban area, 2 = suburban area, 3 = rural area.

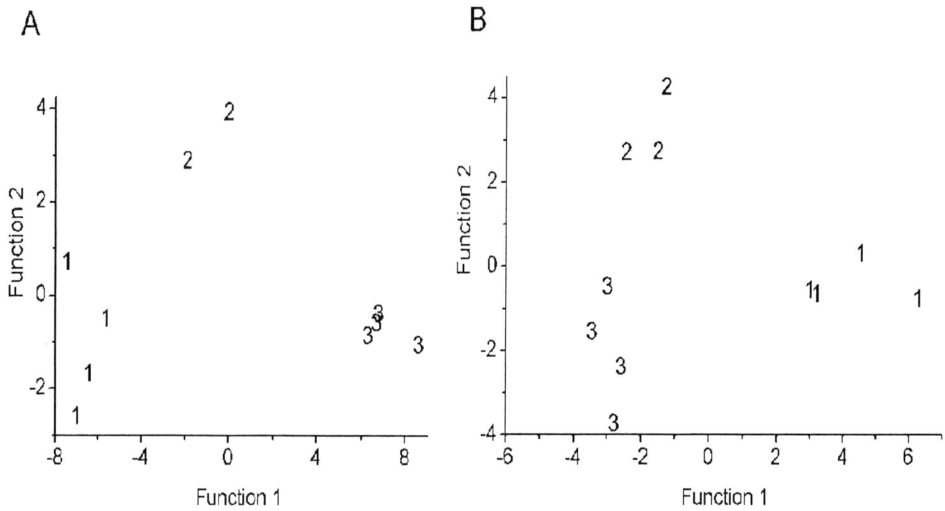

Figure 2. Canonical discriminant analysis of elemental concentrations (mg kg^{-1}) of foliage dust of *Carpinus betulus* (A) and *Quercus robur* (B) along the stable gradient. Notations: 1 = urban area, 2 = suburban area, 3 = rural area.

In the case of the stabile gradient two factors analysis was used for statistical analysis. One factor was the area and the second factor was the tree species. The effect of the area factor was significant on the As ($P < 0.01$), Cu ($P < 0.01$) and Zn ($P < 0.05$) concentration. The effects of interaction of area and tree species on elemental concentration in the cases of following elements were significant: Al, As, Co, Cu, Fe, Pb, S and Zn ($P < 0.05$). In the case of the species factor, significant effect was not found on the elemental concentration ($P > 0.05$). Significant differences were found between the foliage dust concentrations of *Carpinus betulus* according to the different areas. Significantly higher Al, As, Co, Cu, Fe, Pb, S and Zn concentration was found in the rural area than in the urban area ($P < 0.05$). In the cases of Ba and Sr significant differences were not found among areas. Differences between urban and suburban area only occurred for one element - Pb was found in significantly higher concentration in the suburban area than in the urban area. In the case of elemental concentration of foliage dust of *Quercus robur* significant differences were not found in cases of all elements ($P > 0.05$).

Compared to the elemental concentration of dust of *Carpinus betulus* and *Quercus robur* significantly lower elemental concentration was found in dust of *Carpinus betulus* in the cases of all elements than in dust of *Quercus robur* in the urban area ($P < 0.05$). In the suburban and rural areas higher Al, Ba, Cu, Fe, S, Sr and Zn concentration was found in the dust of *Quercus robur* than in the dust of *Carpinus betulus* ($P < 0.05$).

We found that the elemental concentration of foliage dust of *A. pseudoplatanus*, *Carpinus betulus* and *Quercus robur* were significantly different in the urban area compared to the rural area. Wei and Yang (2010) reported that the concentration of lead was wide-spread in urban road dust and our results corroborate that report. Although, we found lower lead concentration in foliage dust in the urban area compared to other studies [Christoforidis and Stamatis 2009, Faiz et al. 2009]. In contrast with earlier studies (Aksoy and Sahin 1999, Baycu et al. 2006, Qiu et al. 2009], significantly higher cadmium concentration was found in the rural area than in the urban area. Margitai and Braun [2005] studied heavy metal concentration of foliage dust in different European cities. According to our present results the zinc concentration in foliage dust is higher in the urban area of Vienna, than in Brussels, Munich, Debrecen and Oradea [Margitai and Braun 2005]. In contrast to the results in other cities, the cadmium concentration of foliage dust was below detection limit (0.4 mg kg^{-1}) in Vienna. We also studied the elemental concentration of sulphur which also indicates the level of air pollution. Remarkably higher sulphur concentration was found in foliage dust in Vienna than in Munich, Debrecen, Oradea and Cluj-Napoca [Margitai and Braun 2005]. In the case of iron, arsenic, copper and zinc also higher concentrations were found in foliage dust in Oradea and Cluj-Napoca than in Vienna [Braun et al. 2007]. In an earlier study, Zechmeister and Riss [2009] used moss species to assess the air pollution in Austria and they reported less cadmium (0.4 μg/g), copper (5.9 μg/g), sulphur (1263 μg/g) and zinc (32.7 μg/g) concentrations compared to our results. The comparison of species demonstrates the effects of urbanization on elemental concentrations in leaves but the results may also depend on the morphological and anatomical parameters of leaves [Kardel et al. 2010].

Conclusion

Plants are especially useful as biological indicators to assess air pollution [Kardel et al. 2010], and several plant species have already been applied as bioindicator [Aksoy and Sahin. 2000, Celik et al. 2005, Baycu et al. 2006, Mingorance and Oliva 2006]. Heavy metals and inorganic contaminants can be emitted into the environment by transportation, industry, and fossil fuels [Celik et al. 2005]. The settling contaminants can deposit on the surface of leaves from air [Salma et al. 2001], and increase their harmful effects on human health [Temesi et al. 2003].

The sources and transport of dust play important role in global climate and biogeochemical cycling [Reheis et al. 2002]. However, the control of dust and other pollution is a complex problem because of identification of sources and emission, evaluation of analytical methods and assessment of risks [Wolterbeek 2002]. But, dust is a useful indicator of heavy metal contamination by anthropogenic activities in urban environment [Zheng et al. 2010]. Our case of study demonstarted that urbanization changed significantly the elemental concentrations of foliage dust and the applied analytical method can be useful for monitoring the environmental load.

References

Abbruzzese, G., Beritognolo, I., Muleo, R., Piazzai, M., Sabatti, M., Mugnozza, G. S. and Kuzminsky, E. (2009). Leaf morphological plasticity and stomatal conductance in three *Populus alba* L. genotypes subjected to salt stress. *Environmental and Experimental Botany*, *66*, 381-388.

Adachi, K. and Tainoshob, Y. (2004). Characterization of heavy metal particles embedded in tire dust. *Environment International, 30*, 1009-1017.

Aksoy, A. and Sahin, U. (1999). *Elaeagnus angustifolia* L. as a biomonitor of heavy metal pollution. *Turkish Journal of Botany, 23*, 83-87.

Alfani, A., Baldantoni, D., Maisto, G., Bartoli, G. and Virzo De Santo, A. (2000). Temporal and spatial variation in C, N, S and trace element contents in the leaves of of Quercus ilex within the urban areas of Naples. *Environmental Pollution, 109*, 119-129.

Al-Khashman, O. A. (2004). Heavy metal distribution in dust, street dust and soils from the work place in Karak Industrial Estate, Jordan. *Atmospheric Environment, 38*, 6803-6812.

Apeagyei, E., Bank, M. S. and Spengler, J. D. (2011). Distribution of heavy metals in road dust along an urban-rural gradient in Massachusetts. *Atmospheric Environment, 45*, 2310-2323.

Baycu, G., Tolunay, D., Özden, H. and Günebakan, S. (2006). Ecophysiological and seasonal variations in Cd, Pb, Zn, and Ni concentrations in the leaves of urban deciduous trees in Istanbul. *Environmental Pollution, 143*, 545-554.

Braun, M., Margitai, Z., Tóth, A. and Leermakers, M. (2007). Environmental monitoring using linden tree leaves as natural traps of atmospheric deposition: A pilot study in Transylvania, Romania. *Acta Geographica Debrecina Landscape and Environment, 1*, 24-35.

Cape, J. N., Fowler, D. and, Davison, A. (2003). Ecological effects of sulfur dioxide, fluorides, and minor air pollutants: recent trends and research needs. *Environment International, 29*, 201–211.

Caroll, J., Miller, P. and Pronos, J. (2003). Historical perspectives on ambient ozone and its effects on the Sierra Nevada. In: Bytnerowicz, A., Aurbaugh, M.J., Alonso, R. (Eds.), Ozone Air Pollution in the Sierra Nevada-Distribution and Effects on Forests. *Developments in Environmental Science*, 2. Elsevier, Amsterdam, pp. 33-54.

Celik, A., Kartal, A. A., Akdogan, A. and Kaska, Y. (2005). Determining the heavy metal pollution in Denizli (Turkey) by using *Robinio pseudo-acacia* L. *Environment International, 31*, 105-112.

Charlesworth, S., Everett, M., McCarthy, R., Ordónez, A. and De Miguel, E. (2003). A comparative study of heavy metal concentration and distribution in deposited street dusts in a large and a small urban area: Birmingham and Coventry, West Midlands, UK. *Environment International, 29*, 563–573.

Christoforidis, A. and Stamatis, N. (2009). Heavy metal contamination in street dust and roadside soil along the major national road in Kavala's region, Greece. *Geoderma, 151*, 257–263.

Cisneros, R., Bytnerowicz, A., Schweizer, D., Zhong, S., Traina, S. and Bennett, D. H. (2010). Ozone, nitric acid, and ammonia air pollution is unhealthy for people and ecosystems in southern Sierra Nevada, California. *Environmental Pollution, 158*, 3261-3271.

De Miguel, E., Llamas, J. F., Chacón, E., Berg, T., Larssen, S., Royset, O. and Vadset, M. (1997). Origin and patterns of distribution of trace elements in street dust: unleaded petrol and urban lead. *Atmospheric Environment, 31*, 2733-2740.

Duong, T. T. T. and Lee, B. K. (2011). Determining contamination level of heavy metals in road dust from busy traffic areas with different characteristics. *Journal of Environmental Management, 92,* 554-562.

Duzgoren-Aydin, N. S., Wong, C. S. C., Aydin, A., Song, Z., You, M. and Li, X. D. (2006). Heavy metal contamination and distribution in the urban environment of Guangzhou, SE China. *Environmental Geochemistry and Health, 28*, 375-391.

Ellenberg, H., Leuschner, C. (eds.) Vegetation Mitteleuropas mit den Alpen in ökologischer, dynamischer und historischer Sicht. 6. Auflage, Ulmer Verlag, Stuttgart; 2010. pp. 1333.

Faiz, Y., Tufail, M., Javed, M. T., Chaudhry, M. M. and Siddique, N. (2009). Road dust pollution of Cd, Cu, Ni, Pb and Zn along Islamabad Expressway, Pakistan. *Microchemical Journal, 92*, 186–192.

Fenn, M. E., Baron, J. S., Allen, E. B., Ruth, H. M., Nydick, K. R., Geiser, L., Bowman, W. D., Sickman, J. O., Meixner, T., Johnson, D. W. and Neitlich, P. (2003). Ecological effects of nitrogen deposition in the western United States. *BioScience, 53*, 404-420.

Kardel, F., Wuyts, K., Babanezhad, M, Vitharana, U.W.A., Wuytack, T., Potters, G. and Samson R. (2010). Assessing urban habitat quality based on specific leaf area and stomatal characteristics of *Plantago lanceolata* L. *Environmental Pollution, 158*, 788-794.

Kim, K. W., Myung, J. H., Ahn, J. S. and Chon, H. T. (1998). Heavy metal contamination in dusts and stream sediments in the Taejon area, Korea. *Journal of Geochemical Exploration, 64*, 409-419.

Krommer, V., Zechmeister, H.G., Roder, I., Scharf, S. and, Hanus-Illnar, A., (2007). Monitoring athmospheric pollutants in the biosphere reserve Wienerald by a combined approach of biomonitoring methods and technical measurements. *Chemosphere, 67*, 1956-1966.

Kumar, A. V., Patil, R. S. and Nambi, K. S. V. (2001). Source apportionment of suspended particulate matter at two traffic junctions in Mumbai, India. *Atmospheric Environment, 35*, 4245–4251.

Lohr, V. I. and Pearson-Mims, C. H. (1996). Particulate matter accumulation on horizontal surfaces in interiors: influence of foliage plants. *Atmospheric Environment, 30*, 2565-2568.

Lu, S. G., Zheng, Y. W. and Bai, S. Q. (2008). A HRTEM/EDX approach to identification of the source of dust particles on urban tree leaves. *Atmospheric Environment, 42*, 6431–6441.

Margitai, Z. and Braun, M. (2005). Nyolc európai város légszennyezettségének vizsgálata falevelekről gyűjtött por elemösszetételének diszkriminancia analízésével. (Air quality assessment via discriminant analysis of elemental composition data dust collected from tree leaves, in Hungarian). *Magyar Kémiai Folyóirat, 111*, 38-41.

Meza-Figueroa, D., De la O-Villanueva, M. and De la Parra, L. M. (2007). Heavy metals distribution in dust from elementary schools in Hermosillo, Sonora, Mexico. *Atmospheric Environment, 41*, 276–288.

Mingorance, M. D. and Oliva, S. R. (2006). Heavy metals content in N. Oleander leaves as urban pollution assessment. *Environmental Monitoring and Assessment, 119*, 57-68.

Morani, A., Nowak, D. J., Hirabayashi, S. and Calfapietra, C. (2011). How to select the best tree planting locations to enhance air pollution removal in the MillionTreesNYC initiative. *Environmental Pollution, 159*, 1040-1047.

Pujadas, M., Plaza, J., Terés, J., Artínano, B. and Millán, M. (2000). Passive remote sensing of nitrogen dioxide as a tool for tracking air pollution in urban areas the Madrid urban plume, a case of study. *Atmospheric Environment, 34*, 3041-3056.

Qiu, Y., Guan, D., Song, W. and Huang, K. (2009). Capture of heavy metals and sulfur by foliar dust in urban Huizhou, Guangdong Province, China. *Chemosphere, 75*, 447–452.

Reheis, M., Budahn, J. and Lamothe, P. (2002). Geochemical evidence for diversity of dust sources in the southwestern United States. *Geochimica et Cosmochimica Acta, 66*, 1569-1587.

Robinson, M., Boardman, J., Evans, R., Heppell, K., Packman, J. and, Leeks, G. Land use change. In: Acreman M, editor. The hydrology of the UK. Routledge Publishers; 2000. pp. 30– 54.

Salma, I., Maenhaut, W., Zemplén-Papp, É. and Záray, G. (2001). Comprehensive characterisation of atmospheric aerosols in Budapest, Hungary: physicochemical properties of inorganic species. *Atmospheric Environment, 35*, 4367–4378.

Schauer, J. J., Lough, G .C., Shafer, M .M., Christensen, W. F., Arndt, M. F., DeMinter, J. T. and Park, J. S. (2006). Characterization of metals emitted from motor vehicles. *Health Effect Institute, 133*, 1-88.

Shi, G., Chen, Z., Xu, S., Zhang, J., Wang, L., Bi, C. and, Teng, J., (2008). Potentially toxic metal contamination of urban soils and roadside dust in Shanghai, China. *Environmental Pollution, 156*, 251–260.

Simon, E., Braun, M., Vidic, A., Bogyó, D., Fábián, I. and Tóthmérész, B. (2011). Air pollution assessment based on elemental concentration of leaves tissue and and foliage dust along an urbanization gradient in Vienna. *Environmental Pollution, 159*, 1229-1233.

Singh, A., Agrawal, S. B. and Rathore, D. (2005). Amelioration of Indian urban air pollution phytotoxicity in *Beta vulgaris* L. by modifying NPK nutrients. *Environmental Pollution, 134*, 385-395.

Temesi, D., Molnár, Á., Mészáros, E. and Feczkó, T. (2003). Seasonal and diurnal variation in the size distribution of fine carbonaceous particles over rural Hungary. *Atmospheric Environment, 37*, 139–146.

Theakston, F. Air quality Guidelines for Europe. World Health Organization Regional Office for Europe Copenhagen. 2000. pp.: 288.

Thomas, V. M. (1995). The elimination of lead in gasoline. *Annual Review of Energy and Environment, 20,* 301-324.

Tsai, Y. I., Kuo, S. C. and Lin, Y. H. (2003). Temporal characteristics of inhalablemercury and arsenic aerosols in the urban atmosphere in southern Taiwan. *Atmospheric Environment, 37*, 3401–3411.

Wang, W., Ren, L., Zhang, Y., Chen, J., Liu, H., Bad, L., Fan, S. and Tang, D. (2008). Aircraft measurements of gaseous pollutants and particulate matter over Pearl River Delta in China. *Atmospheric Environment, 42*, 6187–6202.

Wei, B. and Yang, L. (2010). A review of heavy metal contaminations in urban soils, urban road dusts and agricultural soils from China. *Microchemical Journal, 94*, 99–107.

Wolterbeek, B. (2002). Biomonitoring of trace element air pollution: principles, possibilities and perspectives. *Environmental Pollution, 120*, 11–21.

World Health Organization, 1987. Air Quality Guidelines for Europe. In: Regional Publications European Series, No. 23 Copenhagen, Denmark.

Zechmeister, H.G., Hohenwallner, D., Hanus-Illnar, A., Roder, I., Riss, A., 2009. Schwermetalldepositionen in Österreich. Biomonitoring mit Moosen. Aufsammlung 2005. Report. REP-0201. Wien.

Zechmeister, H.G., Riss, A., 2009. Schwermetalldepositionen in Österreich. Biomonitoring mit Moosen. Aufsammlung 2000. Report. REP-0200. Wien.

Zheng, N., Liu, J., Wang, Q. and Liang, Z. (2011). Health risk assessment of heavy metal exposure to street dust in the zinc smelting district, Northeast of China. *Science of the Total Environment, 408*, 726–733.

INDEX

D

E

F

G

H

I

N

O

P

Q

R

S

X

Y

Z